T0305664

Reliability Engineering

Advanced Research in Reliability and System Assurance Engineering

Series Editor: Mangey Ram, Professor, Graphic Era (Deemed to be University), Dehradun, India

Modeling and Simulation Based Analysis in Reliability Engineering
Edited by Mangey Ram

Reliability Engineering
Theory and Applications
Edited by Ilia Vonta and Mangey Ram

System Reliability Management
Solutions and Technologies
Edited by Adarsh Anand and Mangey Ram

Reliability Engineering
Methods and Applications
Edited by Mangey Ram

For more information about this series, please visit: https://www.crcpress.com/ Reliability-Engineering-Theory-and-Applications/Vonta-Ram/p/book/9780815355175

Reliability Engineering

Methods and Applications

Edited by
Mangey Ram

CRC Press
Taylor & Francis Group
Boca Raton London New York

CRC Press is an imprint of the
Taylor & Francis Group, an **informa** business

CRC Press
Taylor & Francis Group
6000 Broken Sound Parkway NW, Suite 300
Boca Raton, FL 33487-2742

First issued in paperback 2021

ISBN-13: 978-1-138-59385-5 (hbk)
ISBN-13: 978-1-03-217675-8 (pbk)
DOI: 10.1201/9780429488009

Publisher's Note

The publisher has gone to great lengths to ensure the quality of this reprint but points out that some imperfections in the original copies may be apparent.

Library of Congress Cataloging-in-Publication Data

Names: Ram, Mangey, editor.
Title: Reliability engineering : methods and applications / edited by Mangey Ram.
Other titles: Reliability engineering (CRC Press : 2019)
Description: Boca Raton, FL : CRC Press/Taylor & Francis Group, 2018.
Series: Advanced research in reliability and system assurance engineering | Includes bibliographical references and index.
Identifiers: LCCN 2019023663 (print) | LCCN 2019023664 (ebook) | ISBN 9781138593855 (hardback) | ISBN 9780429488009 (ebook)
Subjects: LCSH: Reliability (Engineering)
Classification: LCC TA169 .R439522 2019 (print) | LCC TA169 (ebook) | DDC 620/.00452--dc23
LC record available at https://lccn.loc.gov/2019023663

Visit the Taylor & Francis Web site at
http://www.taylorandfrancis.com

and the CRC Press Web site at
http://www.crcpress.com

Contents

Preface

The theory, methods, and applications of reliability analysis have been developed significantly over the last 60 years and have been recognized in many publications. Therefore, awareness about the importance of each reliability measure of the system and its fields is very important to a reliability specialist.

This book *Reliability Engineering: Methods and Applications* is a collection of different models, methods, and unique approaches to deal with the different technological aspects of reliability engineering. A deep study of the earlier approaches and models has been done to bring out better and advanced system reliability techniques for different phases of the working of the components. Scope for future developments and research has been suggested.

The main areas studied follow under different chapters:

Chapter 1 provides the review and analysis of preventive maintenance modeling issues. The discussed preventive maintenance models are classified into two main groups for one-unit and multi-unit systems.

Chapter 2 provides the literature review on the most commonly used optimal inspection maintenance mode using appropriate inspection strategy analyzing the complexity of the system whether single or multi-stage system etc. depending on the requirements of quality, production, minimum costs, and reducing the frequency of failures.

Chapter 3 presents the application of stochastic processes in degradation modeling to assess product/system performances. Among the continuous stochastic processes, the Wiener, Gamma, and inverse Gaussian processes are discussed and applied for degradation modeling of engineering systems using accelerated degradation data.

Chapter 4 presents a novel approach for analysis of Failure Modes and Effect Analysis (FMEA)-related documents through a semi-automatic procedure involving semantic tools. The aim of this work is reducing the time of analysis and improving the level of detail of the analysis through the introduction of an increased number of considered features and relations among them.

Chapter 5 studies the reliability and availability modeling of a system through Markov chains and stochastic Petri nets.

Chapter 6 talks about the fault tree analysis technique for the calculation of reliability and risk measurement in the transportation of radioactive materials. This study aims at reducing the risk of environmental contamination caused due to human errors.

Chapter 7 surveys the failure rate functions of replacement times, random, and periodic replacement models and their properties for an understanding of the complex maintenance models theoretically.

Chapter 8 highlights the design of accelerated life tests with competing failure modes which give rise to competing risk analysis. This design helps in the prediction of the product reliability accurately, quickly, and economically.

Chapter 9 presents an analysis, classification, and orientation of content to encourage researchers, organizations, and professionals to use IEC standards as applicable procedures and/or as reference guides. These standards provide methods and mathematical metrics known worldwide.

Chapter 10 discusses the time-variant reliability analysis methods for real-life dynamic structures under uncertainties and vibratory systems having high nonlinear performance. These methods satisfy the accuracy requirements by considering the time correlation.

Chapter 11 presents a few reliability or survival analysis models involving latent variables. The latent variable model considers missing information, heterogeneity of observations, measurement of errors, etc.

Chapter 12 highlights the failure mode and effects analysis technique that estimates the system reliability when the components are dependent on each other and there is common cause failure as in redundant systems using the logical algorithm.

Chapter 13 provides an overview of the current state-of-the-art reliability assessment approaches, including testing and probabilistic data analysis approaches, for vehicle components and systems, vehicle exhaust components, and systems. The new concepts include a fatigue S-N curve transformation technique and a variable transformation technique in a damage-cycle diagram.

Chapter 14 is an attempt to develop a semi-Markov model of a ship's electric power generation system and use multi-state systems theory to develop an alternative aspect of maintenance policy, indicating the importance of the human capital management relating to its cost management optimization.

Chapter 15 discusses the quantitative models proposed in the software security literature called vulnerability discovery model for predicting the total number of vulnerabilities detected, identified, or discovered during the operational phase of the software. This work also described the modeling framework of the vulnerability discovery models and vulnerability patching models.

Chapter 16 discusses the signature and its factor such as mean time to failure, expected cost, and Barlow-Proschan index with the help of the reliability function and the universal generating function also using Owen's method for a coherent system, which has independent identically, distributed elements.

Throughout this book, engineers and academician gain great knowledge and help in understanding reliability engineering and its overviews. This book gives a broad overview on the past, current, and future trends of reliability methods and applications for the readers.

Mangey Ram
Graphic Era (Deemed to be University), India

Acknowledgments

The Editor acknowledges CRC Press for this opportunity and professional support. My special thanks to Ms. Cindy Renee Carelli, Executive Editor, CRC Press/ Taylor & Francis Group for the excellent support she provided me to complete this book. Thanks to Ms. Erin Harris, Editorial Assistant to Mrs. Cindy Renee Carelli, for her follow up and aid. Also, I would like to thank all the chapter authors and reviewers for their availability for this work.

Mangey Ram
Graphic Era (Deemed to be University), India

Editor

Dr. Mangey Ram received a PhD degree major in Mathematics and minor in Computer Science from G. B. Pant University of Agriculture and Technology, Pantnagar, India. He has been a Faculty Member for over 11 years and has taught several core courses in pure and applied mathematics at undergraduate, postgraduate, and doctorate levels. He is currently a Professor at Graphic Era (Deemed to be University), Dehradun, India. Before joining Graphic Era, he was a Deputy Manager (Probationary Officer) with Syndicate Bank for a short period. He is Editor-in-Chief of *International Journal of Mathematical, Engineering and Management Sciences* and the guest editor and member of the editorial board of various journals. He is a regular reviewer for international journals, including IEEE, Elsevier, Springer, Emerald, John Wiley, Taylor & Francis, and many other publishers. He has published 150-plus research publications in IEEE, Taylor & Francis, Springer, Elsevier, Emerald, World Scientific, and many other national and international journals of repute and presented his works at national and international conferences. His fields of research are reliability theory and applied mathematics. Dr. Ram is a Senior Member of the IEEE, life member of Operational Research Society of India, Society for Reliability Engineering, Quality and Operations Management in India, Indian Society of Industrial and Applied Mathematics, member of International Association of Engineers in Hong Kong, and Emerald Literati Network in the UK. He has been a member of the organizing committee of a number of international and national conferences, seminars, and workshops. He was conferred with the *Young Scientist Award* by the Uttarakhand State Council for Science and Technology, Dehradun, in 2009. He was awarded the *Best Faculty Award* in 2011; the *Research Excellence Award* in 2015; and the *Outstanding Researcher Award* in 2018 for his significant contribution in academics and research at Graphic Era (Deemed to be University) in, Dehradun, India.

Contributors

Misbah Anjum
Amity Institute of Information
    Technology
Amity University
Noida, India

Laurent Bordes
Laboratory of Mathematics and its
    Applications—IPRA, UMR 5142
University of Pau and Pays
    Adour—CNRS—E2S UPPA
Pau, France

Jose Carpio
Department of Electrical, Electronic
    and Control Engineering
Spanish National Distance Education
    University
Madrid, Spain

Jamilson Ramalho Dantas
Departamento de Ciência da
    Computação Centro de Informática
    da UFPE—CIN Recife
Pernambuco, Brasil

and

Departamento de Ciência da
    Computação Universidade Federal
    do Vale do São Francisco—
    UNIVASF Campus Salgueiro
Salgueiro, Pernambuco, Brasil

Maritza Rodriguez Gual
Department of Reactor Technology
    Service (SETRE)
Centro de Desenvolvimento da
    Tecnologia Nuclear—CDTN
Belo Horizonte, Brazil

Kanchan Jain
Department of Statistics
Panjab University
Chandigarh, India

Rivero Oliva Jesús
Departamento de Engenharia Nuclear
Universidade Federal do Rio de Janeiro
    (UFRJ)
Rio de Janeiro, Brazil

Salomón Llanes Jesús
GAMMA SA
La Habana, Cuba

P. K. Kapur
Amity Centre for Interdisciplinary
    Research
Amity University
Noida, India

Akshay Kumar
Department of Mathematics
Graphic Era Hill University
Dehradun, India

Shah Limon
Industrial & Manufacturing
    Engineering
North Dakota State University
Fargo, North Dakota

Claudio Cunha Lopes
Department of Reactor Technology
    Service (SETRE)
Centro de Desenvolvimento da
    Tecnologia Nuclear—CDTN
Belo Horizonte, Brazil

Paulo Romero Martins Maciel
Departamento de Ciência da
  Computação Centro de Informática
  da UFPE—CIN Recife
Pernambuco, Brasil

Perdomo Ojeda Manuel
Instituto Superior de Tecnologías y
  Ciencias Aplicadas
Universidad de La Habana (UH)
La Habana, Cuba

Thomas Markopoulos
Department of Financial and
  Management Engineering
University of the Aegean
Chios, Greece

Rubens de Souza Matos Júnior
Coordenadoria de Informática Instituto
  Federal de Educação, Ciência e
  Tecnologia de Sergipe, IFS Lagarto
Sergipe, Brasil

Rogerio Pimenta Morão
Department of Reactor Technology
  Service (SETRE)
Centro de Desenvolvimento da
  Tecnologia Nuclear—CDTN
Belo Horizonte, Brazil

Toshio Nakagawa
Department of Business Administration
Aichi Institute of Technology
Toyota, Japan

Miguel Angel Navas
Department of Electrical, Electronic
  and Control Engineering
Spanish National Distance Education
  University
Madrid, Spain

Vagner de Oliveira
Department of Reactor Technology
  Service (SETRE)
Centro de Desenvolvimento da
  Tecnologia Nuclear—CDTN
Belo Horizonte, Brazil

Agapios N. Platis
Department of Financial and
  Management Engineering
University of the Aegean
Chios, Greece

Mangey Ram
Department of Mathematics; Computer
  Science & Engineering
Graphic Era (Deemed to be University)
Dehradun, India

Edson Ribeiro
Centro de Desenvolvimento da
  Tecnologia Nuclear—CDTN
Belo Horizonte, Brazil

Davide Russo
Department of Management,
  Information and Production
  Engineering
University of Bergamo
Bergamo, Italy

Carlos Sancho
Department of Electrical, Electronic
  and Control Engineering
Spanish National Distance Education
  University
Madrid, Spain

Ameneh Forouzandeh Shahraki
Civil & Industrial Engineering
North Dakota State University
Fargo, North Dakota

Avinash K. Shrivastava
Department: QT, IT and Operations
International Management Institute
Kolkata, West Bengal, India

Luiz Leite da Silva
Department of Reactor Technology
   Service (SETRE)
Centro de Desenvolvimento da
   Tecnologia Nuclear—CDTN
Belo Horizonte, Brazil

S. B. Singh
Department of Mathematics,
   Statistics & Computer Science
G. B. Pant University of Agriculture &
   Technology
Pantnagar, India

Christian Spreafico
Department of Management,
   Information and Production
   Engineering
University of Bergamo
Dalmine, Italy

Preeti Wanti Srivastava
Department of Operational Research
University of Delhi
New Delhi, India

Zhonglai Wang
School of Mechanical and Electrical
   Engineering
University of Electronic Science and
   Technology of China
Chengdu, China

Zhigang Wei
Tenneco Inc.
Grass Lake, Michigan

Sylwia Werbińska-Wojciechowska
Department of Operation and
   Maintenance of Logistic,
   Transportation and Hydraulic
   Systems Faculty of Mechanical
   Engineering
Wroclaw University of Science and
   Technology
Wrocław, Poland

Om Prakash Yadav
Civil & Industrial Engineering
North Dakota State University
Fargo, North Dakota

Shui Yu
School of Mechanical and Electrical
   Engineering
University of Electronic Science and
   Technology of China
Chengdu, China

Xufeng Zhao
College of Economics and Management
Nanjing University of Aeronautics and
   Astronautics
Nanjing, China

1 Preventive Maintenance Modeling
State of the Art

Sylwia Werbińska-Wojciechowska

CONTENTS

1.1 INTRODUCTION

Preventive maintenance (PM) is an important part of facilities management in many of today's companies. The goal of a successful PM program is to establish consistent practices designed to improve the performance and safety of the operated equipment. Recently, this type of maintenance strategy is applied widely in many technical systems such as production, transport, or critical infrastructure systems.

Many studies have been devoted to PM modeling since the 1960s. One of the first surveys of maintenance policies for stochastically failing equipment—where PM models are under investigation—is given in [1]. In this work, the author investigated PM for known and uncertain distributions of time to failure. Pierskalla and Voelker [2] prepared another excellent survey of maintenance models for proper scheduling and optimizing maintenance actions, which Valdez-Flores and Feldman [3] updated later. Other valuable surveys summarize the research and practice in this area in different ways (e.g., [4–18]. In turn, the comparison between time-based maintenance and condition-based maintenance is the authors' area of interest, e.g., in works [19,20]).

In this chapter, the author focuses on the review and summary of recent PM policies developed and presented in the literature. The adopted main maintenance models classification is based on developments given in [15–18]. The models classification includes two main groups of maintenance strategies—single- and multi-unit systems. The main scheme for classification of PM models for technical system is presented in Figure 1.1.

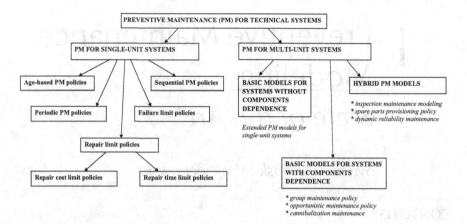

FIGURE 1.1 The classification for preventive maintenance models for technical system. (Own contribution based on Wang, H., *European Journal of Operational Research*, 139, 469–489, 2002; Werbińska-Wojciechowska, S., *Technical System Maintenance,* Delay-time-based modeling, Springer, London, UK, 2019; Werbińska-Wojciechowska, S., Multicomponent technical systems maintenance models: State of art (in Polish), in Siergiejczyk, M. (ed.), *Technical Systems Maintenance Problems: Monograph* (in Polish), Publication House of Warsaw University of Technology, Warsaw, Poland, pp. 25–57, 2014.)

Many well-known research papers focus on PM models dedicated for optimization of single-unit systems performance. The well-known maintenance models for single-unit systems are age-dependent PM and periodic PM models. In these areas, the most frequently used replacement models are based on age replacement and block replacement policies. The basic references in this area are [3,15,22,23]. The maintenance policies comparison is presented, e.g., in works [24–29].

According to Cho and Parlar [4], "multi component maintenance models are concerned with optimal maintenance policies for a system consisting of several units of machines or many pieces of equipment, which may or may not depend on each other." In 1986, Thomas, in his work [30], presents classification of optimal maintenance strategies for multi-unit systems. He focuses on the models that are based on one of three types of dependence that occurs between system elements—economic, failure, and structural. According to the author, economic dependence implies that an opportunity for a group replacement of several components costs less than separate replacements of the individual components. Stochastic dependence, also called failure or probabilistic dependence, occurs if the condition of components influences the lifetime distribution of other components. Structural dependence means that components structurally form a part, so that maintenance of a failed component implies maintenance of working components. These definitions are adopted in this chapter.

Literature reviews are given, e.g., in works [5,31–33] that are compatible with research findings given in [30]. More comprehensive discussion in maintenance from an application point of view can be found in [34,35]. For other recent references, see, e.g., [8,18,23]. A detailed review of the most commonly used PM policies for single- and multi-unit systems is presented in subchapters 1.2 and 1.3.

1.2 PREVENTIVE MAINTENANCE MODELING FOR SINGLE-UNIT SYSTEMS

First, the PM models for single-unit systems are investigated. Here a unit may be perceived as a component, an assembly, a subsystem, or even the whole system (treated as a complex system). The main classification for maintenance models of such systems is given in Figure 1.2. The comparisons concerning different PM policies are given in works [22,24,25,28,29,36–38].

One of the most commonly used PM policies for single-unit systems is an *age replacement policy* (ARP) that was developed in the early 1960s [39]. Under this policy, a unit is always replaced at its age *T* or at failure, whichever occurs first [40].

The issues of ARP modeling have been extensively studied in the literature since the 1990s. The main extensions that are developed for this maintenance policy apply to minimal repair, imperfect maintenance performance, shock modeling, or inspection action implementation. Following this, in the known maintenance models, the PM at *T* and corrective maintenance (CM) at failure might be either minimal, imperfect, or perfect. The main optimization criteria are based on maintenance cost structure. Therefore, in the case of the simple ARP, the expected cost per unit of time for an infinite time span is given as [39,41]:

$$C(T) = \frac{c_r F(T) + c_p \overline{F}(T)}{\displaystyle\int_0^T \overline{F}(t)dt} \tag{1.1}$$

where:

$C(T)$ is the long-run expected cost per unit time
c_p is the cost of preventive replacement of a unit
c_r is the cost of failed unit replacement
$F(t)$ is the probability distribution function of system/unit lifetime: $\overline{F}(t) = 1 - F(t)$

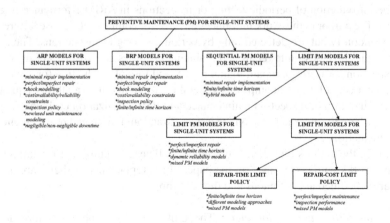

FIGURE 1.2 The classification for PM models for single-unit systems.

The first investigated group of ARP models apply to minimal repair implementation. Minimal repair is defined herein as "the repair that put the failed item back into operation with no significant effect on its remaining life time" [39]. A simple ARP model with minimal repair is given in [42], where the author investigates a one-unit system that is replaced at first failure after age T. All failures that happen before the age T are minimally repaired. The model is based on the optimization of the mean cost rate function. The extension of this model is given in [43,44], where the authors develop the ARP with minimal repair and general random repair cost.

The continuation of this research also is given in [45], where the author introduces the model for determining the optimal number of minimal repairs before replacement. The main assumptions are compatible with [43,44] and incorporate minimal repair, replacement, and general random repair cost.

A similar problem is analyzed later in [46], where the authors investigate PM with Bayesian imperfect repair. In the given PM model, the failure that occurred (for the unit age $T_y < T$) can be either minimally repaired or perfectly repaired with random probabilities. The expected cost per unit time is investigated for the infinite-horizon case and the one-replacement-cycle case.

The implementation of Bayesian approach for determining optimal replacement strategy also is given in [47]. In this paper, the authors present a fully Bayesian analysis of the optimal replacement problem for the block replacement protocol with minimal repair and the simple age replacement protocol. The optimal replacement strategies are obtained by maximizing the expected utility with uncertainty analysis.

The ARP with minimal repair usually is investigated with the use of maintenance costs constraints for optimization performance. However, a few PM models are developed based on availability optimization. For example, in [48] the authors investigate the steady-state availability of imperfect repair model for repairable two-state items. The authors use the renewal theory for providing analytical solutions for single and multi-component systems.

In another work [49], the author introduces an ARP with non-negligible downtimes. In this work, the author develops the sufficient conditions for the ARP in the aspect of the existence of a global minimum to the asymptotic expected cost rate.

The introduction of periodic testing or inspections in ARP performance is given in [50]. The author in this work introduces an ARP for components whose failures can occur randomly but are detected only by periodic testing or inspections. The developed model includes finite repair and maintenance times and cost contributions due to inspection (or testing), repair, maintenance, and loss of production (or accidents). The analytical solution encompasses general cost rate and unavailability equations. The continuation of inspection maintenance and PM optimization problems is given in [51], where the authors focus on the issues of random failure and replacement time implementation.

In [52], the authors introduce replacement policies for a unit that is running successive works with cycle times. In the paper, three replacement policies are defined that are scheduled at continuous and discrete times:

- *Continuous age replacement*: The unit is replaced before failure at a planned time T

- *Discrete age replacement*: The unit is replaced before failure at completion of the N_{wcth} working cycle
- *Age replacement with overtime*: The unit is replaced before failure at the first completion of some working cycle over the planned time T

Analytical equations of the expected cost rate with numerical solutions are provided. The authors also present the comparison of given replacement policies.

Another extension of ARP modeling is given in [53], where the authors investigate the problem of PM uncertainty by assuming that the quality of PM actions is a random variable with a defined probability distribution. Following this, the authors analyze an age reduction PM model and a failure rate PM model. Under the age reduction PM model, it is assumed that each PM reduces operational stress to the existing time units previous to the PM intervention, where the restoration interval is less than or equal to the PM interval. The optimization criteria also is based on maintenance cost structure.

The issues of warranty policy are investigated in [54]. The author in this work investigates a general age-replacement model that incorporates minimal repair, planned replacement, and unplanned replacement for a product under a renewing free-replacement warranty policy. The main assumptions of the ARP are compatible with [43,44]. The authors assume that all the product failures that cause minimal repair can be detected instantly and repaired instantaneously by a user. Thus, it is assumed in this study that the user of the product should be responsible for all minimal repairs before and after the warranty expires. Following this, for the product with an increasing failure rate function, the authors show that a unique optimal replacement age exists such that the long-run expected cost rate is minimized. The authors also compare analytically the optimal replacement ages for products with and without warranty.

The warranty policy problem is analyzed in [55], where the authors propose an age-dependent failure-repair model to analyze the warranty costs of products. In this paper, the authors consider four typical warranty policies (fixed warranty, renewing warranty, mixture of minimal and age-reducing repairs, and partial rebate warranty).

The last group of ARP models applies to PM strategies based on the implementation of shock models. The simple age-based policy with shock model is presented in [56]. In this work, the authors introduce the three main cumulative damage models: (1) a unit that is subjected to shocks and suffers some damage due to shocks, (2) the model includes periodic inspections, and (3) the model assumes that the amount of damage increases linearly with time. For the defined shock models, optimal replacement policies are derived for the expected cost rate minimization.

The extension of the given models is presented in [57], where the authors study the mean residual life of a technical object as a measure used in the age replacement model assessment. The analytical solution is supplied with a new U-statistic test procedure for testing the hypothesis that the life is exponentially distributed against the alternative that the life distribution has a renewal-increasing mean residual property.

Another development of general replacement models of systems subject to shocks is presented in [58], where the authors introduce the fatal and nonfatal shocks occurrence. The fatal shock causes the system total breakdown and the system is replaced, whereas the nonfatal shock weakens the system and makes it more expensive to run.

Following this, the authors focus on finding the optimal T that minimizes the long-run expected cost per unit time.

Another extension of the ARP with shock models is to introduce the minimal repair performance. Following this, in [59] the authors extend the generalized replacement policy given in [58] by introducing minimal repair of minor failures. Moreover, in the given PM model, the cost of minimal repair of the system is age dependent.

Later, in [60], the authors introduce an extended ARP policy with minimal repairs and a cumulative damage model implementation. Under the developed maintenance policy, the fatal shocks are removed by minimal repairs and the minor shocks increase the system failure rate by a certain amount. Without external shocks, the failure rate of the system also increases with age due to the aging process. The optimality criteria also are focused on the long-run expected cost per unit time. This model is extended later in [61], where the authors consider the ARP with minimal repair for an extended cumulative damage model with maintenance at each shock. According to the developed PM policy, when the total damage does not exceed a predetermined failure level, the system undergoes maintenance at each shock. When the total damage has reached a given failure level, the system fails and undergoes minimal repair at each failure. The system is replaced at periodic times T or at Nth failure, whichever occurs first.

To sum up, many authors usually discuss ARPs of single-unit systems analytically. The main models that address this maintenance strategy also should be supplemented by works that investigate the problem of ARP modeling with the use of semi-Markov processes (see, e.g., [62,63]), TTT-plotting (see, e.g., [64]), heuristic models (see, e.g., [65]), or approximate methods implementation (see, e.g., [66]). The authors in [67] introduce the new stochastic order for ARP based on the comparison of the Laplace transform of the time to failure for two different lifetime distributions. The comparison of ARP models for a finite horizon case based on a renewal process application and a negative exponential and Weibull failure-time distribution is presented in [68]. The additional interesting problems in ARP modeling may be connected with spare provisioning policy implementation (see, e.g., [69]) or multi-state systems investigation (see, e.g., [62,70,71]).

The quick overview of the given ARPs is presented in Table 1.1.

Another popular PM policy for single-unit systems is *block replacement policy* (BRP). For the given maintenance policy, it is assumed that all units in a system are replaced at periodic intervals regardless of their individual age in kT time moments, where $k = 1, 2, 3$, and so on. The maintenance problem usually is aimed at finding the optimal cycle length T either to minimize total maintenance and operational costs or to maximize system availability. The simple BRP, when the maintenance times are negligible, is based on the optimization of the expected long-run maintenance cost per unit time as a function of T, given as [72]:

$$C(T) = \frac{c_r N(T) + c_p}{T} \tag{1.2}$$

where:

$N(t)$ is the expected number of failure/renewals for time interval $(0,t)$

TABLE 1.1

Summary of PM Policies for Single-Unit Systems

Type of Maintenance Policy	Planning Horizon	Optimality Criterion	Modeling Method	Typical References
ARP	Infinite (∞)	The long-run expected cost per time unit	Bayesian approach	[47]
ARP	Infinite (∞)	The long-run expected cost per unit time, availability function	Analytical	[38]
ARP	Infinite (∞)	The long-run expected cost per time unit	Analytical	[39,40–42,44,53,54, 60,118]
ARP	Infinite (∞)	The expected cost rate	Analytical	[45,49,51,56,59,61, 66,119]
ARP	Infinite (∞)	The mean cost rate	Analytical	[120]
ARP	Infinite (∞)	The total cost rate, the expected unavailability	Analytical	[50]
ARP	Infinite (∞)	The expected replacement cost rate	Analytical	[52]
ARP	Infinite (∞)	The expected warranty cost	Analytical	[55]
ARP	Infinite (∞)	The steady-state availability function	Analytical	[48]
ARP	Infinite (∞)	The survival function	Analytical	[121]
ARP	Infinite (∞)	The mean time to failure	Analytical (Laplace transform)	[67]
ARP	Infinite (∞)	The long-run expected cost per unit time, availability, lifetime, and reliability functions	Multi-attribute value model	[122]
ARP	Infinite (∞)	The expected long-run cost rate	Heuristic model	[65]
ARP	Infinite (∞)	The expected long-run cost rate	Semi-Markov decision process	[63]

(Continued)

TABLE 1.1 (*Continued*)

Summary of PM Policies for Single-Unit Systems

Type of Maintenance Policy	Planning Horizon	Optimality Criterion	Modeling Method	Typical References
ARP	Infinite (∞)	The expected long-run cost rate	Semi-Markov process	[62]
ARP	Infinite (∞)	The long-run average cost per unit time	Proportional hazard model and TTT-plotting	[64]
ARP	Infinite (∞)	The total system costs	Simulation model	[69]
ARP	Infinite (∞)	State-age-dependent policy	Multi-phase Markovian model	[71]
ARP	Infinite (∞)	Mean residual life	Analytical/simulation	[57]
ARP	Infinite (∞)	The expected cost of operating the system over a time interval	Analytical	[123]
ARP	Infinite (∞)	The expected long-run cost per unit time, the total discounted cost	Analytical	[78]
ARP	Infinite (∞)/finite	The expected cost rate per unit time	Analytical	[46]
ARP	Infinite (∞)/finite	The long-run expected cost per unit time	Analytical	[43,58,124]
ARP	Finite	Expected cumulative cost	Analytical	[68]
ARP	Finite	Customer's expected discounted maintenance cost	Continuous-time Markov process	[70]
BRP	Infinite (∞)	The long-run expected cost per time unit	Analytical	[72,74–80,83, 125–127]
BRP	Infinite (∞)	The long-run expected cost per time unit	Analytical/semi-Markov processes	[81]
BRP	Finite	The long-run expected cost per time unit	Analytical	[7]

(Continued)

TABLE 1.1 (*Continued*)

Summary of PM Policies for Single-Unit Systems

Type of Maintenance Policy	Planning Horizon	Optimality Criterion	Modeling Method	Typical References
Sequential PM policy	Infinite (∞)	Mean maintenance costs	Analytical	[41]
Sequential PM policy	Infinite (∞)	Expected cost rate	Analytical	[88]
Sequential PM policy	Infinite (∞)	Expected costs per unit time	Analytical	[90]
Sequential PM policy	Infinite (∞)	Total expected maintenance costs	Genetic algorithm	[92]
Sequential PM policy	Infinite (∞)	Mean cost rate	Bayesian approach	[93]
Sequential PM policy	Infinite (∞)/finite	Expected cost rate till replacement	Analytical	[89]
Sequential PM policy	Finite	Expected cost till replacement	Analytical	[7]
Sequential PM policy	Finite	Expected profit	Genetic algorithm	[91]
Failure limit policy (Failure rate through wear/accumulated damage or stress)	Infinite (∞)	Total expected long-run cost per unit time	Analytical	[94]
Failure limit policy (Failure rate)	Infinite (∞)	Cost rate	Analytical	[95,96]
Failure limit policy (Failure rate)	Infinite (∞)	Availability function	Analytical	[128]
Failure limit policy (Degradation ratio)	Infinite (∞)	Total expected long-run cost per unit time/availability function	Analytical	[129]
Failure limit policy (Failure rate)	Infinite (∞)	Unit-cost life of a system	Genetic algorithms	[98]
Failure limit policy (Age)	Finite	Total costs function	Analytical (branching algorithm)	[97]

(*Continued*)

TABLE 1.1 (Continued)

Summary of PM Policies for Single-Unit Systems

Type of Maintenance Policy	Planning Horizon	Optimality Criterion	Modeling Method	Typical References
Repair-time limit policy	Infinite (∞)	Expected cost per unit time	Markov renewal process	[101]
Repair-time limit policy	Infinite (∞)	Expected cost per unit time	Analytical	[100]
Repair-time limit policy	Infinite (∞)	The total expected costs per unit time	Graphical approach (TTT)	[102,107,112]
Repair-time limit policy	Infinite (∞)	The expected total discounted cost	Graphical approach	[106]
Repair-time limit policy	Infinite (∞)	The expected cost per unit time	Lorenz curve	[105]
Repair-time limit policy	Infinite (∞)	The long-run average profit rate/the total discounted profit	Analytical/nonparametric algorithms	[104]
Repair-cost limit policy	Infinite (∞)	Cost rate	Analytical	[108,110,111]
Repair-cost limit policy	Infinite (∞)	Mean cost rate	Analytical	[109,115]
Repair-cost limit policy	Infinite (∞)	Mean cost rate	Markov renewal process	[117]
Repair-cost limit policy	Infinite (∞)	The long-term cost per unit time	Analytical	[113]
Repair-cost limit policy	Infinite (∞)	The long-run total maintenance cost rate	Analytical	[114]
Repair-cost limit policy	Infinite (∞)	Total expected cost per unit time	Graphical approach (TTT)	[103]
Repair-cost limit policy	Infinite (∞)	The expected average cost per unit time	Optimal stopping theory	[130]
Repair-cost limit policy	Infinite (∞)	The long-run average expected maintenance cost per unit time	Semi-Markov decision process	[131]
Repair-cost limit policy	Finite	The expected cost of servicing	Analytical	[132]

The main advantage of this policy is its simplicity. However, the main drawback of simple block replacement policy is that at planned replacement times practically new items might be replaced and a major portion of the useful life of these units is wasted. Thus, to overcome this disadvantage, various modifications have been introduced in the literature. The main extensions for the simple BRP include minimal repair implementation, finite/infinite time horizon, shock modeling use, and inspection maintenance performance.

The introduction of minimal repair performance was analyzed first in the 1970s. (see, e.g., [41,73]). Later, in [74], the author considers a BRP with minimal repair at failure for a used unit of age T_{ax}. In the given model, the item is preventively replaced by new ones at times kT, $k = 1, 2, 3$, and so on. If the system fails in $[(k-1)T, kT-\Delta_\delta]$, then the item either is replaced by new ones or is repaired minimally. If the failure occurs in $[kT-\Delta_\delta, kT]$, then the item either is replaced by used ones with age varying from Δ_δ to T or is repaired minimally. The choice is random with age-dependent probability. The cost structure also is age-dependent. For the given assumptions, the author defines the expected long-run cost per unit time function. This maintenance model is extended later in [75] for single and multi-unit cases.

An interesting model is introduced in [76], where the authors investigate optimal maintenance model for repairable systems under two types of failures with different maintenance costs. The model assumes that there are performed periodic visual inspections that detect potential failures of type I. For the given assumptions, the total expected costs are estimated.

The presented models are developed for an infinite time span. In [7] finite replacement models are considered. Taking into account, that the working time of a unit is given by a specified value T_{wo}, the long-run expected costs per unit time are estimated.

Another extension of the simple BRP applies to shock modeling implementation. For example, in [77] the authors investigate the system subjected to shocks, which occur independently and according to a Poisson process with intensity rate λ_s. The occurred shocks either may be nonlethal with probability p_s or lethal with probability $(1-p_s)$. Later, the extension of the given model is presented in [78]. In the given paper, the author analyzes a system subject to shocks that arrive according to an Non-Homogeneous Poisson (NHP) process. As shocks occur, the system has two types of failures:

- *Type I (minor) failure*: Removed by minimal repair
- *Type II (catastrophic) failure*: Removed by unplanned replacement

The probability of the type II failure is dependent on the number of shocks suffered since the last replacement. The author derives the expressions for the expected long-run cost per unit time and the total α-discounted cost for each policy. This model is later extended in [79], where the authors consider a BPR model for a system subjected to shock occurrence and with minimal repair at failure for a used unit of age T_{ax}. The proposed solution was based on assumptions given in [74].

The time-dependent cost structure is investigated in [80], where the authors determine a replacement time for a system with the use of counting process whose jump size is of one unit magnitude.

To sum up, many authors discuss BRPs of single-unit systems due to their simplicity. The main models that address this maintenance strategy also should be supplemented by works that investigate the problem of imperfect maintenance (see, e.g., [81,82]), joint preventive maintenance with production inventory control policy (see, e.g., [83]), risk at failure investigation (see, e.g., [84]), or estimation issues (see, e.g., [72]). The examples of BRP implementation apply to transportation systems maintenance (see, e.g., [85]), aircraft component maintenance (see, e.g., [86]), or preventive maintenance for milling assemblies (see, e.g., [87]). The quick overview of the given BRPs is presented in Table 1.1.

Another PM policy applied in the area of maintenance of single-unit systems is *sequential PM policy*. Under this PM policy a unit is preventively maintained at unequal time intervals. The unequal time interval usually is related to the age of the system or is predetermined as in periodic maintenance policies [15].

One of the first works where the author considers sequential PM policy is [88]. In this work, the sequential preventive maintenance for a system with minimal repair at failure is investigated. The policy assumes that the system is replaced at constant time intervals and at the Nth failure. This model is later investigated in [7], where the author proposes the simple sequential PM policy with imperfect maintenance for a finite time span.

Another interesting model of the sequential PM policy is presented in [89], where the authors introduce a shock model and a cumulative damage model. In this article, two replacement policies are developed—a periodic PM and a sequential PM policy with minimal repair at failure and imperfect PM. The solutions are obtained for finite and infinite time spans. These problems are investigated later in [90], where the authors adopt improvement factors in the hazard rate function for modeling the imperfect PM performance. The model is presented for an infinite time-horizon. The main characteristic of the given model is connected with considering the age-dependent minimal repair cost and the stochastic failure type.

In [91], the authors present a sequential imperfect PM policy for a degradation system. This model extends assumptions given in [88]. The developed model is based on maximal/equal cumulative-hazard rate constraints. The optimization is obtained using a genetic algorithm. Later, the random adjustment-reduction maintenance model with imperfect maintenance policy for a finite time span is presented in [92]. The authors also use the genetic algorithm implementation.

The Bayesian approach implementation in the sequential PM problem is presented in [93]. The authors determine the optimal PM schedules for a hybrid sequential PM policy, where the age reduction PM model and the hazard rate PM model are combined. Under such a hybrid PM model, each PM action reduces the effective age of the system to a certain value and also adjusts the slope of the hazard rate (slows down the degradation process of the maintained system).

Sequential PM policies are practical for most units that need more frequent maintenance with increasing age. The quick overview of the main known sequential PM models is given in Table 1.1.

The last group of PM policies applies to predefined limit level policies. The PM policy depends on the failure model assumed for operated units—*failure limit policy*. Under this policy, PM is performed only when the defined state variable, which describes the state of the unit at age T (e.g., failure rate), reaches a predetermined level and failures that occur are repaired.

One of the first works that investigates the optimal replacement model with the use of the failure limit policy is in [94]. The author in this work presents the replacement policy based on the failure model defined for an operating unit. In this model, a unit state at age T is defined by a random variable. The replacement is performed either at failure or when the unit state reaches or exceeds a given level, whichever occurs first. Model optimization is based on the average long-run cost per unit time estimation. This problem is investigated later in [95]. The author in his work introduces a PM model with the monotone hazard function affected by system degradation. The author develops a hazard model and achieves a cost optimization of system operation.

The imperfect repair in failure limit policy is introduced in [96]. The authors in their work consider two types of PM (simple PM and preventive replacement) and two types of corrective maintenance (minimal repair and corrective replacement). The developed cost-rate model is based on adjustment of the failure rate after simple PM with the use of a concept of improvement factor. The expected costs are the sum of average costs of both types of PM and average cost of downtime. This problem is addressed continued in [97]. The authors in their work propose a cost model for two types of PM (as in [96]) and one type of corrective maintenance (corrective replacement) that considers inflationary trends over a finite time horizon.

The PM scheduling for a system with deteriorated components also is analyzed in [98]. The authors consider a PM policy compatible with those presented in [97], but the degraded behavior of maintained components is modeled by a dynamic reliability equation. The optimal solution, based on unit-cost life estimation, is obtained with the use of genetic algorithms.

Another example of PM modeling under the failure limit policy is presented in [99], where the authors focus on system availability optimization. In the presented model system failure rate is reduced after each PM and depends on age and on the number of performed PM actions.

Maintenance models under the failure limit policy are summarized in the Table 1.1.

The second group of PM policies based on predefined limit levels are *repair limit policies*. In the known literature, there are two types of repair limit policies: a repair cost limit policy and a repair time limit policy [13]. Under the repair cost limit policy, when a unit fails, a repair cost is estimated and repair is undertaken if the estimated cost is less than a predetermined limit. Otherwise, the unit is replaced. For the repair cost limit policy, a decision variable applies to time of repair. If the time of corrective repair is greater than the specified time T_r^{max}, a unit is replaced. Otherwise, the unit is repaired [15,100].

The first models on repair limit policies are presented in [100,101]. The modeling methods are based on Markov renewal process use. Later, in [102], the authors discuss the optimal repair limit replacement policy based on a graphical approach with the use of the Total Time on Test (TTT) concept. This graphical approach is used in [103] to determine the optimal repair limit replacement policy.

Another extension of the simple repair time limit policy is imperfect maintenance implementation. In this implementation, known models are presented in [104–107].

The implemented modeling methods are based on using the TTT concept and Lorenz statistics.

The second type of repair limit policies is repair cost estimations at a system failure and is defined as a repair-cost limit policy. One of the first studies that investigates a general maintenance model with replacements and minimal repair as a base for repair limit replacement policy is [108]. The author presents three basic maintenance policies (based on age-dependent PM and periodic PM) and two basic repair limit replacement policies. In the first repair-cost limit replacement policy, the author assumes that a system is replaced by the new one if the random repair cost exceeds a given repair cost limit; otherwise, it is minimally repaired. This problem is later investigated in [109], in which the minimal repairs follow Non-Homogeneous Poisson Process (NHPP).

The problem of imperfect maintenance is introduced in [110], whereas in [111] the authors investigate the problem of imperfect estimation of repair cost (imperfect inspection case).

The implementation of a graphical method (TTT concept) in the repair-cost limit replacement problem with imperfect repair is presented in [112]. In the presented model, the authors introduce the imperfect repair (according to [110]) and a lead time for failed unit replacement. The solution is based on the assumption of negligible replacement time and uses the renewal reward process.

The cumulative damage model for systems subjected to shocks is presented in [113]. The author introduces a periodical replacement policy with the concept of repair cost limit under a cumulative damage model and solves it analytically for an infinite time span.

Another interesting approach to the repair-cost limit replacement policies is presented in [114]. The author proposes the total repair-cost limit replacement policy, where a system is replaced by the new one as soon as its total repair cost reaches or exceeds a given level. The presented problem is later investigated and extended in [115,116], where the authors introduce two types of failures (repairable and non-repairable) and propose a mixed maintenance policy similar to the one presented in [117].

The current repair limit policies and their extensions are summarized in the Table 1.1.

1.3 PREVENTIVE MAINTENANCE MODELING FOR MULTI-UNIT SYSTEMS

In this subchapter, the PM models for multi-unit systems are investigated. In this research area models can be distinguished for system with component dependence and for systems without that component dependence defined. For systems without component dependence simple age- and block-maintenance models can be implemented. When there is possibility to identify any occurrence of components dependence in a system, three main types of maintenance policies may be used:

- Group maintenance policy
- Opportunistic maintenance policy
- Cannibalization maintenance

First, the group maintenance policies may be used. Under such a policy, a group of items is replaced at the same time to take advantage of economies of scale.

Opportunity-based replacement models is based on the rule that replacement is performed at the time when an opportunity arrives, such as scheduled downtime, planned shutdown of the machines, or failure of a system in close proximity to the item of interest.

In the situation when one machine is inoperative due to lack of components and at the same time one or more other machines are inoperative due to the lack of different components, maintenance personnel may cannibalize operative components from one or more machines to repair the other or others. This practice is common in systems that are composed of sufficiently identical component parts (see, e.g., [34]).

The main classification for these types of PM maintenance models is given in Figure 1.3. Following is a detailed review of the most commonly used maintenance policies.

First, maintenance policies for multi-unit systems without component dependence are reviewed. In these systems two PM policies usually are used—ARP and BRP.

One of the first works that applies the simple age replacement policy implementation is [133]. The author proposes the simple ARP model for an n_k-out-of-n warm stand-by system, where the lifetime of components is exponentially distributed. The optimal maintenance policy for n failure-independent but non-identical machines in series is given in [134]. The solution is obtained with the use of nonlinear programming models.

The maintenance models with the use of ARP for multi-unit systems mostly implement minimal repair, a shock-modeling approach, and hybrid PM.

The minimal repair is introduced in [135]. In this paper, the model assumes that a system is replaced at age T. When the system fails before age T, it is either replaced or minimally repaired depending on the random repair cost at failure. The model considers finite and infinite time spans and is solved with a Bayesian approach implementation.

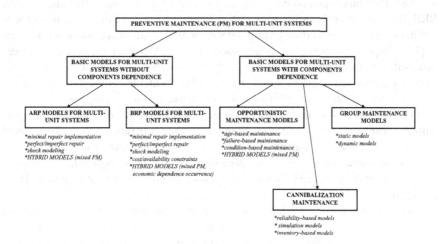

FIGURE 1.3   The classification for PM models for multi-unit systems.

Another interesting extension of the simple ARP is shock-modeling implementation. This problem is investigated in [136,137]. In [136], the authors introduce a maintenance model for a two-unit system subjected to shocks and with a failure rate interaction. The two types of shocks (minor and catastrophic) stem from a non-homogeneous pure birth process and their occurrence is dependent on the number of shocks that have occurred since the last replacement. In [137], this model is extended by a spare parts availability investigation.

The hybrid ARP applies mostly to opportunity-based maintenance implementation. This problem is investigated in [138], where maintenance opportunities arise according to a Poisson process. The problem of opportunity-based ARP also is investigated in [139–141].

In the available literature, ARP models can be found that apply to a repair priority problem (see [142]), a machine repair problem (see, [143]), or production systems maintenance (see [144]). The quick overview of the given ARPs is presented in Table 1.2.

The second group of PM policies for multi-unit systems without economic dependence applies to BRPs. Various BRPs are investigated in [145]. The author analyzes a two-unit system in a series reliability structure.

The maintenance problems of a two-unit parallel system also are investigated in [146]. In this article, the authors introduce a replacement model with minimal repair at minor failure. The analyzed system is based on structural dependence. The significant development of this model is given in [147], where the authors focus on periodic replacement for an n-unit parallel system subject to common cause shock failures. In this model, two types of failures are considered:

• Independent failures of one component in the system
• Failures of many components of the system at the same time, not necessarily independent

The summary of optimum replacement policies for an n-unit system in parallel is given in [148]. The authors compare four replacement policies—a simple BRP and a mixed BRP. This work is the basis for other authors to introduce many extensions of the BRPs for multi-unit systems. The analysis of a system with non-identical components is given in [149]. Imperfect maintenance is introduced in [150]. Moreover, the periodic replacement with minimal repair at failure for a multi-unit system is considered in [151]. In this work, the author investigates a simple model of BRP with minimal repair, when repair costs depend on system age and the number of performed minimal repairs.

The problem of minimal repair performance is investigated in [152], where the authors introduce a periodical inspection for a two-unit parallel system. This model considers the detection capacity of inspections (perfect/imperfect), minimal repairs, and failure interactions to examine dependence between subsystems. The investigation is continued in [153], where the authors examine issues analyzed in [152] and [150].

The main maintenance models focus on optimization of the cycle length T between performance of preventive maintenance actions. A number of research works also deal with the problem of cyclically scheduling maintenance activities assuming a fixed cycle length. In [154], the authors formulate a maintenance scheduling problem to maintain a

TABLE 1.2

Summary of Age and Block Replacement Policies for Multi-unit Systems

Type of Maintenance policy	Planning Horizon	Optimality Criterion	Modeling Method	Typical References
ARP	Infinite (∞)	The expected long-run costs per unit time	Analytical	[133,138–141, 144]
ARP	Infinite (∞)	The expected long-run costs per unit time	Nonlinear programming	[134]
ARP	Infinite (∞)	The expected cost rate	Analytical	[136,137,143]
ARP	Infinite (∞)	Average loss rate	Renewal process/ geometric process/ Markov process	[142]
ARP	Infinite (∞)/ finite	The expected long-run costs per unit time	Renewal reward theory/ Bayesian approach	[119]
BRP	Infinite (∞)	The expected long-run cost per unit time	Analytical/simulation	[145,149]
BRP	Infinite (∞)	The expected long-run cost per unit time	Analytical (hybrid PM)	[152,157]
BRP	Infinite (∞)	The expected long-run cost per unit time	Analytical (expected and critical value models)	[155]
BRP	Infinite (∞)	The expected long-run cost per unit time	Markov processes	[158]
BRP	Infinite (∞)	The expected long-run cost per unit time	Embedded Markov chain	[153]
BRP	Infinite (∞)	The expected long-run cost per unit time	Analytical	[75,146–151]
BRP	Infinite (∞)	The expected long-run cost per unit time, system availability	Analytical	[160]
BRP	Infinite (∞)	System availability	Analytical	[150,161]
BRP	Infinite (∞)	System availability and reliability	Analytical (matrix Laplace transformations)	[156]
BRP	Infinite (∞)	Total operating and servicing cost	Branch and price algorithm	[154]
BRP	Infinite (∞)	System reliability	Simulation	[162]

set of machines for a given determined T. The study presents the completely deterministic approach to decide for each period $t \in T$ which machine to service (if any) such that total servicing costs and operating costs are minimized. The solution is obtained with the use of a branch and price algorithm. Another interesting maintenance problem applies to investigation of uncertain lifetime of system units (see [155]), introduction of repairable and non-repairable failures of a system (see [156]), lives of heterogeneous components of a system (see [157]), implementation of a ergodic Markov environment (see [158], or nearly optimal and optimal PM assessment for real-life systems (see [128,159]). The quick overview of the given BRPs is presented in Table 1.2.

For technical systems, where component dependence can be defined, *group maintenance policies* may be used to optimize system performance. This maintenance policy is based on the performance of a maintenance activity for a group of components. According to [15], the group maintenance is performed either when a fixed time interval is expired or when a fixed number of units have failed, whichever comes first. The main classification of group replacement policies includes two main groups of models—static maintenance models and dynamic maintenance models.

In the group of static maintenance models, four main classes of group replacement policies can be defined. A T-age policy that assumes a system replacement is performed after every T units of time. An m-failure policy that calls for replacing a system at the time of mth failure. The (m, T)-policy combines features of T-age policy and m-failure policy—under such a policy, system replacement is performed at the time of the mth failure or at time T, whichever occurs first. The T-policy refers to the assumptions of the block replacement.

The presented classes of maintenance models are based on the assumption that a failure distribution of a system is known with certainty. However, in practice the failure distribution of a system is usually unknown or known with uncertain parameters. In this case, there are proposed Bayesian group replacement policies.

Considering the planning aspect, group maintenance models can be classified as stationary or dynamic. In stationary models, a long-term stable situation is assumed during which the rules for maintenance do not change over the planning horizon. The models in this overview mostly applies to this type. However, stationary models cannot incorporate dynamically changing information during operational process performance, such as a varying deterioration of components or unexpected opportunities.

To consider such short-term circumstances there are proposed dynamic models that can adapt the long-term plan according to information becoming available in the short term. This situation yields a dynamic grouping policy [163].

The main extensions of the group maintenance apply to minimal repair performance, shock modeling, or periodic inspection implementation.

Additional replacement problems that are investigated in grouping maintenance models apply to risk management (see [164]), continuous deteriorating process implementation (see [165]), or joint optimization of production scheduling (see [166]). In [164], the author analyzes the correlation among potential human error, grouping maintenance, and major accident risk. In [165], the authors introduce the novel stochastic Petri-Net and genetic algorithm-based approach to solve maintenance modeling and optimization problems. The authors in [166] present a Bayesian approach to develop a joint optimization model connecting group PM with production scheduling of a series system.

Group maintenance models are investigated widely in the literature. A review is presented in Table 1.3.

TABLE 1.3

Summary of Group Maintenance Policies for Deteriorating Multi-unit Systems

Planning Horizon	Type of Group Maintenance	Optimality Criterion	Modeling Method	Typical References
Infinite (∞)	Static (T-policy)	The long-run cost per unit time	Analytical	[167,168]
		The expected cost per unit time		[169]
		The expected cost rate		[148]
		System maintenance cost in a unit time		[170]
		Stationary availability		[161]
		Expected discounted cost to go	Control theory of jump process/dynamic programming	[171]
	Static (T-age policy)	The long run expected cost per unit of time	Analytical	[172,173]
	Static (T-age policy, m-failure policy, (m, T)-policy)	The expected cost per unit time		[174,175]
	Static	The long run expected cost per unit of time	Bayesian approach	[164,176,177]
		The long-run average maintenance cost per unit time	Markov processes	[35]
			Discrete-time Markov decision chains/simulation	[178]
		Total maintenance possession time and cost	Petri-net and GA-based approach	[165]
		Total maintenance costs	Random-key genetic algorithm	[166]
Finite rolling horizon	Dynamic	The long-term tentative plan	Dynamic programming	[179]
		The economic profit of group	Heuristic approach based on genetic algorithm and MULTIFIT algorithm	[180]
		The economic profit of group	Heuristic approach based on GA	[181]
		Penalty cost function, total maintenance cost savings over the scheduling interval	Analytical	[163]

Another group of maintenance policies for multi-unit systems with component dependence is opportunity-based maintenance. During performance processes of a multi-unit system, some maintenance opportunities may occur due to breakdowns of units in a series configuration. In most cases opportunities cannot be predicted in advance and, because of their random occurrence, *opportunistic maintenance models* can be used for effective maintenance planning. Types of opportunistic maintenance policies considered in this chapter are based mainly on [182] and include four main groups of maintenance policies:

- Age-based opportunity maintenance models
- Failure-based opportunity maintenance models
- Opportunity and condition-based maintenance models
- Mixed PM models that consider implementation of different types of maintenance policies

The detailed classification and review of the given opportunity-based maintenance policies is presented in Table 1.4.

The main extensions of opportunity-based maintenance models apply to minimal repair performance, imperfect maintenance implementation, data uncertainty investigation, finite horizon case, or shock modeling. The main applications are maintenance of production systems (see [183–185]) or offshore wind turbine systems (see [186]).

A few papers deal with an opportunistic maintenance policy under a multi-criteria perspective. The main research studies apply to production system performance (see [187]) and a power plant (see [188]).

Worth mentioning also is a group of risk-based opportunistic maintenance models. This modeling problem is considered in [189]. The authors develop a reliability model for a system that releases signals as it degrades. These released signals are used to inform opportunistic maintenance. They assume that system vulnerability to shock occurrence is dependent on its deterioration level. The risk-based opportunistic maintenance model also is analyzed in [190]. In [190], the authors present the model that uses risk evaluation of system shutdown caused by component failure. The proposed approach is based on the analysis of fault coupling features of a complex mechanical system considering age and risk factors.

In this research area, the issues of dynamic opportunistic maintenance policy optimization are analyzed. For example, in [191], the authors develop a dynamic opportunistic maintenance policy for a continuously monitored multi-unit series system with imperfect maintenance. The model is based on short-term optimization. It is assumed also that a unit's hazard rate distribution in the current maintenance cycle can be directly derived through condition-based predictive maintenance. This problem is later investigated in [192], where the authors present a dynamic

TABLE 1.4

Summary of Opportunity-Based Maintenance Policies for Deteriorating Multi-unit Systems

Planning Horizon	Maintenance Model	Optimality Criterion	Modeling Method	Typical References
Infinite (∞)	Age-based	Expected total discounted time/ expected total discounted value of good time minus costs, total discounted good time vs cost ratio	Analytical	[203]
		Cost rate		[168]
		Expected long-run cost per unit time	Analytical (deterministic problem)	[204]
		Optimal production stops	Odds algorithm-based approach	[198]
		One-step cost function	discrete-time Markov chain	[205]
		Total expected maintenance cost per unit per day	Simulation	[206]
		Total maintenance cost		[207]
		The expected cost per unit time	Monte Carlo simulation	[184]
			MC simulation and Bootstrap technique	[208]
Finite		Total maintenance cost in a given time period	Shortest path algorithm	[209]
		The total maintenance cost	Linear programming	[194]
		The cumulative maintenance cost in a given time horizon	Monte Carlo simulation	[210]
		The average cost per unit time	Heuristic approach	[211]
Infinite (∞)	Failure-based	Expected system cost rate	Analytical	[212]
		The long-run mean cost rate		[213]
		Long-run expected system maintenance cost per unit time		[214]
		Number of failures	Analytical/coupling technique	[215]
		The total maintenance cost rate	Dynamic simulation	[190]
		Signals of failure state and degradation state of a component	Signal model/simulation	[189]
		System availability	MAM	[188]

(Continued)

TABLE 1.4 (Continued)

Summary of Opportunity-Based Maintenance Policies for Deteriorating Multi-unit Systems

Planning Horizon	Maintenance Model	Optimality Criterion	Modeling Method	Typical References
Finite		The expected total maintenance cost	Analytical	[185]
		The total maintenance cost	Simulation	[216]
			Genetic algorithm	[217]
			MAM-APB model	[187]
		Survival function	Expert judgment	[202]
		The total maintenance cost	Genetic algorithm	[196]
Infinite (∞)	Condition-based	The long-run expected maintenance cost rate	Simulation	[218]
		The long-run average maintenance cost rate	Markov decision process	[219]
		The long-run average maintenance cost per blade and per time unit	Analytical	[220]
Finite		Cumulative OM cost saving		[186]
		The long-term average maintenance cost		[191]
		The expected total cost per unit time		[192]
Infinite (∞)	Mixed PM	Joint stationary probability	Dynamic Bayesian networks	[201]
			A deterioration state space partition method	[182]
Finite		Optimal total cost	Discrete-event simulation model	[200]
		The expected cost incurred in a cycle	Analytical	[221]
		The total maintenance cost per unit time		[193]
		Average net benefit over failure replacement policy	Genetic algorithm	[195]
		The expected maintenance cost	Dynamic programming	[197]
	−	Components proximity measure	Fuzzy approach	[199]

opportunistic condition-based maintenance strategy that is based on real-time predictions of the remaining useful life of components with stochastic and economic dependencies.

In [193], the authors propose a dynamic opportunistic PM optimization policy for multi-unit series systems that integrates two PM techniques: periodic PM and sequential PM policies. Whenever one unit reaches its reliability threshold level, the whole system has to stop and at that time PM opportunities arise for other units of the system. The optimal PM policy is determined by maximizing the cost saving for short-term cumulative opportunistic maintenance of the whole system.

Moreover, some research studies are based on the implementation of linear programming (see [194]), genetic algorithms (see [195,196]), dynamic programming (see [197]), theory of optimal stopping (see [198]), fuzzy modeling approach (see [199]), and simulations (see [200]). A generalized modeling method for maintenance optimization of single- and multi-unit systems is given in [182]. Moreover, a Bayesian perspective in opportunistic maintenance is investigated in [201], where the authors propose a PM policy for multi-component systems based on dynamic Bayesian networks (DBN)—Hazard and Operability Study (HAZOP) model. The use of expert judgment to parameterize a model for degradation, maintenance, and repair is provided in [202].

The last group of PM models for multi-unit systems with component dependence applies to *cannibalization maintenance*. Cannibalization in maintenance occurs "when a failed unit in a system is replaced with a functioning component from another system that is failed for some other reason" [222]. The key issue in cannibalization is how to use the component of failed units to maximize the number of working units. Thus, cannibalization actions often are used in systems with large costs associated with their critical components maintenance and operation (e.g., critical infrastructures, transport systems, and production systems).

In the recent literature, a significant amount of research is available on the use of mathematical modeling to analyze the effects of cannibalization. For a literature survey, see [18,223,224].

Following [222,225], this research can be separated into the three main approaches [18]:

- Reliability-based models
- Inventory-based maintenance models
- Simulation (queuing) maintenance models

The detailed classification and review of the given opportunity-based maintenance policies is presented in Table 1.5.

TABLE 1.5
Summary of Cannibalization Maintenance Policies for Deteriorating Multi-unit Systems

Optimality Criterion	Approach	Modeling Method	Typical References
System minimum condition	Reliability-based	Analytical	[226]
Cannibalized structure function		(allocation model)	[227]
Four measures: expected system state, defectives per failed machine, MTTCF[a], total cannibalizations		Analytical (allocation model)/simulation	[228]
The survival function of number of units of equipment available or use at the end of given time period		Analytical	[229]
System reliability for mission		Nonlinear programming	[225]
Total profit resulting from a component reusing		Simulation	[230]
Reasons for product returns		Case study	[223]
Expected number of inoperative machines		Markov process	[34]
The average total maintenance investments	Simulation-based	A closed-network, discrete-event simulation	[222]
Average total maintenance costs/average fleet readiness			[231]
NORS rate	Inventory-based	NORS model	[232]
Optimal portfolio, optimal stock level		Allocation problem – heuristic approach	[233]
The expected availability objective function		DRIVE model	[224]
Aircraft availability		Analytical (AAM model)	[234]
Cannibalization rates		Analytical	[235]
Cannibalization rates		Performance indicators analysis	[236]
Product cannibalization		Statistical data analysis	[237]
e.g., Inter-Squadron cannibalization		Balanced Scorecard	[238]

[a] MTTCF – Mean time to complete failure

1.4 CONCLUSIONS AND DIRECTIONS FOR FURTHER RESEARCH

In this chapter, the literature is reviewed on the most commonly used preventive maintenance models for single- and multi-unit systems. The literature was selected based on using Google Scholar as a search engine and ScienceDirect, JStor, SpringerLink, and SAGEJournals. The author primarily searched the relevant literature based on keywords, abstracts, and titles. The following main terms and/or a combination of them were used for searching the literature: *preventive maintenance, maintenance model, time-based maintenance.*

The selection methodology was based on searching for the defined keywords, and later choosing the models, that satisfy the main reviewing criteria. For example, when searching for the keyword preventive maintenance in a Google search, there were about 260 million hits. In the ScienceDirect database, this keyword had about 68,440 hits. Comparing the obtained search results to the main required criteria, such as *age-based maintenance model, block-based maintenance model, maintenance optimization for multi-unit system,* and *periodic maintenance,* the author focused on the most frequently used inspection models published from 1964 to 2015.

Preventive maintenance issues have been investigated by various researchers and practitioners for over 60 years. Thus, it is impossible to present all of the known models that appeared during the period under consideration. As a result, just a few of the other problems are presented that are investigated in the literature but omitted in this chapter:

- Spare part optimization issues (see [239,240])
- Data uncertainty (see [241,242])
- Maintenance decision-making issues (see [243]).

Moreover, the given literature overview provided definition for the following main conclusions:

- The most commonly used mathematical methods for analyzing maintenance scheduling problems include applied probability theory, renewal reward processes, and Markov decision theory. When the functional relationship between the system's input and output parameters cannot be described analytically, various maintenance models have been developed that apply linear and nonlinear programming, dynamic programming, simulation processes, genetic algorithms, Bayesian approach, and heuristic approaches, which were only mentioned in the presented overview.
- The investigated maintenance models usually are based on cost criterion to obtain the optimal maintenance parameters. However, maintenance actions focused on improving system dependability. Thus, for complex systems, where various types of components have different maintenance cost and different reliability importance in the system, it is more appropriate to analyze the optimal maintenance policy under cost and reliability constraints simultaneously.
- Many maintenance models consider the grouping of maintenance activities on a long-term basis with an infinite horizon. In practice, planning horizons are usually finite for a number of reasons: information is only available over the short term, a modification of the system changes the maintenance problem completely, and some events are unpredictable.
- In the most existing literature on maintenance theory, the maintenance time is assumed to be negligible. This assumption makes availability modeling impossible or unrealistic.

- Most maintenance models are based on the assumption of fully available logistic support when it is needed. Thus, in the modeling approach, it is assumed that whenever a system component is to be replaced, a new component is immediately available. However, considering real life situations, the number of spare parts is usually limited and the procurement lead-time is non-negligible. This situation implies that the maintenance policy and spare provisioning policy should be modeled and optimized jointly.

- Another problem applies to data availability and reliability. Maintenance and replacement decisions are based on the information available, such as the failure data of the equipment under consideration, maintenance performance times, and type and number of necessary support resources. Sufficient data rarely exist for estimating parameters in a complex model, and if data do exist, they are often unreliable. This situation makes the application of mathematical models to support maintenance and replacement decisions less obvious.

In summary, traditional PM programs often require very time-consuming, manual data and rely heavily on "tribal knowledge" estimates or require in-depth knowledge and analysis of each individual piece of equipment on an ongoing basis to stay up-to-date. Thus, based on the authors main conclusions and following the global trends in maintenance (see [244,245] for recent reports), in the future most likely the main interests will be on more advanced maintenance optimization models that are based on the use of digital technologies.

REFERENCES

1. Mccall, J. J. (1965). Maintenance policies for stochastically failing equipment: A survey. *Management Science* 11(5): 493–524.
2. Pierskalla, W. P. and Voelker, J. A. (1976). A survey of maintenance models: The control and surveillance of deteriorating systems. *Naval Research Logistics Quarterly* 23: 353–388.
3. Valdez-Flores, C. and Feldman, R. (1989). A survey of preventive maintenance models for stochastically deteriorating single-unit systems. *Naval Research Logistics* 36: 419–446.
4. Cho, I. D. and Parlar, M. (1991). A survey of maintenance models for multi-unit systems. *European Journal of Operational Research* 51(1): 1–23.
5. Dekker, R., Wildeman, R. E., and Van Der Duyn Schouten, F. A. (1997). A review of multi-component maintenance models with economic dependence. *Mathematical Methods of Operations Research* 45: 411–435.
6. Mazzuchi, T. A., Van Noortwijk, J. M., and Kallen, M. J. (2007). Maintenance optimization. Technical Report, TR-2007-9.
7. Nakagawa, T. and Mizutani, S. (2009). A summary of maintenance policies for a finite interval. *Reliability Engineering and System Safety* 94: 89–96. doi:10.1016/j.ress.2007.04.004.
8. Nicolai, R. P. and Dekker, R. (2007). A review of multi-component maintenance models. In: Aven, T. and Vinnem, J. M. (eds.) *Risk, Reliability and Societal Safety: Proceedings of European Safety and Reliability Conference ESREL 2007*, Stavanger, Norway, June 25–27, 2007, Leiden, the Netherlands: Taylor & Francis Group: pp. 289–296.

9. Nowakowski, T. and Werbińska, S. (2009). On problems of multi-component system maintenance modelling. *International Journal of Automation and Computing* 6(4): 364–378.
10. Pham, H. and Wang, H. (1996). Imperfect maintenance. *European Journal of Operational Research* 94: 425–438.
11. Pophaley, M. and Ways, R. K. (2010). Plant maintenance management practices in automobile industries: A retrospective and literature review. *Journal of Industrial Engineering and Management* 3(3): 512–541. doi:10.3926/jiem..v3n3.p512-541.
12. Popova, E. and Popova, I. (2014). Replacement strategies. Wiley StatsRef: Statistics Reference Online.
13. Sarkar, A., Behera, D. K., and Kumar, S. (2012). Maintenance policies of single and multi-unit systems in the past and present. *International Journal of Current Engineering and Technology* 2(1): 196–205.
14. Vasili, M., Hond, T. S., Ismail, N., and Vasili, M. (2011). Maintenance optimization models: A review and analysis. In: *Proceedings of the 2011 International Conference on Industrial Engineering and Operations Management*, January 22–24, 2011, Kuala Lumpur, Malaysia: pp. 1131–1138.
15. Wang, H. (2002). A survey of maintenance policies of deteriorating systems. *European Journal of Operational Research* 139(3): 469–489. doi:10.1016/S0377-2217(01)00197-7.
16. Wang, H. and Pham, H. (2003). Optimal imperfect maintenance models. In: Pham, H. (ed.) *Handbook of Reliability Engineering*, London, UK: Springer-Verlag London Limited: pp. 397–414.
17. Wang, H. and Pham, H. (1997). A survey of reliability and availability evaluation of complex networks using Monte Carlo techniques. *Microelectronics Reliability* 37(2): 187–209. doi:10.1016/S0026-2714(96)00058-3.
18. Werbińska-Wojciechowska, S. (2019). *Technical System Maintenance. Delay-Time-Based Modelling*. London, UK: Springer.
19. Ahmad, R. and Kamaruddin, S. (2012). An overview of time-based and condition-based maintenance in industrial application. *Computers and Industrial Engineering* 63: 135–149. doi:10.1016/j.cie.2012.02.002.
20. Geurts, J. H. J. (1983). Optimal age replacement versus condition based replacement: Some theoretical and practical considerations. *Journal of Quality Technology* 15(4): 171–179.
21. Werbińska-Wojciechowska, S. (2014). Multicomponent technical systems maintenance models: State of art (in Polish). In: Siergiejczyk, M. (ed.) *Technical Systems Maintenance Problems: Monograph* (in Polish), Warsaw, Poland: Publication House of Warsaw University of Technology: pp. 25–57.
22. Barlow, R. E. and Proschan, F. (1964). Comparison of replacement policies, and renewal theory implications. *The Annals of Mathematical Statistics* 35(2): 577–589. doi:10.1214/aoms/1177703557.
23. Wang, H. and Pham, H. (2006). *Reliability and Optimal Maintenance*, London, UK: Springer-Verlag.
24. Aven, T. and Dekker, R. (1997). A useful framework for optimal replacement models. *Reliability Engineering and System Safety* 58(1): 61–67. doi:10.1016/S0951-8320(97)00055-0.
25. Block, H. W., Langberg, N.A., and Savits, T.H. (1990). Maintenance comparisons: Block policies. *Journal of Applied Probability* 27: 649–657. doi:10.2307/3214548.
26. Block, H. W., Langberg, N. A., and Savits, T. H. (1990). Comparisons for maintenance policies involving complete and minimal repair. *Lecture Notes-Monograph Series* 16(Topics in Statistical Dependence): 57–68.
27. Christer, A. H. and Keddie, E. (1985). Experience with a stochastic replacement model. *Journal of Operational Research Society* 36(1): 25–34.

28. Frostig, E. (2003). Comparison of maintenance policies with monotone failure rate distributions. *Applied Stochastic Models in Business and Industry* 19: 51–65. doi:10.1002/asmb.485.

29. Langberg, N. A. (1988). Comparisons of replacement policies. *Journal of Applied Probability* 25: 780–788.

30. Thomas, L. C. (1986). A survey of maintenance and replacement models for maintainability and reliability of multi-item systems. *Reliability Engineering* 16(4):297–309.

31. Aboulfath, F. (1995). Optimal maintenance schedules for a fleet of vehicles under the constraint of the single repair facility. MSc Thesis. Toronto, ON: University of Toronto.

32. Nicolai, R. P. and Dekker, R. (2006). Optimal maintenance of multicomponent systems: A review. Economic Institute Report.

33. Lamberts, S. W. J. and Nicolai, R. P. (2008). *Maintenance Models for Systems Subject to Measurable Deterioration*. Rotterdam, the Netherlands: Rozenberg Publishers, University Dissertations.

34. Fisher, W. W. (1990). Markov process modelling of a maintenance system with spares, repair, cannibalization and manpower constraints. *Mathematical Computer Modelling* 13(7): 119–125.

35. Gurler, U. and Kaya, A. (2002). A maintenance policy for a system with multi-state components: An approximate solution. *Reliability Engineering and System Safety* 76: 117–127. doi:10.1016/S0951-8320(01)00125-9.

36. Block, H. W., Langberg, N. A., and Savits, T. H. (1993). Repair replacement policies. *Journal of Applied Probability* 30: 194–206. doi:10.2307/3214632.

37. Park, M. and Pham, H. (2016). Cost models for age replacement policies and block replacement policies under warranty. *Applied Mathematical Modelling* 40(9–10): 5689–5702. doi:10.1016/j.apm.2016.01.022.

38. Scarf, P. A., Dwight, R., and Al-Musrati, A. (2005). On reliability criteria and the implied cost of failure for a maintained component. *Reliability Engineering and System Safety* 89: 199–207. doi:10.1016/j.ress.2004.08.019.

39. Chowdhury, C. H. (1988). A systematic survey of the maintenance models. *Periodica Polytechnica. Mechanical Engineering* 32(3–4): 253–274.

40. Glasser, G. J. (1967). The age replacement problem. *Technometrics* 9(1): 83–91.

41. Rakoczy, A. and Żółtowski, J. (1977). About the issues on technical object renewal principles definition (in Polish). In: *Proceedings of Winter School on Reliability*, Szczyrk, Poland: pp. 175–191.

42. Yun, W. Y. (1989). An age replacement policy with increasing minimal repair cost. *Microelectronics Reliability* 29(2): 153–157.

43. Sheu, S.-H. (1991). A general age replacement model with minimal repair and general random repair cost. *Microelectronics Reliability* 31(5): 1009–1017.

44. Sheu, S.-H. and Liou, C.-T. (1992). An age replacement policy with minimal repair and general random repair cost. *Microelectronics Reliability* 32(9): 1283–1289.

45. Sheu, S.-H. (1993). A generalized model for determining optimal number of minimal repairs before replacement. *European Journal of Operational Research* 69: 38–49.

46. Lim, J. H., Qu, J., and Zuo, M. J. (2016). Age replacement policy based on imperfect repair with random probability. *Reliability Engineering and System Safety* 149: 24–33. doi:10.1016/j.ress.2015.10.020.

47. Mazzuchi, T. A. and Soyer, R. (1996). A Bayesian perspective on some replacement strategies. *Reliability Engineering and System Safety* 51: 295–303.

48. Cha, J. H. and Kim, J. J. (2002). On the existence of the steady state availability of imperfect repair model. *Sankhya: The Indian Journal of Statistics* 64, series B. Pt. 1: 76–81.

49. Dagpunar, J. S. (1994). Some necessary and sufficient conditions for age replacement with non-zero downtimes. *Journal of Operational Research Society* 45(2): 225–229.

50. Vaurio, J. K. (1999). Availability and cost functions for periodically inspected preventively maintained units. *Reliability Engineering and System Safety* 63: 133–140. doi:10.1016/S0951-8320(98)00030-1.

51. Nakagawa, T., Zhao, X., and Yun, W. Y. (2011). Optimal age replacement and inspection policies with random failure and replacement times. *International Journal of Reliability, Quality and Safety Engineering* 18(5): 405–416. doi:10.1142/S0218539311004159.

52. Zhao, X., Mizutani, S., and Nakagawa, T. (2015). Which is better for replacement policies with continuous or discrete scheduled times? *European Journal of Operational Research* 242: 477–486. doi:10.1016/j.ejor.2014.11.018.

53. Wu, S. and Clements-Croome, D. (2005). Preventive maintenance models with random maintenance quantity. *Reliability Engineering and System Safety* 90: 99–105.

54. Chien, Y.-H. (2008). A general age-replacement model with minimal repair under renewing free-replacement warranty. *European Journal of Operational Research* 186: 1046–1058. doi:10.1016/j.ejor.2007.02.030.

55. Dimitrov, B., Chukova, S., and Khalil, Z. (2004). Warranty costs: An age-dependent failure/repair model. *Naval Research Logistics* 51(7): 959–976. doi:10.1002/nav.20037.

56. Ito, K. and Nakagawa, T. (2011). Comparison of three cumulative damage models. *Quality Technology and Quantitative Management* 8(1): 57–66. doi:10.1080/16843703.2011.11673246.

57. Sepehrifar, M. B., Khorshidian, K., and Jamshidian, A. R. (2015). On renewal increasing mean residual life distributions: An age replacement model with hypothesis testing application. *Statistics and Probability Letters* 96: 117–122. doi:10.1016/j.spl.2014.09.009.

58. Sheu, S.-H. (1992). A general replacement of a system subject to shocks. *Microelectronics Reliability* 32(5): 657–662.

59. Sheu, S.-H., Griffith, W. S., and Nakagawa, T. (1995). Extended optimal replacement model with random repair cost. *European Journal of Operational Research* 85: 636–649.

60. Lai, M.-T. and Leu, B.-Y. (1996). An economic discrete replacement policy for a shock damage model with minimal repairs. *Microeconomics Reliability* 36(10): 1347–1355.

61. Qian, C., Nakamura, S., and Nakagawa, T. (2003). Replacement and minimal repair policies for a cumulative damage model with maintenance. *Computers and Mathematics with Applications* 46: 1111–1118.

62. Lam, C. T. and Yeh, R. H. (1994). Optimal replacement policies for multi-state deteriorating systems. *Naval Research Logistics* 41(3): 303–315.

63. Segawa, Y., Ohnishi, M., and Ibaraki, T. (1992). Optimal minimal-repair and replacement problem with age dependent cost structure. *Computers and Mathematics with Applications* 24(1/2): 91–101.

64. Kumar, D. and Westberg, U. (1997). Maintenance scheduling under age replacement policy using proportional hazards model and TTT-ploting. *European Journal of Operational Research* 99: 507–515.

65. Mahdavi, M. and Mahdavi, M. (2009). Optimization of age replacement policy using reliability based heuristic model. *Journal of Scientific and Industrial Research* 68: 668–673.

66. Zhao, X., Al-Khalifa, K. N., and Nakagawa, T. (2015). Approximate method for optimal replacement, maintenance, and inspection policies. *Reliability Engineering and System Safety* 144: 68–73. doi:10.1016/j.ress.2015.07.005.

67. Kayid, M., Izadkhah, S., and Alshami, S. (2016). Laplace transform ordering of time to failure in age replacement models. *Journal of the Korean Statistical Society* 45(1): 101–113.

68. Christer, A. H. (1986). Comments on finite-period applications of age-based replacement models. *IMA Journal of Mathematics in Management* 1: 111–124.

69. Kabir, A. B. M. Z. and Farrash, S. H. A. (1996). Simulation of an integrated age replacement and spare provisioning policy using SLAM. *Reliability Engineering and System Safety* 52: 129–138.

70. Wu, S. and Zuo, M. J. (2010). Linear and nonlinear preventive maintenance models. *IEEE Transactions on Reliability* 59(1): 242–249. doi:10.1109/TR.2010.2041972.

71. Yeh, R. H. (1997). State-age-dependent maintenance policies for deteriorating systems with Erlang sojourn time distributions. *Reliability Engineering and System Safety* 58: 55–60.

72. Crowell, J. I. and Sen, P. K. (1989). Estimation of optimal block replacement policies. Mimeo series/the Institute of Statistics, the Consolidated University of North Carolina, Department of Statistics, available at: stat.ncsu.edu.

73. Rakoczy, A. (1980). Simulation method for technical object's optimal preventive maintenance time assessment (in Polish). In: *Proceedings of Winter School on Reliability*, Szczyrk, Poland: pp. 143–152.

74. Sheu, S.-H. (1994). Extended block replacement policy with used item and general random minimal repair cost. *European Journal of Operational Research* 79(3): 405–416.

75. Sheu, S.-H. (1991). Periodic replacement with minimal repair at failure and general random repair cost for a multi-unit system. *Microelectronics Reliability* 31(5): 1019–1025.

76. Colosimo, E. A., Santos, W. B., Gilardoni, G. L., and Motta, S. B. (2006). Optimal maintenance time for repairable systems under two types of failures. In: Soares, C. G. and Zio, E. (eds.) *Safety and Reliability for Managing Risk: Proceedings of European Safety and Reliability Conference ESREL 2006, Estoril, Portugal*, September 18–22, 2006, Leiden, the Netherlands: Taylor & Francis Group.

77. Lai, M.-T. and Yuan, J. (1993). Cost-optimal periodical replacement policy for a system subjected to shock damage. *Microelectronics Reliability* 33(8): 1159–1168.

78. Sheu, S.-H. (1998). A generalized age and block replacement of a system subject to shocks. *European Journal of Operational Research* 108: 345–362.

79. Sheu, S.-H. and Griffith, W. S. (2002). Extended block replacement policy with shock models and used items. *European Journal of Operational Research* 140: 50–60. doi:10.1016/S0377-2217(01)00224-7.

80. Abdel-Hameed, M. (1986). Optimum replacement of a system subject to shocks. *Journal of Applied Probability* 23: 107–114.

81. Abdel-Hameed, M. (1995). Inspection, maintenance and replacement models. *Computers and Operations Research* 22(4): 435–441. doi:10.1016/0305-0548(94)00051-9.

82. Zhao, X., Qian, C., and Nakagawa, T. (2017). Comparisons of replacement policies with periodic times and repair numbers. *Reliability Engineering and System Safety* 168: 161–170. doi:10.1016/j.ress.2017.05.015.

83. Berthaut, F., Gharbi, A., and Dhouib, K. (2011). Joint modified block replacement and production/inventory control policy for a failure-prone manufacturing cell. *Omega* 39: 642–654. doi:10.1016/j.omega.2011.01.006.

84. Drobiszewski, J. and Smalko, Z. (2006). The equable maintenance strategy. *Journal of KONBiN* 2: 375–383.

85. Pilch, R., Smolnik, M., Szybka, J., and Wiązania, G. (2014). Concept of preventive maintenance strategy for a chosen example of public transport vehicles (in Polish). In: Siergiejczyk, M. (ed.) *Maintenance Problems of Technical Systems*, Warsaw, Poland: Publication House of Warsaw University of Science and Technology: pp. 171–182.

86. Kustroń, K. and Cieślak, Ł. (2012). The optimization of replacement time for non-repairable aircraft component. *Journal of KONBiN* 2(22): 45–58.

87. Pilch, R. (2017). Determination of preventive maintenance time for milling assemblies used in coal mills. *Journal of Machine Construction and Maintenance* 1(104): 81–86.

88. Nakagawa, T. (1986). Periodic and sequential preventive maintenance policies. *Journal of Applied Probability* 23: 536–542.

89. Nakagawa, T. and Mizutani, S. (2008). Periodic and sequential imperfect preventive maintenance policies for cumulative damage models. In: Pham, H. (ed.) *Recent Advances in Reliability and Quality in Design*, London, UK: Springer.

90. Sheu, S.-H., Chang, C. C., and Chen, Y.-L. (2012). An extended sequential imperfect preventive maintenance model with improvement factors. *Communications in Statistics: Theory and Methods* 41(7): 1269–1283. doi:10.1080/03610926.2010.542852.

91. Liu, Y., Li, Y., Huang, H.-Z., and Kuang, Y. (2011). An optimal sequential preventive maintenance policy under stochastic maintenance quality. *Structure and Infrastructure Engineering: Maintenance, Management, Life-Cycle Design and Performance* 7(4): 315–322.

92. Peng, W., Liu, Y., Zhang, X., and Huang, H.-Z. (2015). Sequential preventive maintenance policies with consideration of random adjustment-reduction features. *Eksploatacja i Niezawodnosc: Maintenance and Reliability* 17(2): 306–313.

93. Kim, H. S., Sub Kwon, Y., and Park, D. H. (2006). Bayesian method on sequential preventive maintenance problem. *The Korean Communications in Statistics* 13(1): 191–204.

94. Bergman, B. (1978). Optimal replacement under a general failure model. *Advances in Applied Probability* 10: 431–451.

95. Canfield, R. V. (1986). Cost optimization of periodic preventive maintenance. *IEEE Transactions on Reliability* R-35(1): 78–81. doi:10.1109/TR.1986.4335355.

96. Lie, C. H. and Chun, Y. H. (1986). An algorithm for preventive maintenance policy. *IEEE Transactions on Reliability* R-35(1): 71–75.

97. Jayabalan, V. and Chaudhuri, D. (1992). Cost optimization of maintenance scheduling for a system with assured reliability. *IEEE Transactions on Reliability* 41(1): 21–25. doi:10.1109/24.126665.

98. Tsai, Y.-T., Wang, K.-S., and Teng, H.-Y. (2001). Optimizing preventive maintenance for mechanical components using genetic algorithms. *Reliability Engineering and System Safety* 74: 89–97. doi:10.1016/S0951-8320(01)00065-5.

99. Chan, J.-K. and Shaw, L. (1993). Modeling repairable systems with failure rates that depend on age and maintenance. *IEEE Transactions on Reliability* 42(4): 566–571. doi:10.1109/24.273583.

100. Nakagawa, T. and Osaki, S. (1974). The optimum repair limit replacement policies. *Operational Research Quarterly* 25(2): 311–317.

101. Okumoto, K. and Osaki, S. (1976). Repair limit replacement policies with lead time. *Zeitschrift fur Operations Research* 20: 133–142.

102. Koshimae, H., Dohi, T., Kaio, N., and Osaki, S. (1996). Graphical/statistical approach to repair limit replacement policies. *Journal of the Operations Research* 39(2): 230–246.

103. Dohi, T., Kaio, N., and Osaki, S. (2000). A graphical method to repair-cost limit replacement policies with imperfect repair. *Mathematical and Computer Modelling* 31: 99–106. doi:10.1016/S0895-7177(00)00076-5.

104. Dohi, T., Ashioka, A., Kaio, N., and Osaki, S. (2006). Statistical estimation algorithms for repairs-time limit replacement scheduling under earning rate criteria. *Computers and Mathematics with Applications* 51: 345–356. doi:10.1016/j.camwa.2005.11.004.

105. Dohi, T., Ashioka, A., Kaio, N., and Osaki, S. (2003). The optimal repair-time limit replacement policy with imperfect repair: Lorenz transform approach. *Mathematical and Computer Modelling* 38: 1169–1176. doi:10.1016/S0895-7177(03)90117-8.

106. Dohi, T., Kaio, N., and Osaki, S. (2003). A new graphical method to estimate the optimal repair-time limit with incomplete repair and discounting. *Computers and Mathematics with Applications* 46: 999–1007. doi:10.1016/S0898-1221(03)90114-3.

107. Dohi, T., Matsushima, N., Kaio, N., and Osaki, S. (1996). Nonparametric repair-limit replacement policies with imperfect repair. *European Journal of Operational Research* 96: 260–273.

108. Beichelt, F. (1992). A general maintenance model and its application to repair limit replacement policies. *Microelectronics Reliability* 32(8): 1185–1196. doi:10.1016/0026-2714(92)90036-K.

109. Bai, D. S. and Yun, W. Y. (1986). An age replacement policy with minimal repair cost limit. *IEEE Transactions on Reliability* R-35(4): 452–454.

110. Yun, W. Y. and Bai, D. S. (1987). Cost limit replacement policy under imperfect repair. *Reliability Engineering* 19: 23–28.

111. Yun, W. Y. and Bai, D. S. (1988). Repair cost limit replacement policy under imperfect inspection. *Reliability Engineering and System Safety* 23: 59–64.

112. Dohi, T., Takeita, K., and Osaki, S. (2000). Graphical method for determining/estimating optimal repair-limit replacement policies. *International Journal of Reliability, Quality and Safety Engineering* 7(1): 43–60.

113. Lai, M.-T. (2014). Optimal replacement period with repair cost limit and cumulative damage model. *Eksploatacja i Niezawodnosc: Maintenance and Reliability* 16(2): 246–252.

114. Beichelt, F. (1999). A general approach to total repair cost limit replacement policies. *ORiON* 15(1/2): 67–75.

115. Chang, C.-C., Sheu, S.-H., and Chen, Y.-L. (2013). Optimal replacement model with age-dependent failure type based on a cumulative repair-cost limit policy. *Applied Mathematical Modelling* 37: 308–317. doi:10.1016/j.apm.2012.02.031.

116. Chang, C.-C., Sheu, S.-H., and Chen, Y.-L. (2013) Optimal number of minimal repairs before replacement based on a cumulative repair-cost limit policy. *Computers and Industrial Engineering* 59: 603–610. doi:10.1016/j.cie.2010.07.005.

117. Kapur, P. K. and Garg, R. B. (1989) Optimal number of minimal repairs before replacement with repair cost limit. *Reliability Engineering and System Safety* 26: 35–46.

118. Chien, Y.-H. and Sheu, S.-H. (2006). Extended optimal age-replacement policy with minimal repair of a system subject to shocks. *European Journal of Operational Research* 174: 169–181. doi:10.1016/j.ejor.2005.01.032.

119. Sheu, S.-H. (1999). Extended optimal replacement model for deteriorating systems. *European Journal of Operational Research* 112: 503–516.

120. Chang, C.-C. (2014). Optimum preventive maintenance policies for systems subject to random working times, replacement, and minimal repair. *Computers and Industrial Engineering* 67: 185–194. doi:10.1016/j.cie.2013.11.011.

121. Martorell, S., Sanchez, A., and Serradell, V. (1999). Age-dependent reliability model considering effects of maintenance and working conditions. *Reliability Engineering and System Safety* 64: 19–31.

122. Jiang, R. and Ji, P. (2002). Age replacement policy: A multi-attribute value model. *Reliability Engineering and System Safety* 76: 311–318. doi:10.1016/S0951-8320(02)00021-2.

123. Sheu, S.-H. and Chien, Y.-H. (2004). Optimal age-replacement policy of a system subject to shocks with random lead-time. *European Journal of Operational Research* 159: 132–144. doi:10.1016/S0377-2217(03)00409-0.

124. Legat, V., Zaludowa, A. H., Cervenka, V., and Jurca, V. (1996). Contribution to optimization of preventive replacement. *Reliability Engineering and System Safety* 51: 259–266.

125. Nakagawa, T. and Kowada, M. (1983). Analysis of a system with minimal repair and its application to replacement policy. *European Journal of Operational Research* 12(2): 176–182.

126. Park, D. H., Jung, G. M., and Yum, J. K. (2000). Cost minimization for periodic maintenance policy of a system subject to slow degradation. *Reliability Engineering and System Safety* 68(2): 105–112. doi:10.1016/S0951-8320(00)00012-0.

127. Sheu, S.-H., Chen, Y.-L., Chang, C. H.-C. H., and Zhang, Z. G. (2016). A note on a two variable block replacement policy for a system subject to non-homogeneous pure birth shocks. *Applied Mathematical Modelling* 40(5–6): 3703–3712. doi:10.1016/j.apm.2015.10.001.

128. Bukowski, L. (1980). Optimization of technical systems maintenance policy (case study of metallurgical production line) (in Polish). In: *Proceedings of Winter School on Reliability*. Katowice, Ploand: Centre for Technical Progress: pp. 47–62.

129. Zhao, Y. X. (2003). On preventive maintenance policy of a critical reliability level for system subject to degradation. *Reliability Engineering and System Safety* 79: 301–308. doi:10.1016/S0951-8320(02)00201-6.

130. Jiang, X., Cheng, K., and Makis, V. (1998). On the optimality of repair-cost-limit policies. *Journal of Applied Probability* 35: 936–949.

131. Segawa, Y. and Ohnishi, M. (2000). The average optimality of a repair-limit replacement policy. *Mathematical and Computer Modelling* 31: 327–334.

132. Murthy, D. N. P. and Nguyen, D. G. (1988). An optimal repair cost limit policy for servicing warranty. *Mathematical and Computer Modelling* 11: 595–599.

133. Frees, E. W. (1986). Optimizing costs on age replacement policies. *Stochastic Processes and their Applications* 21: 195–212.

134. Maillart, L. M. and Fang, X. (2006). Optimal maintenance policies for serial, multimachine systems with non-instantaneous repairs. *Naval Research Logistics* 53(8): 804–813.

135. Sheu, S.-H., Yeh, R. H., Lin, Y.-B., and Juang, M.-G. (1999). A Bayesian perspective on age replacement with minimal repair. *Reliability Engineering and System Safety* 65: 55–64.

136. Sheu, S.-H., Sung, C. H.-K., Hsu, T.-S., and Chen, Y.-C. H. (2013a). Age replacement policy for a two-unit system subject to non-homogeneous pure birth shocks. *Applied Mathematical Modelling* 37: 7027–7036. doi:10.1016/j.apm.2013.02.022.

137. Sheu, S.-H., Zhang, Z. G., Chien, Y.-H., and Huang, T.-H. (2013). Age replacement policy with lead-time for a system subject to non-homogeneous pure birth shocks. *Applied Mathematical Modelling* 37: 7717–7725. doi:10.1016/j.apm.2013.03.017.

138. Dekker, R. and Dijkstra, M. C. (1992) Opportunity-based age replacement: Exponentially distributed times between opportunities. *Naval Research Logistics* 39: 175–190.

139. Iskandar, B. P. and Sandoh, H. (2000). An extended opportunity-based age replacement policy. *RAIRO Operations Research* 34: 145–154.

140. Jhang, J. P. and Sheu, S. H. (1999). Opportunity-based age replacement policy with minimal repair. *Reliability Engineering and System Safety* 64: 339–344.

141. Satow, T. and Osaki, S. (2003). Opportunity-based age replacement with different intensity rates. *Mathematical and Computer Modelling* 38: 1419–1426. doi:10.1016/S0895-7177(03)90145-2.

142. Leung, F. K. N., Zhang, Y. L., and Lai, K. K. (2011). Analysis for a two-dissimilar-component cold standby repairable system with repair priority. *Reliability Engineering and System Safety* 96: 1542–1551. doi:10.1016/j.ress.2011.06.004.

143. Armstrong, M. J. (2002). Age repair policies for the machine repair problem. *European Journal of Operational Research* 138: 127–141. doi:10.1016/S0377-2217(01)00135-7.

144. Van Dijkhuizen, G. C. and Van Harten, A. (1998). Two-stage generalized age maintenance of a queue-like production system. *European Journal of Operational Research* 108: 363–378.

145. Scarf, P. A. and Deara, M. (2003). Block replacement policies for a two-component system with failure dependence. *Naval Research Logistics* 50: 70–87. doi:10.1002/nav.10051.

146. Yusuf, I. and Ali, U. A. (2012). Structural dependence replacement model for parallel system of two units. *Journal of Basic and Applied Science* 20(4): 324–326.

147. Lai, M.-T. and Yuan, J. (1991). Periodic replacement model for a parallel system subject to independent and common cause shock failures. *Reliability Engineering and System Safety* 31(3): 355–367.

148. Yasui, K., Nakagawa, T., and Osaki, S. (1988). A summary of optimum replacement policies for a parallel redundant system. *Microelectronic Reliability* 28(4): 635–641.

149. Jodejko, A. (2008). Maintenance problems of technical systems composed of heterogeneous elements. In: *Proceedings of Summer Safety and Reliability Seminars*, June 22–28, 2008, Gdańsk-Sopot, Poland: pp. 187–194.

150. Sheu, S.-H., Lin, Y.-B., and Liao, G.-L. (2006). Optimum policies for a system with general imperfect maintenance. *Reliability Engineering and System Safety* 91(3): 362–369. doi:10.1016/j.ress.2005.01.015.

151. Sheu, S.-H. (1990). Periodic replacement when minimal repair costs depend on the age and the number of minimal repairs for a multi-unit system. *Microelectronics Reliability* 30(4): 713–718.

152. Zequeira, R. I. and Berenguer, C. (2005). A block replacement policy for a periodically inspected two-unit parallel standby safety system. In: Kołowrocki, K. (ed.) *Advances in Safety and Reliability: Proceedings of the European Safety and Reliability Conference (ESREL 2005), Gdynia-Sopot-Gdańsk, Poland*, June 27–30, 2005, Leiden, the Netherlands: A. A. Balkema: pp. 2091–2098.

153. Park, J. H., Lee, S. C., Hong, J. W., and Lie, C. H. (2009). An optimal Block preventive maintenance policy for a multi-unit system considering imperfect maintenance. *Asia-Pacific Journal of Operational Research* 26(6): 831–847. doi:10.1142/S021759590900250X.

154. Grigoriev, A., Van De Klundert, J., and Spieksma, F. C. R. (2006). Modeling and solving the periodic maintenance problem. *European Journal of Operational Research* 172: 783–797. doi:10.1016/j.ejor.2004.11.013.

155. Ke, H. and Yao, K. (2016). Block replacement policy with uncertain lifetimes. *Reliability Engineering and System Safety* 148: 119–124. doi:10.1016/j.ress.2015.12.008.

156. Wells, C. H. E. (2014). Reliability analysis of a single warm-standby system subject to repairable and non-repairable failures. *European Journal of Operational Research* 235: 180–186. doi:10.1016/j.ejor.2013.12.027.

157. Scarf, P. A. and Cavalcante, C. A. V. (2010). Hybrid block replacement and inspection policies for a multi-component system with heterogeneous component lives. *European Journal of Operational Research* 206: 384–394. doi:10.1016/j.ejor.2010.02.024.

158. Anisimov, V. V. (2005). Asymptotic analysis of stochastic block replacement policies for multi-component systems in a Markov environment. *Operations Research Letters* 33: 26–34. doi:10.1016/j.orl.2004.03.009.

159. Caldeira, D. J., Taborda, C. J., and Trigo, T. P. (2012). An optimal preventive maintenance policy of parallel-series systems. *Journal of Polish Safety and Reliability Association Summer Safety and Reliability Seminars* 3(1): 29–34.

160. Duarte, A. C., Craveiro Taborda, J. C., Craveiro, A., and Trigo, T. P. (2005). Optimization of the preventive maintenance plan of a series components system. In: Kołowrocki, K. (ed.) *Advances in Safety and Reliability: Proceedings of the European Safety and Reliability Conference (ESREL 2005), Gdynia-Sopot-Gdańsk, Poland*, June 27–30, 2005, Leiden, the Netherlands: A.A. Balkema.

161. Chelbi, A., Ait-Kadi, D., and Aloui, H. (2007). Availability optimization for multi-component systems subjected to periodic replacement. In: Aven, T. and Vinnem, J. M. (eds.) *Risk, Reliability and Societal Safety: Proceedings of European Safety and Reliability Conference ESREL 2007, Stavanger, Norway*, June 25–27, 2007, Leiden, the Netherlands: Taylor & Francis Group.

162. Okulewicz, J. and Salamonowicz, T. (2008). Preventive maintenance with imperfect repairs of a system with redundant objects. In: *Proceedings of Summer Safety and Reliability Seminars SSARS 2008*, June 22–28, 2008, Gdańsk-Sopot, Poland: pp. 279–286.

163. Do Van, P., Barros, A., Berenguer, C. H., and Bouvard, K. (2013). Dynamic grouping maintenance with time limited opportunities. *Reliability Engineering and System Safety* 120: 51–59. doi:10.1016/j.ress.2013.03.016.

164. Okoh, P. (2015). Maintenance grouping optimization for the management of risk in offshore riser system. *Process Safety and Environmental Protection* 98: 33–39. doi:10.1016/j.psep.2015.06.007.

165. Zhang, T., Cheng, Z., Liu, Y.-J., and Guo, B. (2012). Maintenance scheduling for multiunit system: A stochastic Petri-net and genetic algorithm based approach. *Eksploatacja i Niezawodność: Maintenance and Reliability* 14(3): 256–264.

166. Xiao, L., Song, S., Chen, X., and Coit, D. W. (2016). Joint optimization of production scheduling and machine group preventive maintenance. *Reliability Engineering and System Safety* 146: 68–78. doi:10.1016/j.ress.2015.10.013.

167. Sandve, K. and Aven, T. (1999). Cost optimal replacement of monotone, repairable systems. *European Journal of Operational Research* 116: 235–248.

168. Zequeira, R. I. and Berenguer, C. (2004). Maintenance cost analysis of a two-component parallel system with failure interaction. In: *Proceedings of Reliability and Maintainability, 2004 Annual Symposium*: RAMS, 26-29 Jan. 2004, IEEE, pp. 220–225. doi:10.1109/RAMS.2004.1285451.

169. Sheu, S.-H. and Jhang, J.-P. (1996). A generalized group maintenance policy. *European Journal of Operational Research* 96: 232–247.

170. Bai, Y., Jia, X., and Cheng, Z. (2011) Group optimization models for multi-component system compound maintenance tasks. *Eksploatacja i Niezawodnosc: Maintenance and Reliability* 1: 42–47.

171. Haurie, A. and L'ecuyer, P. L. (1982). A stochastic control approach to group preventive replacement in a multicomponent system. *IEEE Transactions on Automatic Control*, AC-27 2: 387–393.

172. Lai, M.-T. and Chen, Y.-C. H. (2006). Optimal periodic replacement policy for a two-unit system with failure rate interaction. *International Journal of Advanced Manufacturing Technology* 29: 367–371.

173. Shafiee, M. and Finkelstein, M. (2015). An optimal age-based group maintenance policy for multi-unit degrading systems. *Reliability Engineering and System Safety* 134: 230–238. doi:10.1016/j.ress.2014.09.016.

174. Popova, E. and Wilson, J. G. (1999). Group replacement policies for parallel systems whose components have phase distributed failure times. *Annals of Operations Research* 91: 163–189.

175. Ritchken, P. and Wilson, J. G. (1990). (m, T) group maintenance policies. *Management Science* 36(5): 632–639.

176. Popova, E. (2004), Basic optimality results for Bayesian group replacement policies. *Operations Research Letters* 32: 283–287.

177. Sheu, S.-H., Yeh, R. H., Lin, Y.-B., and Juang, M.-G. (2001). A Bayesian approach to an adaptive preventive maintenance model. *Reliability Engineering and System Safety* 71: 33–44. doi:10.1016/S0951-8320(00)00072-7.

178. Dekker, R. and Roelvink, I. F. K. (1995). Marginal cost criteria for preventive replacement of a group of components. *European Journal of Operational Research* 84: 467–480.

179. Wildeman, R. E., Dekker, R., and Smit, A. C. J. M. (1997). A dynamic policy for grouping maintenance activities. *European Journal of Operational Research* 99: 530–551.

180. Do, P., Vu, H. C., Barros, A., and Berrenguer, C. H. (2015). Maintenance grouping for multi-component systems with availability constraints and limited maintenance teams. *Reliability Engineering and System Safety* 142: 56–67. doi:10.1016/j.ress.2015.04.022.

181. Vu, H. C., Do, P., Barros, A., and Berenguer, C. H. (2014). Maintenance grouping strategy for multi-component systems with dynamic contexts. *Reliability Engineering and System Safety* 132: 233–249. doi:10.1016/j.ress.2014.08.002.

182. Zhang, X. and Zeng, J. (2015) A general modelling method for opportunistic maintenance modelling of multi-unit systems. *Reliability Engineering and System Safety* 140: 176–190. doi:10.1016/j.ress.2015.03.030.

183. Zequeira, R. I., Valdes, J. E., and Berenguer, C. (2008). Optimal buffer inventory and opportunistic preventive maintenance under random production capacity availability. *International Journal of Production Economics* 111: 686–696. doi:10.1016/j.ijpe.2007.02.037.

184. Laggoune, R., Chateauneuf, A., and Aissani, D. (2009). Opportunistic policy for optimal preventive maintenance of a multi-component system in continuous operating units. *Computers and Chemical Engineering* 33: 1499–1510.

185. Hou, W. and Jiang, Z. (2013). An opportunistic maintenance policy of multi-unit series production system with consideration of imperfect maintenance. *Applied Mathematics and Information Sciences* 7(1L): 283–290.

186. Shafiee, M., Finkelstein, M., and Berenguer, C. H. (2015). An opportunistic condition-based maintenance policy for offshore wind turbine blades subjected to degradation and environmental shocks. *Reliability Engineering and System Safety* 142: 463–471. doi:10.1016/j.ress.2015.05.001.

187. Xia, T., Jin, X., Xi, L., and Ni, J. (2015). Production-driven opportunistic maintenance for batch production based on MAM-APB scheduling. *European Journal of Operational Research* 240: 781–790. doi:10.1016/j.ejor.2014.08.004.

188. Cavalcante, C. A. V. and Lopes, R. S. (2015). Multi-criteria model to support the definition of opportunistic maintenance policy: A study in a cogeneration system. *Energy* 80: 32–80.

189. Bedford, T., Dewan, I., Meilijson, I., and Zitrou, A. (2011). The signal model: A model for competing risks of opportunistic maintenance. *European Journal of Operational Research* 214: 665–673. doi:10.1016/j.ejor.2011.05.016.

190. Hu, J. and Zhang, L. (2014). Risk based opportunistic maintenance model for complex mechanical systems. *Expert Systems with Applications* 41(6): 3105–3115. doi:10.1016/j.eswa.2013.10.041.

191. Zhou, X., Xi, L., and Lee, J. (2006). A dynamic opportunistic maintenance policy for continuously monitored systems. *Journal of Quality in Maintenance Engineering* 12(3): 294–305. doi:10.1108/13552510610685129.

192. Shi, H. and Zeng, J. (2016). Real-time prediction of remaining useful life and preventive opportunistic maintenance strategy for multi-component systems considering stochastic dependence. *Computers and Industrial Engineering* 93: 192–204. doi:10.1016/j.cie.2015.12.016.

193. Zhou, X., Lu, Z.-Q., Xi, L.-F., and Lee, J. (2010). Opportunistic preventive maintenance optimization for multi-unit series systems with combing multi-preventive maintenance techniques. *Journal of Shanghai Jiaotong University* 15(5): 513–518.

194. Gustavsson, E., Patriksson, M., Stromberg, A.-B., Wojciechowski, A., and Onnheim, M. (2014). Preventive maintenance scheduling of multi-component systems with interval costs. *Computers and Industrial Engineering* 76: 390–400. doi:10.1016/j.cie.2014.02.009.

195. Haque, S. A., Zohrul Kabir, A. B. M., and Sarker, R. A. (2003). Optimization model for opportunistic replacement policy using genetic algorithm with fuzzy logic controller. *Proceedings of the Congress on Evolutionary Computation* 4: 2837–2843.

196. Samhouri, M. S., Al-Ghandoor, A., Fouad, R. H., and Alhaj Ali, S. M. (2009). An intelligent opportunistic maintenance (OM) system: A genetic algorithm approach. *Jordan Journal of Mechanical and Industrial Engineering* 3(4): 246–251.

197. Kececioglu, D. and Sun, F.-B. (1995). A general discrete-time dynamic programming model for the opportunistic replacement policy and its application to ball-bearing systems. *Reliability Engineering and System Safety* 47: 175–185.

198. Iung, B., Levrat, E., and Thomas, E. (2007). Odds algorithm-based opportunistic maintenance task execution for preserving product conditions. *Annals of the CIRP* 56/1: 13–16.

199. Derigent, W., Thomas, E., Levrat, E., and Iung, B. (2009). Opportunistic maintenance based on fuzzy modelling of component proximity. *CIRP Annals – Manufacturing Technology* 58: 29–32.

200. Assid, M., Gharbi, A., and Hajji, A. (2015). Production planning and opportunistic preventive maintenance for unreliable one-machine two-products manufacturing systems. *IFAC-PapersOnLine* 48–43: 478–483. doi:10.1016/j.ifacol.2015.06.127.

201. Hu, J., Zhang, L., and Liang, W. (2012). Opportunistic predictive maintenance for complex multi-component systems based on DBN-HAZOP model. *Process Safety and Environmental Protection* 90: 376–386.

202. Bedford, T. and Alkabi, B. M. (2009). Modelling competing risks and opportunistic maintenance with expert judgement. In: Martorell, S., Guedes Soares, C. and Barnett, J. *Safety, Reliability and Risk Analysis: Theory, Methods and Applications: Proceedings of European Safety and Reliability Conference ESREL 2008, Valencia, Spain,* September 22–25, 2008, Leiden, the Netherlands: Taylor & Francis Group: pp. 515–521.

203. Radner, R. and Jorgenson, D. W. (1963). Opportunistic replacement of a single part in the presence of several monitored parts. *Management Science* 10(1): 70–84.

204. Epstain, S. and Wilamowsky, Y. (1985). Opportunistic replacement in a deterministic environment. *Computers and Operations Research* 12(3): 311–322.

205. Van Der Duyn Schouten, D. A., and Vanneste, S. G. (1990). Analysis and computation of (n, N)-strategies for maintenance of a two-component system. *European Journal of Operational Research* 48: 260–274.

206. Ding, S.-H. and Kamaruddin, S. (2012). Selection of optimal maintenance policy by using fuzzy multi criteria decision making method. In: *Proceedings of the 2012 International Conference on Industrial Engineering and Operations Management,* July 3–6, 2012, Istanbul, Turkey: pp. 435–443.

207. Sarker, B. R. and Ibn Faiz, T. (2016). Minimizing maintenance cost for offshore wind turbines following multi-level opportunistic preventive strategy. *Renewable Energy* 85: 104–113. doi:10.1016/j.renene.2015.06.030.

208. Laggoune, R., Chateauneuf, A., and Aissani, D. (2010). Impact of few failure data on the opportunistic replacement policy for multi-component systems. *Reliability Engineering and System Safety* 95: 108–119. doi:10.1016/j.ress.2009.08.007.

209. Gunn, E. A. and Diallo, C. (2015). Optimal opportunistic indirect grouping of preventive replacements in multicomponent systems. *Computers and Industrial Engineering* 90: 281–291. doi:10.1016/j.cie.2015.09.013.

210. Zhou, X., Huang, K., Xi, L., and Lee, J. (2015). Preventive maintenance modeling for multi-component systems with considering stochastic failures and disassembly sequence. *Reliability Engineering and System Safety* 142: 231–237. doi:10.1016/j.ress.2015.05.005.

211. Hopp, W. J. and Kuo, Y.-L. (1998). Heuristics for multicomponent joint replacement: Applications to aircraft engine maintenance. *Naval Research Logistics* 45: 435–458.

212. Fard, N. and Zheng, X. (1991). An approximate method for non-repairable systems based on opportunistic replacement policy. *Reliability Engineering and System Safety* 33: 277–288.

213. Zheng, X. and Fard, N. (1991). A maintenance policy for repairable systems based on opportunistic failure-rate tolerance. *IEEE Transactions on Reliability* 40(2): 237–244.

214. Pham, H. and Wang, H. (1999). Optimal (τ,T) opportunistic maintenance of a k-out-of-n:G system with imperfect PM and partial failure. *Naval Research Logistics* 47: 223–239.

215. Cui, L. and Li, H. (2006). Opportunistic maintenance for multi-component shock models. *Mathematical Methods of Operations Research* 63(3): 493–511. doi:10.1007/s00186-005-0058-9.

216. Tambe, P. P. and Kularni, M. S. (2013). An opportunistic maintenance decision of a multi-component system considering the effect of failures on quality. In: *Proceedings of the World Congress on Engineering 2013*, Vol. 1, July 3–5, 2013, London, UK: WCE 2013: pp. 1–6.

217. Tambe, P. P., Mohite, S., and Kularni, M. S. (2013). Optimisation of opportunistic maintenance of a multi-component system considering the effect of failures on quality and production schedule: A case study. *International Journal of Advanced Manufacturing Technology* 69(5): 1743–1756.

218. Huynh, T. K., Barros, A., and Berenguer, C.H. (2013). A reliability-based opportunistic predictive maintenance model for k-out-of-n deteriorating systems. *Chemical Engineering Transactions* 33: 493–498.

219. Cheng, Z., Yang, Z., Tan, L., and Guo, B. (2011). Optimal inspection and maintenance policy for the multi-unit series system. In: *Proceedings of 9th International Conference on Reliability, Maintainability and Safety (ICRMS) 2011*, June 12–15, 2011, Guiyang, China: pp. 811–814.

220. Cheng, Z., Yang, Z., and Guo, B. (2013). Optimal opportunistic maintenance model of multi-unit systems. *Journal of Systems Engineering and Electronics* 24(5): 811–817. doi:10.1109/JSEE.2013.00094.

221. Taghipour, S. and Banjevic, D. (2012). Optimal inspection of a complex system subject to periodic and opportunistic inspections and preventive replacements. *European Journal of Operational Research* 220: 649–660. doi:10.1016/j.ejor.2012.02.002.

222. Ormon, S. W. and Cassady, C. R. (2004). Cannibalization policies for a set of parallel machines. In: *Reliability and Maintainability, 2004 Annual Symposium: RAMS*, January 26–29, 2004, Colorado Springs, CO: pp. 540–545.

223. Nowakowski, T. and Plewa, M. (2009). Cannibalization: Technical system maintenance method (in Polish). In: *Proceedings of XXXVII Winter School on Reliability*, Warsaw, Poland: Szczyrk, Publication House of Warsaw University of Technology: pp. 230–238.

224. Sherbrooke, C. C. (2004). *Optimal Modeling Inventory of Systems. Multi-echelon Techniques*. Boston, MA: Kluwer Academic Publishers.

225. Lv, X.-Z., Fan, B.-X., Gu, Y., and Zhao, X.-H. (2013), Selective maintenance model considering cannibalization and its solving algorithm. In: *Proceedings of 2013 International conference on Quality, Reliability, Risk, Maintenance, and Safety Engineering (WR2MSE)*, IEEE: pp. 717–723.

226. Simon, R. M. (1970). Cannibalization policies for multicomponent systems. *SIAM Journal on Applied Mathematics* 19(4): 700–711.

227. Baxter, L. A. (1988). On the theory of cannibalization. *Journal of Mathematical Analysis and Applications* 136: 290–297. doi:10.1016/0022-247X(88)90131-X.

228. Khalifa, D., Hottenstein, M., and Aggarwal, S. (1977). Technical note: Cannibalization policies for multistate systems. *Operations Research* 25(6): 1032–1039.

229. Byrkett, D. L. (1985). Units of equipment available using cannibalization for repair-part support. *IEEE Transactions on Reliability* R-34(1): 25–28.

230. Jodejko-Pietruczuk, A. and Plewa, M. (2012). The model of reverse logistics, based on reliability theory with elements' rejuvenation. *Logistics and Transport* 2(15): 27–35.

231. Salman, S., Cassady, C. R., Pohl, E. A., and Ormon, S. W. (2007). Evaluating the impact of cannibalization on fleet performance. *Quality and Reliability Engineering International* 23: 445–457. doi:10.1002/qre.826.

232. Sherbrooke, C. C. (1971). An evaluator for the number of operationally ready aircraft in a multilevel supply system. *Operations Research* 19(3): 618–635.

233. Shah, J. and Avittathur, B. (2007). The retailer multi-item inventory problem with demand cannibalization and substitution. *International Journal of Production Economics* 106: 104–114. doi:10.1016/j.ijpe.2006.04.004.

234. Gaver, D. P., Isaacson, K. E., and Abell, J. B. (1993). Estimating aircraft recoverable spares requirements with cannibalization of designated items. Santa Monica, CA: RAND Corporation. https://www.rand.org/pubs/reports/R4213.html.

235. Hoover, J., Jondrow, J. M., Trost, R. S., and Ye, M. (2002). A model to study: Cannibalization, FMC, and customer waiting time. Alexandria, VA: CNA.

236. Albright, T. L., Geber, C. A., and Juras, P. (2014). How naval aviation uses the Balanced Scorecard. *Strategic Finance* 10: 21–28.

237. Meenu, G. (2011). Identification of factors affecting product cannibalization in Indian automobile sector. *IJCEM International Journal of Computational Engineering and Management* 12: 2230–7893.

238. Curtin, N. P. (2001). *Military Aircraft: Cannibalizations Adversely Affect Personnel and Maintenance.* Washington, DC: US General Accounting Office.

239. Cheng, Y.-H. and Tsao, H.-L. (2010). Rolling stock maintenance strategy selection, spares parts' estimation, and replacements' interval calculation. *International Journal of Production Economics* 128: 404–412. doi:10.1016/j.ijpe.2010.07.038.

240. Garg, J. (2013). *Maintenance: Spare Parts Optimization.* M2 Research Intern Theses, Ecole Centrale de Paris, Capgemini Consulting.

241. Ondemir, O. and Gupta, S. M. (2014). A multi-criteria decision making model for advanced repair-to-order and disassembly-to-order system. *European Journal of Operational Research* 233: 408–419. doi:10.1016/j.ejor.2013.09.003.

242. Silver, E. A. and Fiechter, C.-N. (1995). Preventive maintenance with limited historical data. *European Journal of Operational Research* 82: 125–144.

243. Nguyen, K.-A., Do, P., and Grall, A. (2015). Multi-level predictive maintenance for multi-component systems. *Reliability Engineering and System Safety* 144: 83–94. doi:10.1016/j.ress.2015.07.017.

244. Predictive maintenance 4.0. Predict the unpredictable. PWC, Mainnovation, Pricewaterhouse Coopers B.V. 2017.

245. Predictive maintenance and the smart factory. Deloitte Development LLC. 2017.

2 Inspection Maintenance Modeling for Technical Systems
An Overview

Sylwia Werbińska-Wojciechowska

CONTENTS

2.1 INTRODUCTION

All equipment breaks down from time to time, requiring materials, tradespeople to repair it, and causing some negative consequences, such as loss in production or transportation delays. To reduce the number of these breakdowns, planned maintenance actions are implemented. One of the most familiar planned maintenance actions is inspection.

Currently, inspection and inspection policy development have an important role in various technical systems, thus they attract a lot of attention in the literature. In many situations there are no apparent systems indicating the forthcoming failure. In such systems with non-self-announcing failures (also called unrevealed faults or latent faults), the typical preventive maintenance policies cannot be used [1]. In maintenance of such systems the inspection actions performance is introduced. Examples of these systems include protective devices, emergency devices, and standby units (see [1,2]).

The main purpose of an inspection is to determine the state of equipment based on the chosen indicators, such as bearing wear, gauge readings, and quality of a product [3]. Following this, the main definition of inspection can be derived.

According to EN 13306:2018 standard [4], inspection is defined as "examination for conformity by measuring, observing, or testing the relevant characteristics of an item." The authors [5] extend this definition, providing that inspection is defined as "measuring, examining, testing, and gauging one or more characteristics of a product or service and comparing the results with specified requirements to determine whether conformity is achieved for each characteristic."

The main benefits obtained from inspection performance include detection and correction of minor defects before major breakdown occurs. Consequently, the inspection maintenance optimization is strictly connected with system's deterioration processes, which are generally stochastic. Thus, the condition of a system is revealed only by its inspection. In other words, inspection models usually assume that the state of the system is completely unknown unless an inspection is performed. Following this, the knowledge about the true status of an inspected system gives the possibility to take appropriate maintenance actions. However, execution of frequent inspections incurs substantial cost. Conversely, infrequent inspections result in a higher cost for system downtime because of longer intervals between performance of these maintenance actions. Following this, to determine an inspection policy, the correct balance between the number of inspections and the resulting output according to the defined optimization criteria (e.g., maximization of profit, minimization of downtime, and maximization of availability) must be sought.

Moreover, inspection schemes may be periodic and non-periodic (sequential) [6]. In this chapter, the focus is on periodic inspection maintenance modeling issues. More information about non-periodic inspection maintenance modeling may be found in [1,7].

Early inspection maintenance models were developed in 1959 by R.E. Barlow and L.C. Hunter in their work *Mathematical models for system reliability* (according to [8]). A standard decision problem includes answering for the question: *An unde-tected failure causes an economic loss which increases in time, whereas inspections are costly too. What is the most cost-efficient way to schedule inspections in time?* Many extensions and modifications of the standard inspection model have been developed and investigated. They have been surveyed in the last five decades.

One of the first research works that surveys inspection models is [9], where the authors focus on the inspection and replacement problems of single and multi-unit systems. The summary of optimal scheduling of replacement and inspection of sto-chastically failing equipment is developed in [10]. Later, in [11] the authors review the research studies that appeared between 1965 and 1976. In this work, the authors present the discrete time maintenance models in which a unit (or units) is monitored and a decision is made to repair, replace, and/or restock the unit(s). In [3], the author gives a state-of-the-art review of the literature related to optimal inspection model-ing of failing systems. The surveyed research papers were published in the 1960s and 1970s. In 1989, the authors in [12] present a survey on the research published after [11]. In this work, the authors focus on single-unit systems (one-unit and com-plex systems), providing a section on inspection models. The authors indicate the main differences between developed models are time horizon, available information, the nature of cost functions, models objective, and system's constraints. The focus on multi-unit systems inspection problems is given in [13]. In [14], the authors present the literature review on inspection maintenance models. The authors focus on the inspection models with different types of inspection information (perfect or not) and different costs of inspections (costly or costless inspection information). The same year, the author in [15] reviews recent developments in the methodology for solving inspection problems. The author focuses on the most important issues that need further development (e.g., fallible tests performance).

In 2002, the authors in work [16] review classical maintenance models including inspection strategies. They focus on the models developed in the 1960s and 1970s that are based on the general inspection policy discussed by R. E. Barlow and F. Proschan in *Mathematical Theory of Reliability*. The author also investigates the standard inspection policies in [17].

Later, in 2012 the authors in [8] review the main inspection models for systems. They present the two main maintenance models—an inspection without replacement and an inspection with replacement. The first group of inspection models includes solutions for three situations: lifetime distribution is known, lifetime distribution is partially known, and lifetime distribution is unknown.

In the second group of maintenance models, the assumption of inspection-replacement process is introduced. The next year, the authors in [18] present the three classes of inspection problems: (1) inspection frequencies for equipment that is in continuous operation and subject to breakdown, (2) inspection intervals for equipment used only in emergency conditions, and (3) condition monitoring of equipment.

The recent literature review on inspection maintenance also is provided in [19], where the author focuses on inspection maintenance for single-unit and multi-unit systems.

Moreover, some recent research works are dedicated to comparing the problems with various maintenance policies. The main comparisons between optimum and nearly optimum inspection policies are given in [20,21], where authors refer to the models developed by R. E. Barlow and F. Proschan as standard optimal policies. In [22], the three sub-optimal inspection polices are proposed and compared—periodic policy, mean residual life policy, and constant hazard policy. The review and comparison of known classical optimum-checking policies is given in [23]. Comparisons for inspection and repair policies are analyzed in [24–26].

In summary, based on the developed literature reviews, the existing inspection models can be classified many ways. One classification is given in [15], where the author defines five main groups of optimal inspection models: imperfect inspection models, inspection with replacement policies, inspection policies with delayed symptoms of failure, inspection models for stand-by systems, and Bayesian models. More general classifications divide existing maintenance models into the inspection models for two-states systems and multi-states systems ([27]), or inspection models for single- and multi-unit systems ([28,29]). According to [1], inspection models are classified considering the type of maintained systems: protective devices (safety systems), or standby units, and operating devices.

In this chapter, classification proposed divides the known models into four main groups of inspection strategies: single-unit systems, multi-unit systems, hybrid inspection models, and models dedicated to solving other maintenance problems (e.g., case studies). Thus, the main scheme for classification of inspection models for technical systems is given in Figure 2.1.

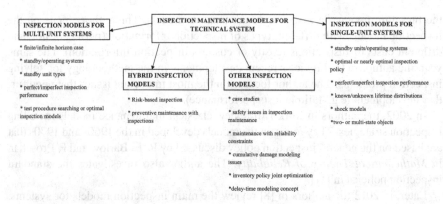

FIGURE 2.1 Inspection maintenance models for technical systems – the main classification. (Own contribution based on Tang, T., Failure finding interval optimization for periodically inspected repairable systems, PhD Thesis, University of Toronto, 2012; Beichelt, F., *Nav. Res. Logist. Q.*, 28, 375–381, 1981; Cazorla, D.M. and R. Perez-Ocon, *Eur. J. Oper. Res.*, 190, 494–508, 2008; Boland, P.J. and E. El-Neweihi, *Comput. Oper. Res.*, 22, 383–390, 1995.)

2.2   INSPECTION MAINTENANCE MODELING FOR SINGLE-UNIT SYSTEMS

In this section, the author investigates a one-unit stochastically failing or deteriorating system in which only actual inspection can detect a system's failure. Following Figure 2.1, inspection models for two-state, single-unit systems are investigated first.

2.2.1   INSPECTION MAINTENANCE FOR TWO-STATE SYSTEMS

The first inspection model formulated by R. E. Barlow and F. Proschan [7] is called a pure inspection model for a system and is characterized by the following assumptions:

- Two-stated system's condition (functioning and failed state)
- The system's condition is known only by inspections
- Inspections are perfect in the sense that a failure will be identified at inspection
- Inspections do not degrade or rejuvenate the system
- System cannot fail or age during inspection performance
- Inspection actions take negligible time

For the given assumptions, the expected total cost is obtained according to the formula:

$$C(T_{in}) = \sum_{n=0}^{\infty} \int_{t_{in}^n}^{t_{in}^{n+1}} \left[c_{in1}(n+1) + c_{in2}(t_{in}^{n+1} - x) \right] dF(x) \qquad (2.1)$$

where:

$C(T_{in})$ Long-run expected cost per unit time
c_{in1} Cost of first inspection action performance
c_{in2} Cost of second (and subsequent) inspection action performance
$F(x)$ Probability distribution function of system/unit lifetime

The main extensions of this pure inspection model of a system applies to perfect/ imperfect inspection process performance, assuming known/unknown system lifetime distribution, cost/reliability optimization criteria use, or shock modeling implementation.

One of the first extensions of the given pure inspection model applies to finite horizon case implementation. In [30], the author analyzes a model that is based on the selection of the best maintenance strategy for the object's reliability state. In [31], the author analyzes the problem of determining an optimum checking schedule over the finite horizon with cost considerations.

In [32,33], a heuristic approach for determining the optimal inspection interval is investigated. The authors in [33] assume that the optimal interval between inspections depends on a likelihood of malfunction, a cost of inspection, and a cost of treatment. The developed model is examined later to analyze the relation of subjects' judgments to the model description. Later, in [32], the author focuses on the development of a mathematical model for determining a periodic inspection schedule in a preventive maintenance program for a single machine.

The second, and very often investigated, extension of the basic inspection model includes the situation when no or only partial information on a lifetime distribution of a system is available. One of the first works that investigates this issue is given in [34]. The author in this work considers that the system lifetime distribution is unknown. To find the optimal inspection policy parameters, the author uses the minimax inspection strategies with respect to cost criterions. This model later is extended in [35] and [36].

Another interesting problem applies to the imperfect inspection performance analysis. For example, in [37] the authors develop an imperfect inspection policy for systems subject to a multiple correlated degradation process. In [38], the author presents a problem of finding the optimum inspection procedure for a system, whose time to failure is exponentially distributed. The problem is considered as a continuous-time Markovian decision process with two states (before and after failure) and provides a basis for the extended model given in [35].

A work worth noting is [39], where the authors introduce an optimal inspection policy that is based on implementation of a failure detection zone. The idea is like a delayed time approach (see [19]) or a Fault Trees with Time Dependencies modeling approach (see [40]). In this model, if inspection is conducted in a pre-specified time zone, a failure will be noticed before it occurs. Otherwise, the failure will remain undetected. The analytical algorithm for searching for the optimal inspection interval is given considering cost and availability criteria.

Another interesting problem is presented in [41], where the authors propose a model in which the ith test increases a remaining failure rate without changing the form of the conditional lifetime distribution. The solution algorithms for finding the best testing times are developed for two cases of uniform and exponential failure time distributions.

The problem of determination of an optimal inspection policy when inspections may be harmful to a maintained unit is continued also in [42]. The author in this work develops a hazardous-inspection model where every performed test may impair the tested unit. The proposed model is developed based on a Markov decision process implementation and the emphasis is put on maximization of the expected lifetime of the inspected unit. A non-Markovian case is analyzed in [43]. The author in this work develops two inspection policies: one-test and two-test. The two-stage inspection procedure is dedicated to expensive devices and is based on performing a fallible test first and an error-free test whenever the first test reports a failure. The models are based on the assumptions of arbitrary failure distributions, general optimality conditions, and algorithms for reduction of the infinite horizon optimization to two dimensions. This inspection problem is continued later in [44].

The problem of imperfect inspections with the implementation of multiple post repair inspections and accidents during inspection is analyzed in [45]. The authors in this model propose an inspection policy for single- and two-unit systems, where a repairman is called immediately to repair a failed unit. The analytical solutions are provided for various measures of reliability such as mean time to system failure, steady-state availability, busy period of repairman for repair, and inspection per unit time by using semi-Markov processes and regenerative point techniques.

Another interesting model is given in [46]. The author in this work considers the problem of the optimal choice of periodic inspection intervals for a renewable equipment without preventive replacement performance. The model is based on two optimization criteria: minimization of maintenance costs and maximization of system availability. The author develops an approximate method for inspection interval calculations and proves that the obtained solutions are very close to the exact ones.

The extended inspection models with imperfect testing also are investigated in [47–50,51]. The continuation of inspection modeling with availability constraints, given in [51], is presented in [52]. The authors in this work analyze the instantaneous availability of a system maintained under periodic inspection with the use of random walk models. Two cases are analyzed: deterministic and stochastic .

Some summary and extensions of the models presented in [52] are given also in [53]. In this work, the authors focus on periodic inspection, developing five basic models with availability requirements. All the inspection models are based on different approaches to the determination of inspection times. In a later work [54], the authors also extend the inspection models given in [51]. The main extension is based on the assumption that periodic inspections take place at fixed time points after repair or replacement in case of failure. The implementation of minimal repairs before replacement or perfect repair is analyzed in [55]. The authors in this work propose a minimal repair model with periodic inspection and constant repair time. The instantaneous availability of the proposed model is derived by a set of recursive formulas, providing the introduction to optimization of system reliability characteristics.

Recently, in [56] the authors focus on the availability of a system under periodic inspection with perfect repair/replacement and non-negligible downtime due to repair/replacement for a detected failure and due to inspection. The model is an extension of the works given in [51,54,57]. The authors in this work analyze a calendar-based inspection policy and an age-based inspection policy.

The last group of inspection policies for two-stated, single-unit systems applies to implementation of shock models. One of the first works focused implementation of random shocks modeling for systems with non-self-announcing failures is given in [85]. The authors in this work consider a periodic inspection model for a system with randomly occurring shocks that follows a Poisson process and cumulatively damages the system. This model is investigated and extended later in [59,60]. The new inspection policy considers random shock magnitudes and times between shock arrivals and focuses on optimization of availability criterion.

Another extension of the model presented in [58] is given in [61]. The authors in this work incorporate a more general deterioration process that includes both shock degradation and graceful degradation (continuous accumulation of damage). With the use of regenerative arguments and considering a constant rate of graceful degradation occurrence, an expression for the limiting average availability is derived.

The maintenance models for systems with two failure modes—type I failure relative to non-maintainable failure mode, and type II failure relative to periodically maintainable failure mode—are developed in [62–65].

In 2006, a model with three types of inspections is introduced in [66]. In this article, the authors assume that a system can fail because of three competing failure types: I, II, and III. Partial inspections detect type I failures without error. Failures of type II can be detected by imperfect inspections. Type III failures are detectable only by perfect inspections. If the system is found to have failed in an inspection, a perfect repair is made.

The summary of the main known models published in the recent literature is presented in Table 2.1. The author considers a few main criteria for summarizing this review:

- The problem category (the main model characteristic that distinguishes it)
- Planning horizon (investigating infinite or finite case)
- Assumption about the quality of performed inspections in a maintained system
- Type of introduced failure modes (for shock modeling)
- Used optimality criterion (cost or reliability constraints)
- Modeling method that is used in order to optimize the inspection policy
- Model's reference with the year of its publication

2.2.2 INSPECTION MAINTENANCE FOR MULTI-STATE SYSTEMS

In some systems, such as critical infrastructure where the safety issues are very important, reliability analysis carried out in relation to two-state technical objects usually is insufficient (see [19] for a review). The solution to this problem is to consider a technical object in terms of a minimum of three reliability states, where a third state is the state of partial failure.

The known inspection models for multi-state deteriorating single unit systems may be classified to the two main groups: models for systems with perfect/imperfect inspection and models for systems subjected to shocks. Following are the main directions of research done in these model groups.

One of the first developed inspection models for multi-state units is given in [79]. In this work, the author presents a Markovian model, which is focused on proper scheduling of inspections and preventive repairs considering minimization of the total expected cost per time unit. The main assumptions include performance of periodic inspections, implementation of perfect repair and inspection actions, and random holding times of systems.

TABLE 2.1

Summary of Inspection Policies for Two-State, Single-Unit Systems

Problem Category	Planning Horizon	Quality of Performed Inspections	Failure Modes	Optimization Criterion	Modeling Method/ Checking Procedures	Type of References	Publication Years
Original algorithm	Infinite	Perfect	n/a	Expected cost per unit of time	Analytical	[67]	1980
Original algorithm	Infinite	Perfect	n/a	Expected cost per unit of time	Analytical/optimal	[68]	1984
Original algorithm	Infinite	Perfect	n/a	Expected profit per unit of time	Analytical/heuristic approach	[32]	1996
Original algorithm	Infinite	Perfect	n/a	Expected cost function	Heuristic approach	[33]	1992
Original algorithm	Infinite	Perfect	n/a	Expected total cost	Analytical	[69]	2005
Original algorithm	Finite	Perfect	n/a	Expected costs of loss	Discrete dynamic programming	[30]	1980
One-parameter optimization model	Infinite	Perfect	n/a	Average total cost per time unit	Analytical	[60]	1998
Model with unknown or partially unknown system lifetime probability[a]	Infinite	Perfect	n/a	Expected loss cost per time unit	Analytical	[34]	1981
Model with known or unknown slp[a]	Infinite	Perfect	n/a	Total expected cost	Analytical	[36]	2006
Model with known slp[a]	Infinite	Perfect/imperfect	n/a	Total expected cost	Analytical	[35]	2001
Model with unknown slp[a]	Infinite	Imperfect	n/a	Cost per unit of time	Analytical	[70]	2002
Model with known slp[a]	Infinite	Imperfect	n/a	Long-run expected cost per unit time/ availability function	Renewal reward process/non linear programming	[71]	1995
Model with known slp[a]	Infinite	Imperfect	n/a	Long-run expected cost per unit time	Renewal reward process	[72]	2003

(Continued)

TABLE 2.1 (*Continued*)

Summary of Inspection Policies for Two-State, Single-Unit Systems

Problem Category	Planning Horizon	Quality of Performed Inspections	Failure Modes	Optimization Criterion	Modeling Method/ Checking Procedures	Type of References	Publication Years
Model with known slp[a]	Infinite	Imperfect	n/a	Long-run expected cost per unit time	Renewal theory, Wiener process	[37]	2016
Model with known slp[a]	Infinite	Imperfect	n/a	Total cost over a lifetime	Continuous-time Markovian decision process	[38]	1982
Model with known slp[a]	Infinite	Imperfect	n/a	Expected cost per time unit	Markovian model	[73]	1998
Model with known slp[a]	Infinite	Fallible/ error-free tests	n/a	Long-run cost per unit time	Dynamic programming	[43]	1993
Model with known slp[a]	Infinite	Fallible tests	n/a	Long-run cost per unit time	Analytical	[44]	1993
Model with known slp[a]	Infinite	Fallible tests	n/a	Mean loss per unit time	Analytical	[41]	1979
Model with known slp[a]	Infinite	Fallible tests	n/a	Expected lifetime of the unit	Markov decision process	[42]	1979
Model with known slp[a]	Infinite/ finite	Failure detection zone	n/a	Long-run cost per unit time	Analytical	[39]	2015
Model with known slp[a]	Finite	Imperfect	n/a	Expected sum of discounted cost	Markov decision process + quasi-Bayes approach + dynamic programming	[74]	2008

(*Continued*)

TABLE 2.1 (*Continued*)
Summary of Inspection Policies for Two-State, Single-Unit Systems

Problem Category	Planning Horizon	Quality of Performed Inspections	Failure Modes	Optimization Criterion	Modeling Method/ Checking Procedures	Type of References	Publication Years
Optimization model	Infinite	Perfect	n/a	Limiting average availability and long-run inspection rate	Analytical	[72]	2000
Optimization model	Infinite	Perfect	n/a	Limiting average availability	Analytical	[51]	2000
Optimization model	Infinite	Perfect	n/a	Long-run average cost per unit time	Analytical	[75]	2012
Optimization model	Infinite	Perfect	n/a	Average availability and the long-run average cost rate	Analytical	[76]	2014
Optimization model	Infinite	Imperfect	n/a	Expected operational readiness of a system	Analytical	[77]	1963
Optimization model	Infinite	Imperfect	n/a	System stationary availability	Analytical	[78]	2008
Optimization model	Infinite	Imperfect	n/a	Measures of system reliability	Semi-Markov process + regenerative point technique	[45]	2005
Optimization model	Infinite	Imperfect	n/a	Stationary availability coefficient and total expected cost per one renewal period	Analytical	[46]	2009
Optimization model	Infinite	Imperfect	n/a	Limiting average availability and the long-run average cost per unit time	Analytical	[49]	2012

(Continued)

TABLE 2.1 (Continued)
Summary of Inspection Policies for Two-State, Single-Unit Systems

Problem Category	Planning Horizon	Quality of Performed Inspections	Failure Modes	Optimization Criterion	Modeling Method/ Checking Procedures	Type of References	Publication Years
Optimization model	Finite/ infinite	Perfect	n/a	Limiting average availability, long-run inspection rate, instantaneous availability, instantaneous inspection rate	Analytical	[53]	2004
Optimization model	Finite/ infinite	Perfect	n/a	Limiting average availability, instantaneous availability	Analytical	[54]	2005
Optimization model	Finite/ infinite	Perfect	n/a	Limiting average availability, instantaneous availability	Analytical	[56]	2013
Optimization model	Finite	Perfect	n/a	Instantaneous availability	Analytical (random walk model)	[52]	2001
Optimization model	Finite	Perfect	n/a	Instantaneous availability	Analytical	[55]	2013
Optimization model	Finite	Imperfect	n/a	Long-run average cost per unit time or cost-rate over the time to retirement	Analytical	[48]	2013
Shock model	Infinite	Perfect	Random shocks arriving according to a Poisson process	Time-stationary availability	Analytical (renewal process)	[58–60]	1994, 1998, 2000
Shock model	Infinite	Perfect	Random shocks (a Poisson process) and graceful degradation	Limiting average availability	Analytical (renewal process)	[61]	2002

(Continued)

TABLE 2.1 (Continued)
Summary of Inspection Policies for Two-State, Single-Unit Systems

Problem Category	Planning Horizon	Quality of Performed Inspections	Failure Modes	Optimization Criterion	Modeling Method/ Checking Procedures	Type of References	Publication Years
Shock model	Infinite	Perfect	Two dependable failure modes: maintainable and non-maintainable	Expected maintenance cost per unit time	Analytical (renewal process)	[64]	2006
Shock model	Infinite	Partial, perfect, and imperfect	Three competing failure modes: I, II, III	Cost rate function	Analytical (renewal process)	[66]	2006
Shock model	Infinite	Perfect	Two failure modes: minor failure and catastrophic failures	Expected cost per unit of time	Analytical (renewal process)	[62,63]	2006
Shock model	Infinite	Perfect	Two failure modes: minor failure and catastrophic failures	Expected net cost rate	Analytical (renewal process)	[65]	2015

[a] slp – information about system lifetime probability

Another implementation of Markovian modeling in multi-state, single-unit systems maintenance problems are given in [80]. The authors in this work use non-homogeneous Markovian techniques to model systems with tolerable down times.

The issues of partially observable process are examined also in [81]. The author in this paper presents a model of a system that deteriorates according to a discrete-time Markov processes and its operation and repair costs increase with system deterioration state number. He proposes a monotonic four-region policy with cost considerations, where the decision process adopts a countable state space and a finite action space. The continuation of this problem is given in [82], where the authors propose a semi-Markov decision algorithm operating on the class of control-limit rules. This problem is extended later in [83], where the authors allow for delayed replacement performance and investigate the discounted cost structure.

The semi-Markov processes are applied in [84]. The author in this work develops a maintenance model for systems with five states that constitute all possible cycles, which begin with inspections. The solution is based on reliability characteristics assessment (asymptotic availability, reliability function).

Moreover, the maintenance inspection issues of production multi-state systems and processes are analyzed in [85–88].

The second investigated problem regards to shock modeling. One of the first works that considers inspection policies for multi-state, single-unit systems with shock modeling is given in [89]. The given model is extended later in work [90], where the author determines an optimal inspection policy for a system with deterioration process assumed to be an increasing pure jump Markov process. Later, in work [91] the authors develop an optimal inspection-replacement policy for an item subject to cumulative damage. In this model, a unit fails depending on the accumulated damage caused by gradual damage. The authors calculate the optimal damage limit according to the long-run expected cost rate criterion using the renewal reward theory.

The problem of imperfect inspections and imperfect repairs is investigated in [92]. A model considers a system submitted to external and internal failures whose deterioration level is known by means of inspections. Moreover, the authors assume the performance of two types of repairs—minimal and perfect—depending on the deterioration level and following a different phase-type distribution. The solutions are based on implementation of a generalized Markov process and the use of a phase-type renewal process as a special case.

Another extension of [89] is given in [93], where the authors propose a state-dependent maintenance policy for a multi-state continuous-time Markovian deteriorating system subject to aging and fatal shocks. The model incorporates the assumptions of state-dependent cost structure, imperfect repair, and perfect inspections, and is based on implementation of periodic inspections.

The availability of periodically inspected systems subjected to shocks is analyzed in [94]. In this model, the authors analyze a system whose deterioration process is modulated by a continuous-time Markov chain and additional damage is induced by a Poisson shock process.

The summary of the main known models published in the recent literature is presented in Table 2.2. The author applies the same classification criteria as in Section 2.2.1.

TABLE 2.2
Summary of Inspection Policies for Multi-state, Single-Unit Systems

Problem Category	Planning Horizon	Quality of Performed Inspections	Failure Modes	Optimization Criterion	Modeling Method/ Checking Procedures	Type of References	Publication Years
Optimization model	Infinite	Perfect	n/a	Discounted and average cost	Discrete-time Markov process	[81]	1976
Optimization model	Infinite	Perfect	n/a	Discounted and average cost	Markov decision process	[88]	1978
Optimization model	Infinite	Perfect	n/a	Total expected cost per time unit	Markovian model	[79]	1976
Optimization model	Infinite	Perfect	n/a	Long-run expected average cost per unit time	Markov renewal theory	[85]	1997
Optimization model	Infinite	Perfect	n/a	Expected long-run discounted cost	Semi-Markov decision process	[83]	1992
Optimization model	Infinite	Imperfect	n/a	Long-run expected cost per unit time	Analytical	[76]	2014
Optimization model	Infinite	Imperfect	n/a	Expected total discounted cost	Discrete-time Markov chain	[86]	1986
Optimization model	Infinite	Imperfect	n/a	Reliability function	Semi-Markov processes	[77]	1962
Inspection with CBM modeling	Infinite	Imperfect	n/a	Operational reliability	Analytical	[50]	2013
Optimization model	Finite	Perfect	n/a	Average cost	Semi-Markov decision model	[82]	1984

(Continued)

TABLE 2.2 (Continued)
Summary of Inspection Policies for Multi-state, Single-Unit Systems

Problem Category	Planning Horizon	Quality of Performed Inspections	Failure Modes	Optimization Criterion	Modeling Method/ Checking Procedures	Type of References	Publication Years
Shock model	Infinite	Perfect	Cumulative damage attributed to shocks occurrence (Poisson process)	Long-run average cost per unit time	Analytical (renewal reward theorem)	[89]	1980
Shock model	Infinite	Perfect	Deterioration level assumed as increasing pure jump Markov process	Long-run average cost per unit time	Markov process/ control-limit policy	[90]	1987
Shock model	Infinite	Perfect	Cumulative damage caused by gradual damage	Expected long-run cost rate	Analytical (renewal reward theorem)	[91]	1997
Shock model	Infinite	Perfect	Poisson shock process	Limiting average availability	Continuous-time Markov chain	[94]	2006
Shock model	Infinite	Perfect	Fatal shocks occurrence	Expected long-run cost rate	Continuous-time Markov process	[93]	2001
Shock model	Infinite	Perfect/imperfect	Internal and external failures occurrence	Total costs per unit time	Generalized Markov process	[92]	2008

2.3   INSPECTION MAINTENANCE MODELING
FOR MULTI-UNIT SYSTEMS

The general classification of the main investigated inspection policies for multi-component systems considers the type of hidden failures. According to [39], there are two types of hidden failures:

- *Type I*: protective devices or standby unit. The function of these devices is to protect the main system in case of failures.
- *Type II*: operating devices. They are operating systems, and their failure will cause direct loss.

At the beginning models are investigated for protective devices and standby units.

2.3.1   INSPECTION MAINTENANCE FOR STANDBY SYSTEMS

The standby units are characteristic for many engineering systems. Spare components, or systems, that are not in continuous operation are the examples of this sort of unit [129]. The main function of the spare unit is to replace the component in use when the latter fails so that the system is restored to operating condition as soon as possible. However, the standby units also deteriorate and fail with its failures remaining undiscovered until the next attempt to use them, unless some test or inspection is carried out (unrevealed failures).

Many inspection models dedicated to the inspection of standby systems were developed in the 1970s and 1980s. For example, a two-unit repairable system is analyzed in [95]. In this work, the first unit is operative and the other is in cold standby. The author in this work considers two types of failure situations: (1) a failure of an active element is detected instantaneously but a failure of a standby unit is revealed at inspection epochs only and (2) a failure of both the active and the standby units is revealed at the time of an inspection only. The extension of this model is presented also in [96], where the authors discuss a two-unit cold standby redundant system with repair, inspection, and preventive maintenance. The model is based on the assumption of arbitrary distributions of failure time, inspection time, repair, and preventive repair times.

The reliability analysis of a two-unit cold standby system with the consideration of single repair facility performance is given in [97]. In this work [97], the authors assume that a single repair facility facilitates inspection, replacement, preparation, and repair. Moreover, failure, delivery, replacement, and inspection times have exponential distributions, whereas all other time distributions are general.

A similar problem is analyzed in [98], where the authors investigate a two-unit warm standby system with minor (internal) and major (external) repair. Another extension of these works applies to the analysis of two non-identical units. Using the regenerative point technique, various pointwise and steady-state reliability characteristics of system effectiveness are obtained.

Later, a warm standby *n*-system with operational and repair times following phase-type distributions is considered in [99]. The analyzed system is governed by a level-dependent quasi-birth-and-death process and the general Markov model is provided. The main reliability characteristics that are calculated include availability and rate of occurrence of failures.

Another extension of the inspection model developed in [97] is given in [100]. In this work, the authors consider a reliability model for a two-unit cold standby system with a single server. In the work, various reliability measures of system effectiveness are obtained by using a semi-Markov process and a regenerative point technique. Later, this model is extended in [101], where the authors investigate two non-identical units, where the first unit goes for repair, inspection, and post repair (when needed), whereas the second unit is as good as new after repair. The priority in operation is given to the first unit (lower running costs), while the priority in repair is given to the second unit (less time consuming). The model also is based on various calculations of reliability characteristics with the use of regenerative point technique and Monte Carlo simulation.

Moreover, the extension of [100] is given in [102]. The authors in this work study two dissimilar (automatic and manual) cold standby systems. An inspection policy is introduced for an automatic machine to detect this kind of a failure. The model solution is based on the estimation of various measures of reliability and profit incurred to the system using a semi-Markov process and a regenerative point technique.

The problem of time-dependent unavailability of periodically tested aging components under various testing and repair policies is analyzed in [103,104].

The investigation of maintenance for multi-component systems, which may be either in operating condition or in the standby mode is presented in [70]. The authors in this work define an inspection policy along with a preventive maintenance (PM) procedure and imperfect testing for a series system. The cost optimization is performed based on the renewal theory use.

The shock model implementation is considered in [105]. The authors in this work consider a parallel redundant system consisting of n components. Considering the assumption that the arrival rate of shocks and the failure probabilities of components may depend on an external Markovian environment, the authors propose several state-dependent maintenance policies based on system availability and cost functions.

The components failure interaction is considered in [106]. The authors in this work investigate a two-component cold standby system under periodic inspections. They assume that a failure of one component can modify the failure probability of a component still operating with a constant probability and obtain the system reliability function for the case of staggered inspections. The failure interaction scheme is like the shock model used in studies of common cause failures (known as a β-Factor model).

The continuation of research studies about testing policies for two-unit parallel standby systems without identical components is presented in [107]. The authors in this work propose an optimal testing policy for a system under the criteria of availability and maintenance costs. The analytical solution is provided in the context of recognition of common cause failure.

Moreover, the comparison of various inspection models for redundant systems is given in [108]. In this work, the authors provide the comparison of four models of two- and three-component systems using discrete Markov chains. The first model applies to active redundancy without component repair, the second model includes active redundancy with component repair, the third and fourth models analyze standby redundancy without and with component repair.

2.3.2 INSPECTION MAINTENANCE FOR OPERATING SYSTEMS

Inspection models for multi-unit operating systems include two main groups of research works: test procedure searching models and optimal inspection models. The first group of models is focused on the development of the best maintenance scheduling order, answering the question: *In what order the components should be tested to satisfy the time requirements?* The second group of inspection models focuses on optimal maintenance policy searching considering cost and/or reliability criteria.

One of the first research works on optimum test procedure models is given in [109]. The author in this work focuses on searching for test procedures that maximize the probability of locating a failed component within the given time. The solution is provided using renewal theory and dynamic programming. Later, the authors in [110] study the problem of scheduling activities of several types under time constraints. The developed model is focused on finding an optimal schedule that specifies the periods to execute each of the activity types to minimize the long-run average cost per period. The discrete time maintenance problem of n machines is solved for finite and infinite time horizon cases.

The implementation of an imperfect inspection case into a maintenance management model is presented in [111]. The authors in this work analyze a two-stage inspection process that considers detection and sizing activities. The purpose of this study is to develop a method that simulates deterioration, inspection, repair, and failure of structures over time using Markov matrices.

Another inspection model that includes an imperfect inspection problem is given in [112]. The authors present a model for determining optimal inspection plans for critical multi-characteristic components. The inspection is performed in stages by inspectors who may make mistakes—errors of false acceptance and false rejection occurrence possibility. This problem is continued later in [113] and the extension of this model is given in [114]. The model is focused on finding the optimal number of inspections necessary to minimize the total cost per accepted component.

The issues of imperfect inspections performance are analyzed in [115,116]. In [116], the authors investigate an imperfect inspection model focused on processes of testing and estimation of model parameters. The probability of failure detection is a constant variable and the solution is based on a Markov chain and use of simulation modeling. In [115], the authors develop a maintenance policy for pipelines subjected to corrosion, including predictive degradation modeling, time-dependent reliability assessment, inspection uncertainty, and expected cost optimization. The solution is obtained with the use of Bayesian modeling. The influence of the type I and type II inspection errors on maintenance costs is investigated in [117].

The second group of models applies to the problem of optimization of inspection policy parameters. In this area, one of the preliminary models is given in [118]. The author in this work develops an optimal inspection and replacement model for a coherent system with components having exponential life-time distributions. The solution is based on the implementation of a semi-Markov decision process framework.

One of the extensions of this model is presented in [119], where the author develops an optimal inspection strategy under two optimality criteria: the long-run average net income and the total expected discounted net income. The author considers a multi-unit machine in a series-reliability structure, if along the inspection process only one unit can be tested. This problem later is investigated in [120], where the author gives an example to demonstrate that the previously presented characterization of the optimal inspection policy for series systems is not correct in the discounted case.

Another extension of the optimal inspection model given in [118] applies to the investigation of reliability characteristics. For example, in [121] the author presents an analytical method that gives upper and lower bounds for the reliability in a case of systems subject to inspections at Poisson random times. This model later is extended in [122] by providing the exact expression of the reliability function, its Laplace transform, and the Mean Time To Failure (MTTF) of the system.

Later, perfect and minimal repair policies in a reliability model are considered in [123]. The author in this work considers two-unit systems with stochastic dependence and two types of failures (soft and hard failures), providing analytical reliability and cost models. The practical application is based on the optimization of steam turbine system maintenance.

The issues of structural reliability are considered in [124], where the authors analyze the optimal time interval for inspection and maintenance of offshore structures. The structural reliability is expressed here by means of closed-form mathematical formulas that are incorporated into the cost-benefit analysis.

Moreover, in the literature inspection maintenance policies for multi-state systems can be found. For example, in [125] the authors focus on a periodic inspection maintenance model for a system with several multi-state components over a finite time horizon. The degradation process of the components is modeled by the non-homogeneous continuous-time Markov chain, and the particle swarm optimization is used to optimize the maintenance threshold and inspection intervals under cost constraints. Later, in [126] an optimization model of an inspection-based PM policy is developed for three-state mechanical components subject to competing failure modes, which integrates continuous degradation and discrete shock effects. Periodic inspection of series systems with revealed and unrevealed failures is considered in [127]. This model extends the one given in [118] by introducing the probability of failure revealing. The simple maintenance model for n independent components in series is based on renewal theory.

Series-parallel systems are considered in [128]. The authors propose a general preventive maintenance model used to optimize the maintenance cost. The model is developed using a simulation approach and a parallel simulation algorithm for availability analysis. A special ratio-criterion is based on a Birnbaum importance factor. The optimization is performed using a genetic algorithm technique.

The summary of the main known models published in the recent literature is presented in Table 2.3. The author considers the same classification criteria as in the previous sections.

TABLE 2.3

Summary of Inspection Policies for Multi-unit Systems

System Type	Stand by Unit Type	Planning Horizon	Quality of Performed Inspections	Optimization Criterion	Modeling Method/ Checking Procedures	Type of References	Publication Years
Standby system	Cold standby	Infinite	Perfect	Main unreliability characteristics	Analytical (regenerative point technique)	[104]	1997
Standby system	Cold standby	Infinite	Perfect	Reliability function, MTTF	Analytical (renewal theory)	[95]	1970
Standby system	Cold standby	Infinite	Perfect	Expected loss due to system unavailability per time unit, the average system unavailability per cycle	Analytical (renewal theory)	[107]	2012
Standby system	Cold standby	Infinite	Perfect	Main reliability characteristics, the expected total profit per unit of time	Semi-Markov process and regenerative point technique	[102]	2016
Standby system	Cold standby	Infinite	Perfect	Main reliability characteristics, the profit function	Semi-Markov process and regenerative point technique	[100]	2011
Standby system	Cold standby	Infinite	Perfect	Main reliability characteristics, the expected total profit per unit of time	Regenerative point technique, MC simulation, Bayesian setup	[101]	2012

(Continued)

TABLE 2.3 (*Continued*)
Summary of Inspection Policies for Multi-unit Systems

System Type	Stand by Unit Type	Planning Horizon	Quality of Performed Inspections	Optimization Criterion	Modeling Method/ Checking Procedures	Type of References	Publication Years
Standby system	Warm standby	Infinite	Perfect	Main reliability characteristics, the total cost of a system per unit of time	Generalized Markov process	[99]	2008
Standby system	Warm standby	Infinite	Perfect/ imperfect	Total cost per unit of time	Analytical (renewal theory)	[129]	2002
Standby system	Cold/warm standby	Infinite	Perfect	Limiting average availability, the expected cost rate	Analytical (renewal theory), Markov jump process	[105]	2009
Standby system	Cold standby	Finite/infinite	Perfect	Main reliability characteristics	Analytical (regenerative point technique)	[97]	1995
Standby system	Warm standby	Finite/infinite	Perfect	Main reliability characteristics, the expected total profit in (0,t] and per unit of time	Analytical (regenerative point technique)	[98]	1995
Standby system	Cold standby	Finite/infinite	Perfect	Main unreliability characteristics	Analytical (regenerative point technique)	[103]	1999
Standby system	Cold standby	Finite	Perfect	Distribution function of time to the first system down and the mean time to the first system down	Analytical (renewal theory)	[96]	1970

(Continued)

TABLE 2.3 (Continued)
Summary of Inspection Policies for Multi-unit Systems

System Type	Stand by Unit Type	Planning Horizon	Quality of Performed Inspections	Optimization Criterion	Modeling Method/ Checking Procedures	Type of References	Publication Years
Standby system	Warm standby	Finite	Perfect	Average unavailability in inspection interval	Analytical	[106]	2005
Operating system	n/a	Infinite	Perfect	Long-run expected cost per unit time	Semi-Markov decision framework	[118]	1987
Operating system	n/a	Infinite	Perfect	Long-run average net income and total expected discounted net income	Renewal theory	[119]	1989
Operating system	n/a	Infinite	Perfect	Expected cost of operation per unit of time	Renewal theory	[126]	2016
Operating system	n/a	Infinite	Perfect	Average total cost of maintenance for unit of time	Renewal theory	[127]	2009
Operating system	n/a	Infinite	Perfect	Total expected discounted net income	Analytical	[120]	1991
Operating system	n/a	Finite/infinite	Perfect	Long-run average cost per period	Analytical	[110]	1998
Operating system	n/a	Finite	Perfect	Probability that the failed component is checked out before given time period	Renewal theory and dynamic programming	[109]	1964
Operating system	n/a	Finite	Imperfect	Total cost of inspection	Analytical	[114]	2008

(Continued)

TABLE 2.3 (Continued)
Summary of Inspection Policies for Multi-unit Systems

System Type	Stand by Unit Type	Planning Horizon	Quality of Performed Inspections	Optimization Criterion	Modeling Method/ Checking Procedures	Type of References	Publication Years
Operating system	n/a	Finite	Imperfect	Expected annual total cost	Markov model and Event based decision theory	[111]	2010
Operating system	n/a	Finite	Imperfect	Expected total cost per accepted component	Analytical and Bayes theorem	[112,113]	1995, 2002
Operating system	n/a	Finite	Perfect	Expected total cost	Analytical	[24]	1995
Operating system	n/a	Finite	Perfect	Failure distribution parameters	Nonlinear programming	[130]	2014
Operating system	n/a	Finite	Perfect	Total inspection cost	Particle swarm optimization algorithm	[131]	2012
Operating system	n/a	Finite	Perfect	Availability function	Analytical	[132]	2011
Operating system	n/a	Finite	Perfect	Sum of inspection, repair and risk cost	Simulation modeling	[133]	1999
Operating system	n/a	Finite	Perfect	Reliability characteristics	Renewal theory	[121]	1999
Operating system	n/a	Finite	Perfect	Reliability function, MTTF	Analytical	[122]	2002
Operating system	n/a	Finite	Perfect	Expected cost incurred in the inspection for each cycle	Analytical	[123]	2016

(Continued)

TABLE 2.3 (Continued)
Summary of Inspection Policies for Multi-unit Systems

System Type	Stand by Unit Type	Planning Horizon	Quality of Performed Inspections	Optimization Criterion	Modeling Method/ Checking Procedures	Type of References	Publication Years
Operating system	n/a	Finite	Perfect	System availability function, inspection cost	GA and MC simulation	[128]	2003
Operating system	n/a	Finite	Perfect	Maintenance cost rate in a renewal cycle	Non-homogeneous continuous Markov chain	[125]	2015
Operating system	n/a	Finite	Perfect/imperfect	Total expected social cost	Markov decision process and quasi-Bayes approach	[74]	2008
Operating system	n/a	Finite	Imperfect	Expected total cost function	Analytical (cost-benefit analysis)	[124]	2014
Operating system	n/a	Finite	Imperfect	Expected cost incurred in a cycle	Analytical (and Bayes theory)	[115]	2013
Operating system	n/a	Finite	Imperfect	Probability functions	Analytical and three-state Markov chain	[116]	1993

2.4 HYBRID INSPECTION MODELS

In the investigation of hybrid inspection models, two main groups of models can be defined:

- Risk-based inspection models (RBI)
- Inspection models with preventive maintenance policy implementation

The first group of models focuses on "designing and optimization of an inspection scheme based on the performance of a risk assessment progress using historical database, analytical methods, experience and engineering judgment" [134]. In this approach, risk assessment is used as a valuable tool to assign priorities among inspection and maintenance activities by analyzing the likelihood of failure and its consequences [135,136]. This approach is predominantly used in the oil and gas industries (see [134,136–139]), but some implementations also may be found for marine systems (see [135]), nuclear power plants (see [140–142]), or railway systems (see [143]). A basic overview on RBI is given in [6].

The second group of the maintenance models is based on different types of problem investigations. For example, in the literature maintenance models can be found that are based on the implementation of maintenance-free operating periods in the development of inspection policy (see [73]). The maintenance model as a mixture of a standard age replacement policy (ARP) and a maintenance procedure for unrevealed failures is given in [70]. The maintenance policy for a unit as inspected and maintained preventively at periodic intervals is given in [144]. The author in this work develops two maintenance models as an extension of the well-known ARP and an inspection model with constant checking time.

The introduction of an inspection-repair-replacement (IRR) policy is given in [71,72]. In these works, the authors assume that a system is inspected at preassigned times to distinguish between the up and down states. If the system is identified as being in the down state during the inspection, then a repair action (perfect repair according to [71] or minimal repair (according to [72]) will be taken. Moreover, periodic preventive replacement is performed. The focus is to determine an optimal IRR policy so that the availability of the system is high enough at any time considering the minimization of cost criterion. The models are based on the renewal reward process use.

Simple and hybrid inspection policies focused on guaranteeing a high level of availability are investigated in [175]. First, the simple periodical inspection is analyzed. To overcome its weaknesses and consider the information about remaining life of a system, the quantile-based inspections are introduced. This inspection policy is valid for increasing failure rate of the system. Later, a hybrid inspection policy is developed that considers performance of maintenance actions (periodic inspections or quantile-based inspections) according to the type of lifetime distributions: increasing failure rate or decreasing failure rate. Analytical solutions and numerical examples are provided for the limiting average availability and the long-run inspection rate assumptions.

A randomly failing single unit system whose failures may be self-announcing or not self-announcing is considered later in [78]. The authors in this work consider a randomly failing single unit system that is submitted to inspection when its age reaches T_{yin} units of time. The model includes imperfect inspection and preventive replacement performance. The proposed model is based on the implementation of the basic strategy of an ARP for the case of self-announcing failures. The objective is to determine the inspection and preventive maintenance interval that maximizes the stationary availability of the system.

The hybrid inspection models are developed for maintenance of multi-unit systems. The block inspection and replacement policy is presented in [106], where the authors introduce a periodical inspection for a two-unit parallel system. This model considers the detection capacity of inspections (perfect/imperfect), minimal repairs, and failure interactions to consider dependence between subsystems.

An interesting model is developed in [146], where the authors continue investigation of issues analyzed in [106] and [147]. The authors consider a multi-unit system composed of identical units having periodic imperfect PM and periodic inspection carried out every T_{in} time units. During the performance of inspection actions, units are checked to ascertain whether they are working or not. Failed units are replaced by new ones at inspection time. Assuming negligible PM times, the authors estimate an average cost per unit time function.

Another interesting problem is presented in [148], where the authors consider periodic and opportunistic inspections of a system with hard-type and soft-type components. Failures of soft-type components can be detected only at inspections. Thus, a system can operate with a soft failure, but its performance may be reduced. The hard-type component failures are self-announcing and create an opportunity for additional inspection (opportunistic inspection) of all soft-type components. Moreover, the system also is inspected periodically. Based on this assumption, the two optimization models are discussed using the simulation modeling approach and cost criteria. This problem also is continued in [149].

The problem of opportunistic inspection performance is considered in [150]. The authors in this work investigate an n_k-out-of-n system with hidden failures and under periodic inspection. The developed model is based on the assumption that every system failure presents an additional opportunity for inspection. The objective is to find the optimal periodic inspection policy and the optimal maintenance action at each inspection for the entire system. Moreover, three types of maintenance are considered: minimal repair, preventive replacement, and corrective replacement. The inspection maintenance model is based on implementation of a genetic algorithm and on cost criteria. The extensions of this model is presented in [151], where the authors focus on an n_k-out-of-n system with components whose failures follow a Non-Homogeneous Poisson Process (NHPP). This model does not optimize the maintenance action, which is based on the components state (age dependent). However, the model considers an inventory policy that focuses on supporting the inspection policy to ensure the required spares when necessary (at inspection times). The modeling approach is based on development of the simulation model.

2.5 OTHER INSPECTION MAINTENANCE MODELS

When analyzing and reviewing the literature on inspection maintenance, other issues (not mentioned in the previous subsections) also are noticeable. To the most commonly investigated issues we may include:

- Production planning and quality control (see [152–155])
- Cumulative damage modeling (see [156,157])
- Joint optimization of inventory policy with inspection maintenance modeling (see [158,159])
- Safety and reliability in maintenance (see [6,160–165])

Some examples of case studies can be found on optimization of inspection schedules for different systems. For example, in the literature optimization of inspection policy can be found for railway carriers (see [166]), nuclear power plants (see [161,167,168]), tunnel lighting systems (see [169]), a scale that weighs products in the final stage of the manufacturing process (see [170,171]), sewing machines (see [172]), or wooden poles structures (see [173]). Other inspection problems that are investigated apply to optimization of the periodic inspection of aircraft (see [130]), maintenance of transport systems with a subjective estimation approach (see [174]), investigations of system reliability structure (see [175]), inspection frequency of safety-related control systems of machinery (see [132,176]), optimization of inspection and maintenance decisions for infrastructure facilities (see [74]), inspection issues of hydraulic components (see [133]), safety-related control systems (see [132]), or multi-stage inspection problems (see [131]). Simulation modeling is investigated in [177].

A widely investigated inspection of production process/systems and the maintenance issues is worth noting. Research in this area focuses mostly on computer-aidediInspection planning systems (see [178] for state of the art) or maintenance and inspection models for production inventory systems (see [179–183]). In this research area, authors are interested in development of inspection policies for systems in storage to provide high reliability (see [184–189]).

2.6 CONCLUSIONS AND DIRECTIONS FOR FURTHER RESEARCH

In this chapter, the author provides a literature review on the most commonly used optimal inspection maintenance models. The literature was selected using Google Scholar as a search engine and ScienceDirect, JStor, SpringerLink, and SAGEJournals. The author primarily searched the relevant literature based on keywords, abstracts, and titles. Moreover, also articles were searched for relevant references. The following main terms and/or a combination of them were used for searching the literature: *inspection maintenance, inspection model,* and *inspection maintenance optimization.*

The selection methodology was based on searching for the defined keywords, and later choosing the models that satisfy the main reviewing criteria. For example,

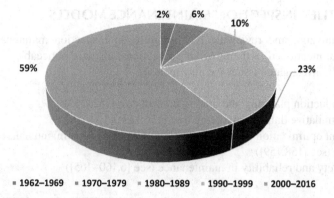

■ 1962–1969   ■ 1970–1979   ■ 1980–1989   ■ 1990–1999   ■ 2000–2016

FIGURE 2.2   Models distribution in relation to the period of their publication.

when searching for the keyword "inspection maintenance" in Google search, there were about 260 million hits. In the ScienceDirect database, this keyword had about 98,500 hits. Comparing the obtained search results to the main required criteria such as *periodic inspection*, *maintenance optimization*, and *technical system*, 122 inspection models published from 1962 to 2016 (see Figure 2.2) were the focus of this chapter.

Due to the plethora of available publications on inspection maintenance, there was no possibility to present all the known models from this research area. The most investigated ones that are not included in this chapter apply to:

- Sequential inspection maintenance modeling (see [17,23,57])
- Condition-based maintenance with inspection modeling issues (see [190])
- Delay-time modeling (see [19])

This literature overview lets the author draw the following main conclusions:

- The most commonly used mathematical methods applied for analysis of inspection maintenance scheduling problems include applied probability theory, renewal theory, Markov decision theory, and Genetic Algorithms (GA) technique. However, there are a lot of inspection maintenance problems that are too complex (e.g., shocks modeling and information uncertainty) to be solved in an analytical way. Thus, in practice, simulation processes and Bayesian approaches can be used widely.
- Most research on periodic inspections for hidden failures assumes that the times for inspection are negligible. However, in some cases the inspection time cannot be ignored due to its influence on system reliability characteristics. Thus, the optimal inspection policy is not obtained using this assumption.
- Many inspection maintenance models are based on simplified assumptions of infinite planning horizon, the steady-state conditions, perfect repair policy, available spare parts, and so on. These assumptions often are not valid for performance of real-life systems.

- Due to the complexity of models developed for inspection maintenance, in many cases there are problems with optimal computation of checking procedures. Thus, in such situations, the nearly optimal methods or algorithms should be implemented. Such algorithms usually are developed for the single-unit case.
- The widely known inspection maintenance models focus on performance of the inspection action that only gives the information about the state of the tested system (up state or down state). There are no models developed that give additional information about the signals of forthcoming failures (some defects occurrence); thus, this type of maintenance models is not enough for systems in which such symptoms may be diagnosed.

REFERENCES

1. Tang T (2012) Failure finding interval optimization for periodically inspected repairable systems. PhD Thesis, University of Toronto.
2. Keller JB (1982) Optimum inspection policies. *Management Science* 28(4): 447–450.
3. Sheriff YS (1982) Reliability analysis: Optimal inspection & maintenance schedules of failing equipment. *Microelectronics Reliability* 22(1): 59–115.
4. PN-EN 13306:2018 Maintenance—Maintenance terminology, The Polish Committee for Standardization, Warsaw.
5. Gulati R, Kahn J, Baldwin R (2010) The professional's guide to maintenance and reliability terminology. Reliabilityweb.com.
6. Peters R (2014) *Reliable, Maintenance Planning, Estimating, and Scheduling*. Gulf Professional Publishing.
7. Barlow RE, Hunter LC, Proschan F (1963) Optimum checking procedures. *Journal of the Society for Industrial and Applied Mathematics* 11(4): 1078–1095. https://www.jstor.org/stable/2946496.
8. Beichelt F, Tittmann P (eds.) (2012) *Reliability and Maintenance. Networks and Systems*. CRC Press.
9. Radner R, Jorgenson DW (1962) Optimal replacement and inspection of stochastically failing equipment. In: Arrow KJ, Karlin S, Scarf H (eds.) *Studies in Applied Probability and Management Science*, Stanford University Press: 184–206.
10. Jorgenson DW, Mccall JJ (1963) Optimal scheduling of replacement and inspection. *Operations Research* 11(5): 732–746.
11. Pierskalla WP, Voelker JA (1976) A survey of maintenance models: The control and surveillance of deteriorating systems. *Naval Research Logistics Quarterly* 23: 353–388.
12. Valdez-Flores C, Feldman R (1989) A survey of preventive maintenance models for stochastically deteriorating single-unit systems. *Naval Research Logistics* 36: 419–446.
13. Cho ID, Parlar M (1991) A survey of maintenance models for multi-unit systems. *European Journal of Operational Research* 51(1): 1–23.
14. Thomas LC, Gaver DP, Jacobs PA (1991) Inspection models and their application. *IMA Journal of Mathematics Applied in Business and Industry* 3: 283–303.
15. Parmigiani G (1991) Scheduling inspections in reliability. Institute of Statistics and Decision Sciences Discussion Paper no. 92–A11:1–21, Duke University. https://stat.duke.edu/research/papers/1992-11 (accessed 17 October 2018).
16. Osaki S (ed.) (2002) *Stochastic Models in Reliability and Maintenance*, Springer-Verlang, Berlin, Germany.
17. Nakagawa T (2005) *Maintenance Theory of Reliability*. Springer.

18. Jardine AKS, Tsang AHC (2013) Maintenance, replacement and reliability. *Theory and Applications*. CRC Press.
19. Werbińska-Wojciechowska S (2019) *Technical System Maintenance. Delay-Time-Based Modeling*. Springer.
20. Kaio N, Osaki S (1989) Comparison of inspection policies. *Journal of Operations Research Society* 40(5): 499–503. Palgrave Macmillan Journals.
21. Kaio N, Osaki S (1988) Inspection policies: Comparisons and modifications. Revenue française d'automatique, d'informatique et de recherché opérationnelle. *Recherche opérationnelle* 22(4): 387–400.
22. Munford AG (1981) Comparison among certain inspection policies. *Management Science* 27(3): 260–267.
23. Jiang R, Jardine AKS (2005) Two optimization models of the optimum inspection problem. *The Journal of the Operational Research Society* 56(10): 1176–1183. doi:10.1057/palgrave.jors.2601885.
24. Boland PJ, El-Neweihi E (1995) Expected cost comparisons for inspection and repair policies. *Computers and Operations Research* 22(4): 383–390. doi:10.1016/0305-0548(94)00047-C.
25. Hu T, Wei Y (2001) Multivariate stochastic comparisons of inspection and repair policies. *Statistics and Probability Letters* 51: 315–324.
26. Mccall JJ (1963) Operating characteristics of opportunistic replacement and inspection policies. *Management Science* 10(1): 85–97.
27. Choi KM (1997) Semi-Markov and delay time models of maintenance. PhD thesis, University of Salford, UK.
28. Chelbi A, Ait-Kadi D (2009) Inspection strategies for randomly failing systems. In: Ben-Daya M, Duffuaa SO, Raouf A, Knezevic J, Ait-Kadi D (eds.) *Handbook of Maintenance Management and Engineering*. Springer, London, UK.
29. Lee C (1999) Applications of delay time theory to maintenance practice of complex plant. PhD thesis, University of Salford, UK.
30. Bobrowski D (1980) Optimisation of technical object maintenance with inspections (in Polish). In: *Proceedings of Winter School on Reliability*, Center for Technical Progress, Katowice, Poland: 31–46.
31. Viscolani B (1991) A note on checking schedules with finite horizon. *Operations Research* 25(2): 203–208. doi:10.1051/ro/1991250202031.
32. Hariga MA (1996) A maintenance inspection model for a single machine with general failure distribution. *Microelectronics Reliability* 36(3): 353–358.
33. Klatzky RL, Messick DM, Loftus J (1992) Heuristics for determining the optimal interval between checkups. *Psychological Science* 3(5): 279–284.
34. Beichelt F (1981) Minimax inspection strategies for single unit systems. *Naval Research Logistics Quarterly* 28(3): 375–381.
35. Leung FKN (2001) Inspection schedules when the lifetime distribution of a single-unit system is completely unknown. *European Journal of Operational Research* 132: 106–115. doi:10.1016/S0377-2217(00)00115-6.
36. Okumura S (2006) Determination of inspection schedules of equipment by variational method. Mathematical Problems in Engineering, Hindawi Publishing Corporation, Article ID 95843: 1–16.
37. Liu B, Zhao X, Yeh R-H, Kuo W (2016) Imperfect inspection policy for systems with multiple correlated degradation processes. *IFAC-PapersOnLine* 49–12: 1377–1382.
38. Senegupta B (1982) An exponential riddle. *Journal of Applied Probability* 19(3): 737–740.
39. Guo H, Szidarovszky F, Gerokostopoulos A, Niu P (2015) On determining optimal inspection interval for minimizing maintenance cost. In: *Proceedings of 2015 Annual Reliability and Maintainability Symposium (RAMS)*, IEEE: 1–7.

40. Magott J, Nowakowski T, Skrobanek P, Werbinska-Wojciechowska S (2010) Logistic system modeling using fault trees with time dependencies—Example of tram network. In: Bris R, Guedes Soares C, Martorell S (eds.) *Reliability, Risk and Safety: Theory and Applications*. Vol. 3, Taylor & Francis, London, UK: 2293–2300.

41. Wattanapanom N, Shaw L (1979) Optimal inspection schedules for failure detection in a model where tests hasten failures. *Operations Research* 27(2): 303–317.

42. Butler DA (1979) A hazardous-inspection model. *Management Science* 25(1): 79–89.

43. Parmigiani G (1993) Optimal inspection and replacement policies with age-dependent failures and fallible tests. *The Journal of the Operational Research Society* 44(11): 1105–1114.

44. Parmigiani G (1993) Optimal scheduling of fallible inspections. DP no. 92–38: 1–30, https://stat.duke.edu/research/papers/1992-38 (accessed 17 October 2018).

45. Rizwan SM, Chauhan H, Taneja G (2005) Stochastic analysis of systems with accident and inspection. *Emirates Journal for Engineering Research* 10(2): 81–88.

46. Hryniewicz O (2009) Optimal inspection intervals for maintainable equipment. In: Martorell S, Guedes-Soares C, Barnett J (eds.) *Safety, Reliability and Risk Analysis: Theory, Methods and Applications*, Taylor & Francis Group, London.

47. Berrade MD (2012) A two-phase inspection policy with imperfect testing. *Applied Mathematical Modelling* 36: 108–114. doi:10.1016/j.apm.2011.05.035.

48. Berrade MD, Cavalcante CAV, Scarf PA (2013) Modelling imperfect inspection over a finite horizon. *Reliability Engineering and System Safety* 111: 18–29. doi:10.1016/j.ress.2012.10.003.

49. Berrade MD, Cavalcante CAV, Scarf PA (2012) Maintenance scheduling of a protection system subject to imperfect inspection and replacement. *European Journal of Operational Research* 218: 716–725. doi:10.1016/j.ejor.2011.12.003.

50. Berrade MD, Scarf PA, Cavalcante CAV, Dwight RA (2013) Imperfect inspection and replacement of a system with a defective state: A cost and reliability analysis. *Reliability Engineering and System Safety* 120: 80–87. doi:10.1016/j.ress.2013.02.024.

51. Sarkar J, Sarkar S (2000) Availability of a periodically inspected system under perfect repair. *Journal of Statistical Planning and Inference* 91: 77–90.

52. Cui L, Xie M (2001) Availability analysis of periodically inspected systems with random walk model. *Journal of Applied Probability* 38: 860–871. doi:10.1017/S0021900200019082.

53. Cui L, Xie M, Loh H-T (2004) Inspection schemes for general systems. *IIE Transactions* 36: 817–825. doi:10.1080/07408170490473006.

54. Cui L, Xie M (2005) Availability of a periodically inspected system with random repair or replacement times. *Journal of Statistical Planning and Inference* 131: 89–100. doi:10.1016/j.jspi.2003.12.008.

55. Yang J, Gang T, Zhao Y (2013) Availability of a periodically inspected system maintained through several minimal repairs before a replacement of a perfect repair. Hindawi Publishing Corporation, Abstracts and Applied Analysis, Article ID 741275: 1–6.

56. Tang T, Lin D, Banjevic D, Jardine AKS (2013) Availability of a system subject to hidden failure inspected at constant intervals with non-negligible downtime due to inspection and downtime due to repair/replacement. *Journal of Statistical Planning and Inference* 143: 176–185. doi:10.1016/j.jspi.2012.05.011.

57. Luss H, Kander Z (1974) Inspection policies when duration of checkings is non-negligible. *Operational Research Quarterly* 25(2): 299–309.

58. Wortman MA, Klutke G-A, Ayhan A (1994) A maintenance strategy for systems subjected to deterioration governed by random shocks. *IEEE Transactions on Reliability* 43(3): 439–445.

59. Chelbi A, Ait-Kadi D (2000) Generalized inspection strategy for randomly failing systems subjected to random shocks. *International Journal of Production Economics* 64: 379–384. doi:10.1016/S0925-5273(99)00073-0.

60. Chelbi A, Ait-Kadi D (1998) Inspection and predictive maintenance strategies. *International Journal of Computer Integrated Manufacturing* 11(3): 226–231. doi:10.1080/095119298130750.
61. Klutke G-A, Yang Y (2002) The availability of inspected systems subject to shocks and graceful degradation. *IEEE Transactions on Reliability* 51(3): 371–374.
62. Badia FG, Berrade MD (2006) Optimal inspection of a system with two types of failures under age dependent minimal repair. *Monografias del Seminario Matematico Garcia de Galdeano* 33: 207–214.
63. Badia FG, Berrade MD (2006) Optimum maintenance of a system under two types of failure. *International Journal of Materials and Structural Reliability* 4(1): 27–37.
64. Zequeira RI, Berenguer C (2006) An inspection and imperfect maintenance model for a system with two competing failure modes. In: *Proceedings of the 6th IFAC Symposium: Supervision and Safety of Technical Processes*: 932–937.
65. Sheu S-H, Tsai H-N, Wang F-K, Zhang ZG (2015) An extended optimal replacement model for a deteriorating system with inspections. *Reliability Engineering and System Safety* 139: 33–49. doi:10.1016/j.ress.2015.01.014.
66. Zequeira RI, Berenguer C (2006) Optimal scheduling of non-perfect inspections. *IMA Journal of Management Mathematics* 17: 187–207. doi:10.1093/imaman/dpi037.
67. Nakagawa T, Yasui K (1980) Approximate calculation of optimal inspection times. *Journal of Operational Research Society* 31: 851–853.
68. Aven T (1984) Optimal inspection when the system is repaired upon detection of failure. *Microelectronics Reliability* 24(5): 961–963.
69. Yeh RH, Chen HD, Wang C-H (2005) An inspection model with discount factor for products having Weibull lifetime. *International Journal of Operations Research* 2(1): 77–81.
70. Badia FG, Berrade MD, Campos CA (2002) Optimal inspection and preventive maintenance of units with revealed and unrevealed failures. *Reliability Engineering and System Safety* 78: 157–163.
71. Yeh L (1995) An optimal inspection-repair-replacement policy for standby systems. *Journal of Applied Probability* 32(1): 212–223.
72. Yang Y, Klutke G-A (2000) Improved inspection schemes for deteriorating equipment. *Probability in the Engineering and Informational Sciences* 14(4): 445–460.
73. Dagg RA, Newby M (1998) Optimal overhaul intervals with imperfect inspection and repair. *IMA Journal of Mathematics Applied in Business and Industry* 9: 381–391.
74. Durango-Cohen PL, Madanat SM (2008) Optimization of inspection and maintenance decisions for infrastructure facilities under performance model uncertainty: A Quasi-Bayes approach. *Transportation Research Part A: Policy and Practice* 42(8): 1074–1085. doi:10.1016/j.tra.2008.03.004.
75. Cheng GQ, Li L (2012) A geometric process repair model with inspections and its optimisation. *International Journal of Systems Science* 43(9): 1650–1655. doi:10.1080/00207721.2010.549586.
76. Wang W, Zhao F, Peng R (2014) A preventive maintenance model with a two-level inspection policy based on a three-stage failure process. *Reliability Engineering and System Safety* 121: 207–220. doi:10.1016/j.ress.2013.08.007.
77. Weiss GH (1963) Optimal periodic inspection programs for randomly failing equipment. *Journal of Research of the National Bureau of Standards—B. Mathematics and Mathematical Physics* 67B(4): 223–228.
78. Chelbi A, Ait-Kadi D, Aloui H (2008) Optimal inspection and preventive maintenance policy for systems with self-announcing and non-self-announcing failures. *Journal of Quality in Maintenance Engineering* 14(1): 34–45, doi:10.1108/13552510810861923.
79. Luss H (1976) Maintenance policies when deterioration can be observed by inspections. *Operational Research* 24(2): 359–366.

80. Becker G, Camarinopoulos L, Ziouas G (1994) A Markov type model for systems with tolerable down times. *The Journal of the Operational Research Society* 45(10): 1168–1178. doi:10.2307/2584479.
81. Rosenfield D (1976) Markovian deterioration with uncertain information. *Operations Research* 24(1): 141–155.
82. Tijms HC, Van Der Duyn Schouten FA (1984) A Markov decision algorithm for optimal inspections and revisions in a maintenance system with partial information. *European Journal of Operational Research* 21: 245–253. Elsevier.
83. Kawai H, Koyanagi J (1992) An optimal maintenance policy of a discrete time Markovian deterioration system. *Computers Mathematics with Applications* 24(1/2): 103–108.
84. Weiss GH (1962) A problem in equipment maintenance. *Management Science* 8(3): 266–277.
85. Fung J, Makis V (1997) An inspection model with generally distributed restoration and repair times. *Microelectronics Reliability* 37(3): 381–389.
86. Ohnishi M, Kawai H, Mine H (1986) An optimal inspection and replacement policy for a deteriorating system. *Journal of Applied Probability* 23(4): 973–988.
87. Wang GJ, Zhang YL (2014) Geometric process model for a system with inspections and preventive repair. *Computers and Industrial Engineering* 75: 13–19. doi:10.1016/j.cie.2014.06.007.
88. White III ChC (1978) Optimal inspection and repair of a production process subject to deterioration. *The Journal of the Operational Research Society* 29(3): 235–243.
89. Zuckerman D (1980) Inspection and replacement policies. *Journal of Applied Probability* 17(1): 168–177.
90. Abdel-Hameed M (1987) Inspection and maintenance policies of devices subject to deterioration. *Advances in Applied Probability* 19(4): 917–931.
91. Kong MB, Park KS (1997) Optimal replacement of an item subject to cumulative damage under periodic inspections. *Microelectronics Reliability* 37(3): 467–472.
92. Delia M-C, Rafael P-O (2008) A maintenance model with failures and inspection following Markovian arrival processes and two repair modes. *European Journal of Operational Research* 186: 694–707. doi:10.1016/j.ejor.2007.02.009.
93. Chiang JH, Yuan J (2001) Optimal maintenance policy for a Markovian system under periodic inspection. *Reliability Engineering and System Safety* 71: 165–172. doi:10.1016/S0951-8320(00)00093-4.
94. Kharoufer JP, Finkelstein DE, Mixon DG (2006) Availability of periodically inspected systems with Markovian wear and shocks. *Journal of Applied Probability* 43(2): 303–317. doi:10.1239/jap/1152413724.
95. Mazumdar M (1970) Reliability of two-unit redundant repairable systems when failures are revealed by inspections. *SIAM Journal on Applied Mathematics* 19(4): 637–647.
96. Osaki S, Asakura T (1970) A two-unit standby redundant system with repair and preventive maintenance. *Journal of Applied Probability* 7(3): 641–648.
97. Mahmoud MAW, Mohie El-Din MM, El-Said Moshref M (1995) Reliability analysis of a two-unit cold standby system with inspection, replacement, proviso of rest, two types of repair and preparation time. *Microelectronics Reliability* 35(7): 1063–1072.
98. Pandey D, Tyagi SK, Jacob M (1995) Profit evaluation of a two-unit system with internal and external repairs, inspection and post repair. *Microelectronics Reliability* 35(2): 259–264.
99. Cazorla DM, Perez-Ocon R (2008) An LDQBD process under degradation, inspection, and two types of repair. *European Journal of Operational Research* 190: 494–508. doi:10.1016/j.ejor.2007.04.056.
100. Kumar J (2011) Cost-benefit analysis of a redundant system with inspection and priority subject to degradation. *IJCSI International Journal of Computer Science Issues* 8(6/2): 314–321.

101. Kishan R, Jain D (2012) A two non-identical unit standby system model with repair, inspection and post-repair under classical and Bayesian viewpoints. *Journal of Reliability and Statistical Studies* 5(2): 85–103.

102. Bhatti J, Chitkara AK, Kakkar MK (2016) Stochastic analysis of dis-similar standby system with discrete failure, inspection and replacement policy. *Demonstratio Mathematica* 49(2): 224–235.

103. Vaurio JK (1999) Availability and cost functions for periodically inspected preventively maintained units. *Reliability Engineering and System Safety* 63: 133–140. doi:10.1016/S0951-8320(98)00030-1.

104. Vaurio JK (1997) On time-dependent availability and maintenance optimization of standby units under various maintenance policies. *Reliability Engineering and System Safety* 56: 79–89. doi:10.1016/S0951-8320(96)00132-9.

105. Kenzin M, Frostig E (2009) M out of n inspected systems subject to shocks in random environment. *Reliability Engineering and System Safety* 94: 1322–1330. doi:10.1016/j.ress.2009.02.005.

106. Zequeira RI, Berenguer C (2005) On the inspection policy of a two-component parallel system with failure interaction. *Reliability Engineering and System Safety* 88: 99–107. doi:10.1016/j.ress.2004.07.009.

107. Lee BL, Wang M (2012) Approximately optimal testing policy for two-unit parallel standby systems. *International Journal of Applied Science and Engineering* 10(3): 263–272.

108. Mendes AA, Coit DW, Duarte Ribeiro JL (2014) Establishment of the optimal time interval between periodic inspections for redundant systems. *Reliability Engineering and System Safety* 131: 148–165. doi:10.1016/j.ress.2014.06.021.

109. Greenberg H (1964) Optimum test procedure under stress. *Operations Research* 12(5): 689–692.

110. Anily S, Glass CA, Hassin R (1998) The scheduling of maintenance service. *Discrete Applied Mathematics* 82(1–3): 27–42. doi:10.1016/S0166-218X(97)00119-4.

111. Sheils E, O'connor A, Breysse D, Schoefs F, Yotte S (2010) Development of a two-stage inspection process for the assessment of deteriorating infrastructure. *Reliability Engineering and System Safety* 95: 182–194. doi:10.1016/j.ress.2009.09.008.

112. Duffuaa S, Al-Najjar HJ (1995) An optimal complete inspection plan for critical multicharacteristic components. *Journal of the Operational Research Society* 46(8): 930–942.

113. Duffuaa S, Khan M (2002) An optimal repeat inspection plan with several classifications. *Journal of the Operational Research Society* 53(9): 1016–1026. doi:10.1057/palgrave.jors.2601392.

114. Duffuaa S, Khan M (2008) A general repeat inspection plan for dependent multicharacteristic critical components. *European Journal of Operational Research* 191: 374–385. doi:10.1016/j.ejor.2007.02.033.

115. Sahraoui Y, Khelif R, Chateauneuf A (2013) Maintenance planning under imperfect inspections of corroded pipelines. *International Journal of Pressure Vessels and Piping* 104: 76–82. doi:10.1016/j.ijpvp.2013.01.009.

116. Srivastava MS, Wu Y (1993) Estimation and testing in an imperfect-inspection model. *IEEE Transactions on Reliability* 42(2): 280–286. IEEE, doi: 10.1109/24.229501.

117. Godziszewski J (2001) The impact of errors of the first and second types made during inspections on the costs of maintenance of a homogeneous equipment park (in Polish). In: *Proceedings of XIX Winter School on Reliability—Computer Aided Dependability Analysis, Publishing House of Institute for Sustainable Technologies*, Radom: 89–100.

118. Aven T (1987) Optimal inspection and replacement of a coherent system. *Microelectronics Reliability* 27(3): 447–450. doi:10.1016/0026-2714(87)90460-4.

119. Zuckerman D (1989) Optimal inspection policy for a multi-unit machine. *Journal of Applied Probability* 26: 543–551.

120. Qiu Y (1991) A note on optimal inspection policy for stochastically deteriorating series systems. *Journal of Applied Probability* 28: 934–939.

121. Dieulle L (1999) Reliability of a system with Poisson inspection times. *Journal of Applied Probability* 36(4): 1140–1154.

122. Dieulle L (2002) Reliability of several component sets with inspections at random times. *European Journal of Operational Research* 139: 96–114.

123. Rezaei E (2017) A new model for the optimization of periodic inspection intervals with failure interaction: A case study for a turbine rotor. *Case Studies in Engineering Failure Analysis* 9: 148–156. doi:10.1016/j.csefa.2015.10.001.

124. Tolentino D, Ruiz SE (2014) Influence of structural deterioration over time on the optimal time interval for inspection and maintenance of structures. *Engineering Structures* 61: 22–30. doi:10.1016/j.engstruct.2014.01.012.

125. Lu Z, Chen M, Zhou D (2015) Periodic inspection maintenance policy with a general repair for multi-state systems. In: *Proceedings of Chinese Automation Congress (CAC)*: 2116–2121.

126. Zhang J, Huang X, Fang Y, Zhou J, Zhang H, Li J (2016) Optimal inspection-based preventive maintenance policy for three-state mechanical components under competing failure modes. *Reliability Engineering and System Safety* 152: 95–103. doi:10.1016/j.ress.2016.02.007.

127. Carvalho M, Nunes E, Telhada J (2009) Optimal periodic inspection of series systems with revealed and unrevealed failures. In: *Safety, Reliability and Risk Analysis: Theory, Methods and Applications—Proceedings of the Joint Esrel and SRA-Europe Conference*, CRC Press: 587–592.

128. Bris R, Chatelet E, Yalaoui F (2003) New method to minimize the preventive maintenance cost of series-parallel systems. *Reliability Engineering and System Safety* 82: 247–255. doi:10.1016/S0951-8320(03)00166-2.

129. Badia FG, Berrade MD, Campos CA (2002) Maintenance policy for multivariate standby/operating units. *Applied Stochastic Models in Business and Industry* 18: 147–155.

130. Huang J, Song Y, Ren Y, Gao Q (2014) An optimization method of aircraft periodic inspection and maintenance based on the zero-failure data analysis. In: *Proceedings of 2014 IEEE Chinese Guidance, Navigation and Control Conference*, Yantai, China: 319–323.

131. Azadeh A, Sangari MS, Amiri AS (2012) A particle swarm algorithm for inspection optimization in serial multi-stage process. *Applied Mathematical Modelling* 36: 1455–1464. doi:10.1016/j.apm.2011.09.037.

132. Dzwigarek M, Hryniewicz O (2011) Frequency of periodical inspections of safety-related control systems of machinery—Practical recommendations for determining methods. In: *Proceedings of Summer Safety and Reliability Seminars*, SSARS 2011, Gdańsk-Sopot, Poland: 17–26.

133. Alfares H (1999) A simulation model for determining inspection frequency. *Computers and Indus-trial Engineering* 36: 685–696. doi:doi.org/10.1016/S0360-8352(99)00159-X.

134. Bai Y, Bai Q (2014) *Subsea Pipeline Integrity and Risk Management*. Elsevier. doi:10.1016/C2011-0-00113-8.

135. Bai Y, Jin W-L (2015) *Marine Structural Design*. Elsevier.

136. Zhaoyang T, Jianfeng L, Zongzhi W, Jianhu Z, Weifeng H (2011) An evaluation of maintenance strategy using risk-based inspection. *Safety Science* 49: 852–860. doi:10.1016/j.ssci.2011.01.015.

137. Hagemeijer PM, Kerkveld G (1998) A methodology for risk-based inspection of pressurized systems. *Proceedings of the Institution of Mechanical Engineers, Part E: Journal of Process Mechanical Engineering* 212(1): 37–47. SAGE Journals.

138. Hagemeijer PM, Kerkveld G (1998) Application of risk-based inspection for pressurized HC production systems in a Brunei petroleum company. *Proceedings of the Institution of Mechanical Engineers, Part E: Journal of Process Mechanical Engineering* 212(1): 49–54.

139. Wang J, Matellini B, Wall A, Phipps J (2012) Risk-based verification of large offshore systems. *Proceedings of the Institution of Mechanical Engineers, Part M: Journal of Engineering for the Maritime Environment* 226(3): 273–298. doi:10.1177/1475090211430302.

140. Jovanovic A (2003) Risk-based inspection and maintenance in power and process plants in Europe. *Nuclear Engineering and Design* 226: 165–182.

141. Kallen MJ, Van Noortwijk JM (2005) Optimal maintenance decisions under imperfect inspection. *Reliability Engineering and System Safety* 90: 177–185. doi:10.1016/j.ress.2004.10.004.

142. You J-S, Kuo H-T, Wu W-F (2006) Case studies of risk-informed inservice inspection of nuclear piping systems. *Nuclear Engineering and Design* 236: 35–46.

143. Podofillini L, Zio E, Vatn J (2006) Risk-informed optimisation of railway tracks inspection and maintenance procedures. *Reliability Engineering and System Safety* 91: 20–35. doi:10.1016/j.ress.2004.11.009.

144. Nakagawa T (1980) Replacement models with inspection and preventive maintenance. *Microelectronics and Reliability* 20: 427–433.

145. Yeh L (2003) An inspection-repair-replacement model for a deteriorating system with unobservable state. *Journal of Applied Probability* 40: 1031–1042.

146. Park JH, Lee SC, Hong JW, Lie CH (2009) An optimal block preventive maintenance policy for a multi-unit system considering imperfect maintenance. *Asia-Pacific Journal of Operational Research* 26(6): 831–847.

147. Sheu S-H, Lin Y-B, Liao G-L (2006) Optimum policies for a system with general imperfect maintenance. *Reliability Engineering and System Safety* 91(3): 362–369.

148. Taghipour S, Banjevic D (2012) Optimum inspection interval for a system under periodic and opportunistic inspections. *IIEE Transactions* 44: 932–948. doi:10.1080/07408 17X.2011.618176.

149. Taghipour S, Banjevic D (2012) Optimal inspection of a complex system subject to periodic and opportunistic inspections and preventive replacements. *European Journal of Operational Research* 220: 649–660. doi:10.1016/j.ejor.2012.02.002.

150. Babishin V, Taghipour S (2016) Joint optimal maintenance and inspection for a k-out-of-n system. *International Journal of Advanced Manufacturing Technology* 87(5): 1739–1749. doi:10.1109/RAMS.2016.7448039.

151. Bjarnason ETS, Taghipour S (2014) Optimizing simultaneously inspection interval and inventory levels (s, S) for a k-out-of-n system. In: *2014 Reliability and Maintainability Symposium*, Colorado Springs, CO: 1–6. doi:10.1109/RAMS.2014.6798463.

152. Chen C-T, Chen Y-W, Yuan J (2003) On dynamic preventive maintenance policy for a system under inspection. *Reliability Engineering and System Safety* 80: 41–47. doi:10.1016/S0951-8320(02)00238-7.

153. Chen Y-C (2013) An optimal production and inspection strategy with preventive maintenance error and rework. *Journal of Manufacturing Systems* 32: 99–106. doi:10.1016/j.jmsy.2012.07.010.

154. Duffuaa S, El-Ga'aly A (2013) A multi-objective mathematical optimization model for process targeting using 100% inspection policy. *Applied Mathematical Modelling* 37: 1545–1552. doi:10.1016/j.apm.2012.04.008.

155. Wang H, Wang W, Peng R (2017) A two-phase inspection model for a single component system with three-stage degradation. *Reliability Engineering and System Safety* 158: 31–40.

156. Feng Q, Peng H, Coit DW (2010) A degradation-based model for joint optimization of burn-in, quality inspection, and maintenance: A light display device application. *International Journal of Advanced Manufacturing Technology* 50: 801–808. doi:10.1007/s00170-010-2532-7.

157. Tsai H-N, Sheu S-H, Zhang ZG (2016) A trivariate optimal replacement policy for a deteriorating system based on cumulative damage and inspections. *Reliability Engineering and System Safety* 160: 122–135. doi:10.1016/j.ress.2016.10.031.

158. Bjarnason ETS, Taghipour S, Banjevic D (2014) Joint optimal inspection and inventory for a k-out-of-n system. *Reliability Engineering and System Safety* 131: 203–215. doi:10.1016/j.ress.2014.06.018.

159. Panagiotidou S (2014) Joint optimization of spare parts ordering and maintenance policies for multiple identical items subject to silent failures. *European Journal of Operational Research* 235: 300–314. doi:10.1016/j.ejor.2013.10.065.

160. Bukowski JV (2001) Modeling and analyzing the effects of periodic inspection on the performance of safety-critical systems. *IEEE Transactions on Reliability* 50(3): 321–329. doi:10.1109/24.974130.

161. Ellingwood BR, Mori Y (1997) Reliability-based service life assessment of concrete structures in nuclear power plants: Optimum inspection and repair. *Nuclear Engineering and Design* 175: 247–258.

162. Estes AC, Frangopol DM (2000) An optimized lifetime reliability-based inspection program for deteriorating structures. In: *Proceedings of the 8th ASCE Joint Specialty Conference on Probabilistic Mechanics and Structural Reliability*, Notre Dame, IN.

163. Faber MH, Sorensen JD (2002) Indicators for inspection and maintenance planning of concrete structures. *Structural Safety* 24: 377–396. doi:10.1016/S0167-4730(02)00033-4.

164. Onoufriou T, Frangopol DM (2002) Reliability-based inspection optimization of complex structures: A brief retrospective. *Computers and Structures* 80: 1133–1144.

165. Woodcock K (2014) Model of safety inspection. *Safety Science* 62: 145–156.

166. Ten Wolde M, Ghobbar AA (2013) Optimizing inspection intervals—Reliability and availability in terms of a cost model: A case study on railway carriers. *Reliability Engineering and System Safety* 114: 137–147. doi:10.1016/j.ress.2012.12.013.

167. Ali SA, Bagchi G (1998) Risk-informed in service inspection. *Nuclear Engineering and Design* 181: 221–224.

168. Garnero M-A, Beaudouin F, Delbos J-P (1998) Optimization of bearing-inspection intervals. *IEEE Proceedings of Annual Reliability and Maintainability Symposium*: 332–338.

169. Aoki K, Yamamoto K, Kobayashi K (2007) Optimal inspection and replacement policy using stochastic method for deterioration prediction, In: *Proceedings of 11th World Conference on Transport Research*, Berkeley CA:1–13.

170. Sandoh H, Igaki N (2003) Optimal inspection policies for a scale. *Computers and Mathematics with Applications* 46: 1119–1127.

171. Sandoh H, Igaki N (2001) Inspection policies for a scale. *Journal of Quality in Maintenance Engineering* 7(3): 220–231.

172. Guduru RKR, Shaik SH, Yaramala S (2018) A dynamic optimization model for multiobjective maintenance of sewing machine. *International Journal of Pure and Applied Mathematics* 118(20): 33–43.

173. Gravito FM, Dos Santos Filho N (2003) Inspection and maintenance of wooden poles structures. *Global ESMO 2003*, Orlando, Florida: 151–155.

174. Jazwinski J, Zurek J (2000) Principles of determining the maintenance set of the condition of the transport system with the use of expert opinions (in Polish). In: *Proceeding of XXVIII Winter School on Reliability—Decision Problems in Dependability Engineering*, Publishing House of Institute for Sustainable Technologies, Radom, Poland: 118–125.

175. Salamonowicz T (2007) Maintenance strategy for systems in k-out-of-n reliability structure (in Polish). In: *Proceedings of XXXV Winter School on Reliability—Problems of Systems Dependability*, Publishing House of Institute for Sustainable Technologies, Radom: 414–420.

176. Dzwigarek M, Hryniewicz O (2012) Periodical inspection frequency of protection systems of machinery—Case studies (in Polish). *Journal of KONBiN* 3(23): 109–120.

177. Landowski B, Woropay M (2003) Simulation of exploitation processes of technical objects preventively maintained (in Polish). In: *Proceedings of XXXI Winter School on Reliability—Forecasting Methods in Dependability Engineering*, Publishing House of Institute for Sustainable Technologies, Radom: 297–308.

178. Zhao F, Xu X, Xie SQ (2009) Computer-aided inspection planning—The state of the art. *Computers in Industry* 60: 453–466. doi:10.1016/j.compind.2009.02.002.

179. Ballou DP, Pazer HL (1982) The impact of inspector fallibility on the inspection policy in serial production systems. *Management Science* 28(4): 387–399. doi:10.1287/mnsc.28.4.387.

180. Darwish MA, Ben-Daya M (2007) Effect of inspection errors and preventive maintenance on a two-stage production inventory system. *International Journal of Production Economics* 107: 301–313. doi:10.1016/j.ijpe.2006.09.008.

181. Lee HL, Rosenblatt MJ (1987) Simultaneous determination of production cycle and inspection schedules in a production system. *Management Science* 33(9): 1125–1136.

182. Meyer RR, Rothkopf MH, Smith SA (1979) Reliability and inventory in a production-storage system. *Management Science* 25(8): 799–807.

183. Tirkel I (2016) Efficiency of Inspection based on out of control detection in wafer fabrication. *Computers and Industrial Engineering* 99: 458–464. doi:10.1016/j.cie.2016.05.022.

184. Ito K, Nakagawa T (2000) Optimal inspection policies for a storage system with degradation at periodic tests. *Mathematical and Computer Modelling* 31: 191–195.

185. Ito K, Nakagawa T (1995) An optimal inspection policy for a storage system with high reliability. *Microelectronics Reliability* 36(6): 875–882.

186. Ito K, Nakagawa T (1995) An optimal inspection policy for a storage system with three types of hazard rate functions. *Journal of the Operations Research Society of Japan* 38(4): 423–431.

187. Ito K, Nakagawa T, Nishi K (1995) Extended optimal inspection policies for a system in storage. *Mathematical and Computer Modelling* 22(10–12): 83–87.

188. Martinez EC (1984) Storage reliability with periodic test. *IEEE Proceedings of Annual Reliability and Maintainability Symposium*: 181–185.

189. Su Ch, Zhang Y-J, Cao B-X (2012) Forecast model for real time reliability of storage system based on periodic inspection and maintenance data. *Eksploatacja i Niezawodnosc – Maintenance and Reliability* 14(4): 342–348.

190. Neves ML, Santiago LP, Maia CA (2011) A condition-based maintenance policy and input parameters estimation for deteriorating systems under periodic inspection. *Computers and Industrial Engineering* 61: 503–511. doi:10.1016/j.cie.2011.04.005.

3 Application of Stochastic Processes in Degradation Modeling
An Overview

*Shah Limon, Ameneh Forouzandeh Shahraki,
and Om Prakash Yadav*

CONTENTS

3.1 INTRODUCTION

Most engineering systems experience the aging phenomena during their life cycle. The operating conditions and external stresses further expedite the aging process of these systems. The aging process reflects the propagation of the failure mechanism, which ultimately results in a decline of product performances and finally product failure. To reduce the downtime and ensure safe operations, it is desirable to identify the product's lifetime and reliability measure accurately so that appropriate maintenance policies can be executed. Therefore, the knowledge of product deterioration characteristics and fundamental root causes is a great source of information to assess the product performance and reliability using the degradation modeling (Limon et al. 2017a; Shahraki et al. 2017). In degradation modeling, a predefined threshold value is considered to identify the time-to-failure. Further, the degradation approach provides more accurate reliability estimates compared to the traditional failure time approaches.

In traditional deterministic models, system behavior is defined by a set of equations that can describe with certainty how the system performance will evolve over the period of time. However, in a reality, there exists variation or uncertainty in system performance that causes probabilistic behavior of the system. This situation led to the increasing importance of the stochastic processes for modeling the probabilistic degradation behavior of the engineering systems. A stochastic process is defined by a collection of random variables that are associated with a set of numbers that represent the random changes of a system over time. It can be divided into two broad categories: discrete and continuous state stochastic process.

The continuous state stochastic processes, mostly the members of the Levy family, such as the Wiener process, Gamma process, and Inverse Gaussian process are being successfully used in modeling degradation processes of the system (Ye et al. 2013; Limon et al. 2017b; Limon et al. 2018). These processes have the independent increment referred to as a Markov property that is very applicable to many engineering degradation phenomena. Further, time-to-failure's explicit expression by the first passage of time concept provides clear advantages of continuous stochastic processes in degradation modeling for reliability assessment.

On the other hand, the discrete state stochastic processes are used to model the degradation process where the overall status of the degradation process can be divided into a finite number of discrete levels ranging from perfect functioning to complete failure. Each state can correspond to a certain level of performance of a system under operation. The discrete state stochastic processes are used in degradation modeling because of the simplicity associated with dealing with only a limited number of states and their practical applications in degradation modeling (Moghaddass and Zuo 2014; Shahraki and Yadav 2018). The change of the system state may happen at the discrete or continuous time that leads to different models. Moreover, in some applications that the system's history and age may influence the future state of the system, the aging Markovian and semi-Markov processes are used as an extension of Markov processes.

The remainder of this chapter is organized as follows. Section 3.2 presents the different types of continuous state stochastic processes, degradation modeling with those processes, and selection of appropriate stochastic process. Section 3.3 describes the discrete state stochastic processes with case examples. Finally, Section 3.4 summarizes the application of stochastic processes in degradation modeling to evaluate the system reliability.

3.2   CONTINUOUS STATE STOCHASTIC PROCESSES

The continuous state stochastic process represents the continuity of the system changes as a function of time and implies a well-behaved sample path property to further analysis. The commonly used continuous state stochastic processes are members of the Levy processes such as the Wiener process, Gamma process, and Inverse Gaussian process. The fundamental idea of using the Levy processes in degradation modeling is based on the assumption that every degradation process is a cumulative result of the small and independent degradation increments. Besides capturing the temporal variation of the degradation processes, these members of the Levy processes also have well-established mathematical properties useful for explaining the degradation behavior. Further, the members of the Levy processes also have a strong Markov property with the following mathematical expression:

$$\Pr(X_{t_i} \mid X_{t_{i-1}}, X_{t_{i-2}}, X_{t_{i-3}} \ldots \ldots \ldots X_{t_1}) = \Pr(X_{t_i} \mid X_{t_{i-1}})$$

This implies that the next degradation increment is only dependent on the current state of the degradation and independent of the past degradation increments. This property is also intuitive and practical for many deterioration processes. The following sections provide the details of each stochastic processes for degradation modeling.

3.2.1 WIENER PROCESS

The basic Wiener process can be expressed as:

$$Y(t) = \mu \Lambda(t) + \sigma B\big(\Lambda(t)\big) \tag{3.1}$$

Here $B(.)$ is the standard Brownian motion, μ and σ represents the drift and volatility parameter respectively, $\Lambda(.)$ indicates the timescale function, and $Y(t)$ is the characteristic indicator that represents the system behavior. Suppose, a random variable $Y(t)$ follows the Wiener stochastic process, then it has the following mathematical properties:

1. $y(0) = 0$
2. $y(t)$ follows a normal distribution with $N \sim (\mu \Lambda(t), \sigma^2 \Lambda(t))$
3. $y(t)$ has an independent increment for every time interval Δt $(\Delta t = t_i - t_{i-1})$
4. The independent increment $\Delta y(t) = y_i - y_{i-1}$ follows the normal distribution $N \sim \big(\mu \Delta \Lambda(t), \sigma^2 \Delta \Lambda(t)\big)$ with probability density function (PDF):

$$f_{\Delta y(t)} = \frac{1}{\sigma \sqrt{2\pi \Delta \Lambda(t)}} e^{\left\{ -\frac{[\Delta y - \mu \Delta \Lambda(t)]^2}{2\sigma^2 \Delta \Lambda(t)} \right\}} \tag{3.2}$$

The Wiener process is known also as the standard Brownian motion that is the random movement of particles suspended in a fluid environment resulting from their collision. This random movement of small particles is very analogous to the random increment of the deterioration path. Besides, the Wiener process has many other attractive properties that are well suited to model the degradation behavior. For example, the degradation process can be viewed as an integration of small environmental effects in a cumulative form. The increment process of these small effects can be approximated by a normal distribution according to the central limit theorem. The environmental effects such as temperature, shocks, and humidity are most often independent, and resulting degradation are also independent in the time interval. Considering this aspect, the Wiener process is a good versatile model to describe many degradation phenomena. In a Wiener process, the drift parameter μ represents the degradation rate and timescale function $\Lambda(.)$ captures the nonlinearity in the degradation process.

The manufacturer often uses the accelerated degradation test (ADT) to quickly analyze the reliability matrices during the product design stages. In ADT, to expedite the degradation process, product samples are subjected to higher stress levels than the normal operating conditions. The effect of stress on product degradation as well as the lifetime can be explained by several existing physics or empirical-based reaction rate models. For example, the temperature or any thermal effect on a product deterioration can be captured easily by the Arrhenius model. Following are several other well-established reaction rate models where $d(s)$ represents the rate of degradation at stress level s, and a_1 and a_2 are the constant coefficients that depend on material or product types (Nelson 2004):

$$d(s) = a_1 e^{-\frac{a_2}{T}} \; ; \; \text{Arrhenius model}\left(s = T\right)$$

$$= a_1 V^{a_2}; \quad \text{Power law model}\left(s = V\right) \tag{3.3}$$

$$= a_1 e^{a_2 W}; \quad \text{Exponential model}\left(s = W\right)$$

Since the magnitude of stress measurement units may differ significantly in the multi-stress scenario, it is important to use standardized transform stresses to disregard the influence of stress measurement units. The transformed stress level is given as (Park and Yum 1997):

$$S_k = \frac{1/S_0' - 1/S_k'}{1/S_0' - 1/S_M'} , \qquad \text{for Arrhenius model}$$

$$= \frac{\log\left(S_k'\right) - \log\left(S_0'\right)}{\log\left(S_M'\right) - \log\left(S_0'\right)} , \quad \text{for Power law model} \tag{3.4}$$

$$= \frac{S_k' - S_0'}{S_M' - S_0'} , \qquad \text{for Exponential law model}$$

where S_0', S_k', and S_M' represent the operational, applied accelerated, and maximum stress level in their original form, whereas S_k represents corresponding transformed stress. It is considered the multiple stress degradation test with possible interaction effect between stresses. The nonlinear behavior of the degradation is described by the power law function $(\Lambda(t) = t^c, c$ is a constant). Considering both the Wiener parameter is stress dependent, the log-likelihood function can be written as:

$$L(\hat{\theta}) = \prod_{i=1}^{n} \prod_{j=1}^{m} \prod_{k=1}^{p} -\frac{1}{2}\log 2\pi \left(\left(t_{ijk}^c - t_{(i-1)jk}^c\right)\right) - \frac{1}{2}\log(\sigma^2) - \frac{\left[\Delta y_{ijk} - \mu(s)\left(t_{ijk}^c - t_{(i-1)jk}^c\right)\right]^2}{2\sigma^2\left(t_{ijk}^c - t_{(i-1)jk}^c\right)}$$

$$\tag{3.5}$$

The maximum likelihood estimation (MLE) method can be applied to estimate the model parameter of the previous function. The time to failure according to the Wiener process is defined when the first passage of time reaches the threshold degradation D and it follows the inverse Gaussian (IG) distribution with the PDF:

$$f_{IG,\ (y,a,b)} = \left(\frac{b}{2\pi y^3}\right)^{\frac{1}{2}} e^{\left[-\frac{b(y-a)^2}{2a^2 y}\right]} \tag{3.6}$$

Here, a and b are the *IG* distribution parameters. The mean time to failure than can be written as:

$$\xi_w = \left[\frac{D - y_0}{\mu(s)}\right]^{\frac{1}{c}} \tag{3.7}$$

The reliability function can be approximated with:

$$R(t) \approx \Phi\left(\frac{D - y_0 - \mu(s)t^c}{\sqrt{\sigma^2(s)t^c}}\right) \tag{3.8}$$

3.2.2 GAMMA PROCESS

The gamma process represents the degradation behavior in a form of cumulative damage where the deterioration occurs gradually over the period of time. Assuming a random variable $Y(t)$ represents the deterioration, then the gamma process that is a continuous-time stochastic process has the following mathematical properties (O'Connor 2012):

1. $y(0) = 0$
2. $y(t)$ follow a gamma distribution with $Ga \sim (\alpha t, \beta)$
3. $y(t)$ has an independent increment in a time interval Δt $(\Delta t = t_i - t_{i-1})$
4. The independent increment $\Delta y(t) = y_i - y_{i-1}$ also follows the gamma distribution $Ga \sim (\alpha \Delta t, \beta)$ with PDF:

$$f_{\Delta y(t)} = \frac{\beta^{\alpha(t_i^c - t_{i-1}^c)}}{\Gamma(\alpha(t_i^c - t_{i-1}^c))} \Delta y^{\alpha(t_i^c - t_{i-1}^c)-1} e^{-(\beta \Delta y)} \tag{3.9}$$

where $\alpha > 0$ and $\beta > 0$ represent the gamma shape and scale parameters, respectively, c is a nonlinearity parameter, and $\Gamma(.)$ is a gamma function with $\Gamma(a) = \int_0^\infty x^{a-1} e^{-(x)} dx$.

Now, considering the accelerated test and both gamma parameter dependent on stresses with interaction effect, the log-likelihood function can be written as:

$$L(\hat{\theta}) = \prod_{i=1}^{n}\prod_{j=1}^{m}\prod_{k=1}^{p} \frac{[\alpha(s)]^{\beta(s)\left(t_{ijk}^c - t_{(i-1)jk}^c\right)}}{\Gamma\left[\alpha(s)\left(t_{ijk}^c - t_{(i-1)jk}^c\right)\right]}\Delta y_{ijk}^{\left[\alpha(s)\left(t_{ijk}^c - t_{(i-1)jk}^c\right)-1\right]}e^{-\Delta y_{ijk}\beta(s)} \qquad (3.10)$$

The MLE method with advanced optimization software can be used to solve this complex equation. Now assuming that a failure occurs while the degradation path reaches the threshold D, then the time to failure ξ is defined as the time when the degradation path crosses the threshold D and the reliability function at time t will be:

$$R(t) = P(t < t_D) = 1 - \frac{\Gamma\left(\alpha t^c, D_\beta\right)}{\Gamma\left(\alpha t^c\right)} \qquad (3.11)$$

where $D_\beta = (D - y_0)\beta$ and y_0 is the initial degradation value. The cumulative distribution function (CDF) of t_D is given as:

$$F(t) = \frac{\Gamma\left(\alpha t^c, D_\beta\right)}{\Gamma\left(\alpha t^c\right)} \qquad (3.12)$$

Because of the gamma function, the evaluation of the CDF becomes mathematically intractable. To deal with this issue, Park and Padgett (2005) proposed an approximation of time-to-failure ξ with a Birnbaum-Saunders (BS) distribution having the following CDF:

$$F_{BS}(t) \approx \phi\left[\frac{1}{a}\left(\sqrt{\frac{t^c}{b}} - \sqrt{\frac{b}{t^c}}\right)\right] \qquad (3.13)$$

where $a = 1/\sqrt{(\omega_\beta)}$ and $b = \omega_\beta/\alpha$. Considering BS approximation, the expected failure time can be estimated as:

$$\xi_G = \left(\frac{\omega_\beta}{\alpha} + \frac{1}{2\alpha}\right)^{\frac{1}{c}} \qquad (3.14)$$

3.2.3 INVERSE GAUSSIAN PROCESS

Consider a system's behavior is represented by the IG process. If $Y(t)$ indicates the system's performance characteristic at time t, then the IG process has the following properties (Wang and Xu 2010):

1. $y(0) = 0$ with probability one
2. $y(t)$ has an independent increment in each time interval Δt ($\Delta t = t_i - t_{i-1}$)
3. The independent increment $\Delta y(t) = y_i - y_{i-1}$ follows the IG distribution $IG \sim \left(\mu \Delta \Lambda(t), \lambda \Delta \Lambda(t)^2 \right)$ with PDF:

$$f_{\left(\Delta y \mid \mu \Delta \Lambda(t),\ \lambda \Delta \Lambda(t)^2\right)} = \left(\frac{\lambda \Lambda(t)^2}{2\pi \Delta y^3} \right)^{1/2} e^{\left[-\frac{\lambda(\Delta y - \mu \Lambda(t))^2}{2\mu^2 \Delta y} \right]} \tag{3.15}$$

Here μ and λ denote the mean and scale parameter and $\Lambda(t)$ represents the shape function. The mean of $Y(t)$ is defined by $\mu\Lambda(t)$ and the variance is $\mu^3 \Lambda(t)/\lambda$. The shape function is nonlinear, and a power law is chosen in this work to represent the nonstationary process ($\Lambda(t) = t^c$). By the properties of the IG process and Equation 3.15, the likelihood function of the degradation increment can be given as:

$$L(\hat{\theta}) = \prod_{i=1}^{n} \prod_{j=1}^{m} \prod_{k=1}^{p} \sqrt{\frac{\lambda_{ijk} \left(t_{ijk}^c - t_{ijk}^c \right)^2}{2\pi \Delta y_{ijk}^3}} \, e^{-\lambda_{ijk} \frac{\left(\Delta y_{ijk} - \mu_{ijk} \left(t_{ijk}^c - t_{ijk}^c \right) \right)^2}{2\mu_{ijk}^2 \Delta y_{ijk}}} \tag{3.16}$$

Suppose $Y(t)$ is a monotonic degradation process and the lifetime ξ_D is defined by the first passage of time where degradation reaches the threshold value D. If the initial degradation is indicated by y_0, then $y(t) - y_0$ follows the IG distribution. Therefore, the CDF of ξ_D can be written as:

$$F\left(\xi_D \mid D, \mu t^c, \lambda(t^c)^2 \right) = \Phi \left[\sqrt{\frac{\lambda}{D - y_0}} \left(t^c - \frac{D - y_0}{\mu} \right) \right] - e^{\left(\frac{2\lambda t^c}{\mu} \right)} \Phi \left[-\sqrt{\frac{\lambda}{D - y_0}} \left(t^c + \frac{D - y_0}{\mu} \right) \right] \tag{3.17}$$

where $\Phi(.)$ is the CDF of the standard normal distribution. However, when $\mu\Lambda(t)$ and t are large, $Y(t)$ can be approximated by the normal distribution with mean $\mu\Lambda(t)$ and variance $\mu^3 \Lambda(t)/\lambda$. Therefore, the CDF of ξ_D also can be approximated by the following equation (Ye and Chen 2014):

$$F\left(\xi_{IG} \mid D, \mu t^c, \lambda(t^c)^2 \right) = \Phi \left[\frac{D - \mu(s)t^c}{\sqrt{\mu(s)^3 t^c / \lambda}} \right] \tag{3.18}$$

And the approximated mean lifetime expression is:

$$\xi_{IG} = \left(\frac{D}{\mu(s)} \right)^{1/c} \tag{3.19}$$

3.2.4 CASE EXAMPLE: DEGRADATION ANALYSIS WITH A CONTINUOUS STATE STOCHASTIC PROCESS

To demonstrate the proposed method, light emitting diodes (LEDs) are taken as a case study example. Recently, LEDs have become very popular due to their very low energy consumption, low costs, and long life (Narendran and Gu 2005). As a solid-state lighting source, the use of LEDs is increasing in many sectors such as communications, medical services, backlighting, sign-post, and general lighting purposes. LEDs produce illumination and unlike the traditional lamp light instead of catastrophic failure, the output light of LEDs is usually degraded over the useful time and experiences soft failure modes. Therefore, it is reasonable to consider the light intensity of LEDs as a degradation of performance characteristics in this study.

The experiment data on degradation of LEDs are taken from the literature (Chaluvadi 2008). Table 3.1 provides the details of experimental set up of the LED

TABLE 3.1

Accelerated Degradation Test Dataset of LEDs

Stress Level	Sample/time (hrs)	Degradation Measurement (lux)					
		0	50	100	150	200	250
	1	1	0.866	0.787	0.76	0.716	0.68
	2	1	0.821	0.714	0.654	0.617	0.58
	3	1	0.827	0.703	0.64	0.613	0.593
	4	1	0.798	0.683	0.623	0.6	0.59
	5	1	0.751	0.667	0.628	0.59	0.54
	6	1	0.837	0.74	0.674	0.63	0.613
40 mA	7	1	0.73	0.65	0.607	0.583	0.58
	8	1	0.862	0.676	0.627	0.6	0.597
	9	1	0.812	0.65	0.606	0.593	0.573
	10	1	0.668	0.633	0.593	0.573	0.565
	11	1	0.661	0.642	0.594	0.58	0.553
	12	1	0.765	0.617	0.613	0.597	0.56
	1	1	0.951	0.86	0.776	0.7	0.667
	2	1	0.933	0.871	0.797	0.743	0.73
	3	1	0.983	0.924	0.89	0.843	0.83
	4	1	0.966	0.882	0.851	0.814	0.786
	5	1	0.958	0.89	0.84	0.81	0.8
	6	1	0.94	0.824	0.774	0.717	0.706
35 mA	7	1	0.882	0.787	0.75	0.7	0.693
	8	1	0.867	0.78	0.733	0.687	0.673
	9	1	0.89	0.8	0.763	0.723	0.713
	10	1	0.962	0.865	0.814	0.745	0.742
	11	1	0.975	0.845	0.81	0.75	0.741
	12	1	0.924	0.854	0.8	0.733	0.715

Source: Chaluvadi, V.N.H., Accelerated life testing of electronic revenue meters, PhD dissertation, Clemson University, Clemson, SC, 2008.

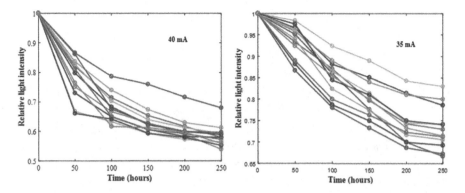

FIGURE 3.1 LED degradation data at a different stress level.

and degradation data from the test. Two different combinations of constant acceler-ated stresses were used to accelerate the lumen degradation of LEDs. At each stress level, twelve samples are assigned, and the light intensity of each sample LED was measured at room temperature every 50 hours up to 250 hours. The operating stress is defined as 30 mA and 50 percent degradation of the initial light intensity is con-sidered to be the failure threshold value.

Figure 3.1 shows the nonlinear nature of the LEDs degradation path that justifies our assumption of the non-stationary continuous state stochastic pro-cess. The nonlinear likelihood function with multiple model parameters makes a greater challenge to estimate parameter values. The MLE method with an advanced optimization software R has been used to solve these complex equa-tions. The built-in "mle" function that uses the Nelder-Mead algorithm (optim) to optimize the likelihood function is used to estimate model parameters. After the model parameters for each stochastic process have been estimated, the lifetime and reliability under any given set of operating conditions can be estimated. Now, considering the different stochastic process models, the parameter and lifetime estimates are provided in Table 3.2.

The results show that the Wiener process has deviated (larger) lifetime estimates compared to the Gamma and IG process. Figure 3.2 illustrates the reliability estimates considering different stochastic process models. Similar to lifetimes, reliability plots also show deviated (higher) estimate by the Wiener process.

TABLE 3.2

Parameter and Lifetime Estimates with Different Degradation Model

Model	$\hat{\gamma}_0$	$\hat{\gamma}$	$\hat{\delta}_0$	$\hat{\delta}_1$	$\hat{c}$	Lifetime
Weibull	−4.3516	0.9483	−3.8413	0.1570	0.4569	3002.26
Gamma	−0.7636	0.0954	4.2685	−.08528	0.5802	1812.28
IG	−5.1956	0.9481	−6.1025	0.1185	0.6097	1611.15

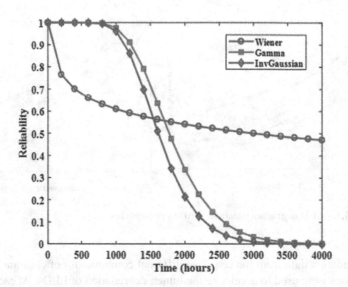

FIGURE 3.2   Reliability estimates using various continuous stochastic processes.

3.2.5   SELECTION OF APPROPRIATE CONTINUOUS STATE STOCHASTIC PROCESS

The appropriate selection of the stochastic process is very important because effective degradation modeling depends on the appropriate choice of the process. The reliability estimation and its accuracy also are dependent on the appropriate stochastic process selection. From the LED case study example, it is observed that the lifetime and reliability estimates differ among three continuous state stochastic processes. There are several criteria to choose an appropriate stochastic process for specific degradation cases which are discussed next.

The graphical analysis is a very common method to check the data patterns and behavior. Figure 3.3 illustrates the histogram and CDF graphs to compare the fitness of three different stochastic processes. The histogram and the CDF graphs suggest that the Gamma process provides the best fit for LED degradation data. On the other hand, the Wiener process is the least fitted degradation model for LED data. Besides, quantile-quantile (Q-Q) plot and probability plots are also a very useful graphical technique to check the model fitness. These plots also provide the same conclusion for the LED data (see Figure 3.4).

Besides graphical methods, there are other stronger statistical methods that are used to check the model fitness such as goodness-of-fit tests. Several parametric or nonparametric methods are available to compare the model fitness such as KS (Kolmogorov-Smirnov) statistic, CVM (Cramer-von Mises) statistic, AD (Anderson-Darling) statistic, AIC (Akaike's Information Criterion), and BIC (Bayesian Information Criterion). All these statistics and criteria are used to select the best-fitted model. Table 3.3 provides the goodness-of-fit statistic value to compare the fitness to the stochastic processes for LED data. It is observed that the Gamma process has the least statistic value in all cases and Wiener has the highest statistic value. This observation implies that the Gamma process is the most suitable and

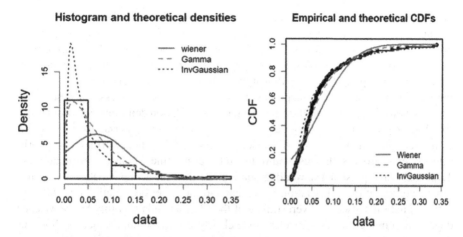

FIGURE 3.3 Graphical model fitness of LED degradation data.

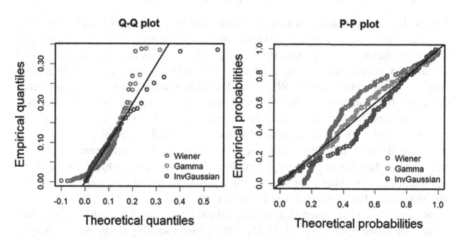

FIGURE 3.4 Q-Q and probability plots of degradation data.

TABLE 3.3

Goodness-of-fit Statistics for Stochastic Processes

Goodness-of-fit Statistic	Wiener	Gamma	Inverse Gaussian
KS statistic	0.1802	0.0708	0.1590
CVM statistic	1.27821	0.1159	0.5977
AD statistic	7.1927	0.6224	2.9771
AIC	−315.2034	−407.427	−389.4947
BIC	−309.6285	−401.852	−383.9197

Wiener is the least suitable model for the LED degradation data. This result explains the huge discrepancy between the lifetime and reliability estimates of the Wiener process compared to other two degradation models. The physical degradation phenomena also is intuitive to this fitness checking criteria. As LEDs are monotonically degraded over a period of time, thus it basically follows the assumption of a monotonic and nonnegative Gamma process most and then an IG process. Because of the clear monotonic behavior of the LED data, the degradation definitely does not follow the Wiener process. All the model fitness test statistic and criteria also indicate an ill-fitted degradation behavior of Wiener process for LED data. Further, this poorly fitted Wiener process also resulted in much lower nonlinear constant estimates (see Table 3.2) that represent a slower degradation rate than the actual situation. This misrepresentation of the degradation increment and the lower degradation rate than the actual situation causes the overestimate of the lifetime and reliability by the Wiener degradation modeling. This case example clearly shows the importance of choosing the right stochastic process for assessing the system's degradation behavior.

3.3   DISCRETE STATE STOCHASTIC PROCESSES

This section presents and discusses different stochastic processes used to model the discrete state degradation process. Unlike the Wiener process, Gamma process, and IG process models, a finite state stochastic process evolves through a finite number of states. In a continuous state degradation process, the degradation process is modeled as a continuous variable. When the degradation process exceeds a predefined threshold, the item is considered failed. However, most engineering systems consist of components that have a range of performance levels from perfect functioning to complete failure. In the discrete-state space, the overall status of the degradation process is divided into several discrete levels with different performances ranging from perfect functioning to complete failure. It is important to highlight here that when a number of states approach to infinity, the discrete-state space and continuous-state space become equivalent to each other.

In general, it is assumed that the degradation process $\{X(t), t \geq 0\}$ evolves on a finite state space $S = \{0, 1, \ldots, M-1, M\}$ with 0 corresponding to the perfect healthy state, M representing the failed state of the monitored system, and others are intermediate states. At time $t = 0$, the process is in the perfect state and as time passes it moves to degraded states. A state transition diagram used for modeling the degradation process is shown in Figure 3.5. Each node represents the state of the degradation process and each branch between two nodes represents the transition between the states corresponding to the nodes. A system can degrade according to three types of transitions: transition to the neighbor state (Type 1), transition to any intermediate state (Type 2), and transition to the failure state (Type 3). Type 1 transitions from one state to the next degraded state are typical of degradation mechanisms driven by cumulative damage and is called minor degradation. Type 2 and Type 3 transitions are called major degradation.

In the context of modeling degradation process, this section focuses on cases in which there is no intervention in the degradation process; i.e., once the process transits to a degradation state, the previous state is not visited again.

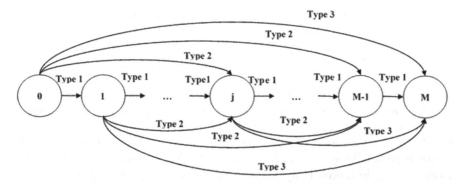

FIGURE 3.5 A multi-state degradation process with minor and major degradation.

The discrete state stochastic process used to model the degradation process can be divided into different categories depending on the continuous or discrete nature of the time variable, and Markovian and non-Markovian property (Moghaddass and Zuo 2014).

From a time viewpoint, the multistate degradation process can evolve according to a discrete-time stochastic process or a continuous-time stochastic process. In the discrete-time type, the transition between different states occurs only at a specific time; however, transitions can occur at any time for the continuous-time stochastic process. With respect to the dependency of degradation transitions to the history of the degradation process, the multistate degradation process can be divided into Markovian degradation process and non-Markovian degradation process. When the degradation transition between two states depends only on its current states, that is, the degradation process is independent of the history of the process, the degradation model follows the Markovian structure. On the other hand, in a multistate degradation process with a non-Markovian structure, the transition between two states may depend on other factors like previous states, the age of the system, and on how long the system has been in its current state. The following sections provide a detailed discussion on Markovian structure and semi-Markov process with suitable examples.

3.3.1 Markovian Structure

A stochastic process $\{X(t)|t \geq 0\}$ is called a Markov process if for any $t_0 < t_1 < t_2 < \cdots < t_{n-1} < t_n < t$ the conditional distribution of $X(t)$ for given values of $X(t_0), X(t_1), \ldots, X(t_n)$ depends only on $X(t_n)$:

$$\Pr\{X(t) \leq x \mid X(t_n) = x_n, X(t_{n-1}) = x_{n-1}, \ldots, X(t_1) = x_1,$$

$$X(t_0) = x_0)\} = \Pr\{X(t) \leq x \mid X(t_n) = x_n\} \tag{3.20}$$

This applies to a Markov process with discrete-state space or continuous-state space. A Markov process with discrete-state space is known as a Markov chain. If the time space is discrete, then it is a discrete-time Markov chain otherwise it is a continuous-time Markov chain.

A discrete-time Markov chain is a sequence of random variables $X_0, X_1, \ldots, X_n, \ldots$ that satisfy the following equation for every n ($n = 0, 1, 2, \ldots$):

$$\Pr\left(X_n = x_n \mid X_0 = x_0, X_1 = x_1, \ldots, X_{n-1} = x_{n-1}\right) = \Pr\left(X_n = x_n \mid X_{n-1} = x_{n-1}\right) \quad (3.21)$$

If the state of the Markov chain at time step n is x_n, we denote it as $X_n = x_n$. Equation 3.21 implies that the chain behavior in the future depends only on its current state and it is independent of its behavior in the past. Therefore, the probability that the Markov chain is going from state i into state j in one step, which is called one-step transition probability, is $p_{ij} = \Pr\left(X_n = j \mid X_{n-1} = i\right)$. For time a homogeneous Markov chain, the transition probability between two states does not depend on the n, i.e., $p_{ij} = \Pr\left(X_n = j \mid X_{n-1} = i\right) = \Pr\left(X_1 = j \mid X_0 = i\right) = \text{constant}$. The one-step transition probabilities can be condensed into a transition probability matrix for a discrete-time Markov chain with $M + 1$ states as follows:

$$P = \begin{pmatrix} p_{00} & p_{01} & \cdots & p_{0M} \\ p_{10} & p_{11} & \cdots & p_{1M} \\ \cdots & \cdots & \cdots & \cdots \\ p_{M0} & p_{M1} & \cdots & p_{MM} \end{pmatrix} \quad (3.22)$$

The sum of each row in P is one and all elements are non-negative. As the discrete-time Markov chain is used to model the degradation process of an item, the transition probability matrix P is in upper-triangular form ($p_{ij} = 0$ for $i > j$) to reflect the system deterioration without considering maintenance or repair. Moreover, for the failure state M, which is also known as an absorbing state, $p_{MM} = 1$ and $p_{Mj} = 0$ for $j = 0, 1, \ldots, M - 1$.

Having the transition probability matrix P and the knowing the initial conditions of the Markov chain, $p(0) = \left[p_0(0), p_1(0), \ldots, p_M(0)\right]$, we can compute the state probabilities at step n, $p(n) = \left[p_0(n), p_1(n), \ldots, p_M(n)\right]$. $p_j(n) = \Pr\{X_n = j\}, j = 1, \ldots, M$, which is the probability that the chain is in state j after n transitions. For many applications such as reliability estimation and prognostics, state probabilities are of utmost interest.

Based on the Chapman-Kolmogorov equation, the probability of a process moving from state i to state j after n steps (transitions) can be calculated by multiplying the matrix P by itself n times (Ross 1995). Thus, assuming that $p(0)$ is the initial state vector, the row-vector of the state probabilities after the nth step is given as:

$$p(n) = p(0).P^n \quad (3.23)$$

For most of the systems, as the system is in the perfect condition at the beginning of its mission, the initial state vector is given as $p(0) = [1, 0, 0, \ldots, 0]$.

When the transition from the current state i to a lower state j takes place at any instant of the time, the continuous-time Markov chain is used to model the degradation process. In analogy with discrete-time Markov chains, a stochastic process

$\{X(t)\,|\,t \geq 0\}$ is a continuous-time Markov chain if the following equation holds for every $t_0 < t_1 < \ldots < t_{n-1} < t_n$ (n is a positive integer):

$$\Pr\left(X(t_n) = x_n \mid X(t_0) = x_0, \ldots, X(t_{n-1}) = x_{n-1}\right) = \Pr\left(X(t_n) = x_n \mid X(t_{n-1}) = x_{n-1}\right) \quad (3.24)$$

Equation 3.24 is analogous to Equation 3.21. Thus, most of the properties of the continuous-time Markov process are similar to those of the discrete-time Markov process. The probability of the continuous-time Markov chain going from state i into state j during Δt, which is called transition probability, is $\Pr\left(X(t + \Delta t) = j \mid X(t) = i\right) = \pi_{ij}(t, \Delta t)$. They satisfy: $\pi_{ij}(t, \Delta t) \geq 0$ and $\sum_{j=0}^{M} \pi_{ij}(t, \Delta t) = 1$.

For time homogeneous continuous-time Markov chain, the transition probability between two states does not depend on the t but depends only on the length of the time interval Δt. Moreover, the transition rate $(\lambda_{ij}(t))$ from state i to state $j\,(i \neq j)$ at time t is defined as: $\lambda_{ij}(t) = \lim_{\Delta t \to 0} \frac{\pi_{ij}(t, \Delta t)}{\Delta t}$, which does not depend on t and is constant for a homogeneous Markov process.

Like the discrete-time case, it is important to get the state probabilities for calculating the availability and reliability measures for the system. The state probabilities of $X(t)$ are:

$$p_j(t) = \Pr\{X(t) = j\}, j = 0,1,\ldots, M \text{ for } t \geq 0 \quad \text{and} \quad \sum_{j=0}^{M} p_j(t) = 1 \quad (3.25)$$

Knowing the initial condition and based on the theorem of total probability and Chapman-Kolmogorov equation, the state probabilities are obtained using the system of differential equations as (Trivedi 2002; Ross 1995):

$$\dot{p}_j(t) = \frac{dp_j(t)}{dt} = \sum_{\substack{i=0 \\ i \neq j}}^{M} p_i(t)\lambda_{ij} - p_j(t) \sum_{\substack{i=0 \\ i \neq j}}^{M} \lambda_{ji}, \quad j = 0,1,\ldots, M \quad (3.26)$$

Equation 3.26 can be written in the matrix notation as:

$$\frac{d\mathbf{p}(t)}{dt} = \mathbf{p}(t)\lambda, \ \mathbf{p}(t) = \left[p_0(t), p_1(t), \ldots, p_M(t)\right], \lambda = \begin{pmatrix} \lambda_{00} & \lambda_{01} & \cdots & \lambda_{0M} \\ \lambda_{10} & \lambda_{11} & \cdots & \lambda_{1M} \\ \cdots & \cdots & \cdots & \cdots \\ \lambda_{M0} & \lambda_{M1} & \cdots & \lambda_{MM} \end{pmatrix}$$

$$(3.27)$$

In the transition rate matrix, $\lambda_{jj} = -\sum_{i \neq j} \lambda_{ji}$ and $\sum_{j=0}^{M} \lambda_{ij} = 0$ for $0 \leq i \leq M$. As the continuous-time Markov chain is used to model the degradation process, the transition rate matrix λ is in upper-triangular form ($\lambda_{ij} = 0$ for $i > j$) to reflect the degradation process without considering maintenance or repair. Since state M is an absorbing state, all the transition rates from this state are equal to zero, $\lambda_{Mj} = 0$ for $j = 0,1,\ldots, M-1$.

Regarding the method to solve the system of Equation 3.27, there are several methods including numerical and analytical methods such as enumerative method

(Liu and Kapur 2007), recursive approach (Sheu and Zhang 2013), and Laplace-Stieltjes transform (Lisnianski and Levitin 2003).

Example 3.3.1.1

Consider a system that can have four possible states, $S = \{0,1,2,3\}$, where state 0 indicates that the system is in as good as new condition, states 1 and 2 are intermediate degraded conditions, and state 3 is the failure state. The system has only minor failures; i.e., there is no jump between different states without passing all intermediate states. The transition rate matrix is given as:

$$\lambda = \begin{pmatrix} \lambda_{00} & \lambda_{01} & \lambda_{02} & \lambda_{03} \\ \lambda_{10} & \lambda_{11} & \lambda_{12} & \lambda_{13} \\ \lambda_{20} & \lambda_{21} & \lambda_{22} & \lambda_{23} \\ \lambda_{30} & \lambda_{31} & \lambda_{32} & \lambda_{33} \end{pmatrix} = \begin{pmatrix} -3 & 3 & 0 & 0 \\ 0 & -2 & 2 & 0 \\ 0 & 0 & -1 & 1 \\ 0 & 0 & 0 & 0 \end{pmatrix}$$

The $\lambda_{33} = 0$ shows that the state 3 is an absorbing state. If the system is in the best state at the beginning ($p(0) = [p_0(0), p_1(0), p_2(0), p_3(0)] = [1,0,0,0]$), the goal is to compute the system reliability at time $t > 0$.

Solution 3.3.1.1: For the multi-state systems, the reliability measure can be based on the ability of the system to meet the customer demand W (required performance level). Therefore, the state space can be divided into two subsets of acceptable states in which their performance level is higher than or equal to the demand level and unacceptable states. The reliability of the system at time t is the summation of probabilities of all acceptable states. All the unacceptable states can be regarded as failed states, and the failure probability is a sum of probabilities of all the unacceptable states.

First, find the state probabilities at time t for each state solving the following differential equations:

$$\begin{cases} \dfrac{dp_0(t)}{dt} = -\lambda_{01} p_0(t) \\ \dfrac{dp_1(t)}{dt} = -\lambda_{01} p_0(t) - \lambda_{12} p_1(t) \\ \dfrac{dp_2(t)}{dt} = -\lambda_{12} p_1(t) - \lambda_{23} p_2(t) \\ \dfrac{dp_3(t)}{dt} = -\lambda_{23} p_2(t) \end{cases}$$

Using the Laplace-Stieltjes transforms and inverse Laplace-Stieltjes transforms (Lisnianski et al. 2010), the state probabilities at time t are found as:

$$\begin{cases} p_0(t) = e^{-\lambda_{43}t} \\ p_1(t) = \dfrac{\lambda_{01}}{\lambda_{01} - \lambda_{12}}(e^{-\lambda_{12}t} - e^{-\lambda_{01}t}) \\ p_2(t) = \dfrac{-\lambda_{12}\lambda_{01}[(\lambda_{01} - \lambda_{12})e^{-\lambda_{23}t} + (\lambda_{23} - \lambda_{01})e^{-\lambda_{12}t} + (\lambda_{12} - \lambda_{23})e^{-\lambda_{01}t}]}{(\lambda_{12} - \lambda_{21})(\lambda_{01} - \lambda_{12})(\lambda_{23} - \lambda_{01})} \\ p_3 = 1 - p_2(t) - p_1(t) - p_0(t) \end{cases}$$

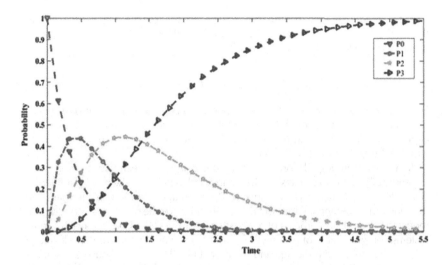

FIGURE 3.6   System state probabilities: Example 3.3.1.1.

The plot of the state probabilities is shown in Figure 3.6. As shown, the probability of being in state 0 is decreasing with time and the probability of being in state 3 is increasing with time.

Then the reliability of the system at time t is calculated based on the demand level by summation of the probabilities of all acceptable states as:

$$\begin{cases} \text{If acceptable states are}: 0,1,2 \rightarrow R_1(t) = p_0(t) + p_1(t) + p_2(t) \\ \text{If acceptable states are}: 0,1 \rightarrow R_2(t) = p_0(t) + p_1(t) \\ \text{If acceptable states are}: 0 \rightarrow R_3(t) = p_0(t) \end{cases}$$

The plots of the system reliability for all three cases are shown in Figure 3.7.

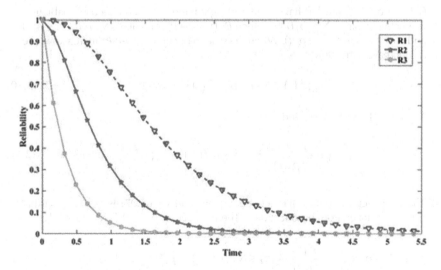

FIGURE 3.7   System reliability for various cases.

Let τ_i denote the time that the degradation process spent in state i. According to the Markov property in Equation 3.24, i does not depend on the past state of the process, so the following equation holds:

$$P\left(\tau_i > t + \Delta t \mid \tau_i > t\right) = h\left(\Delta t\right) \tag{3.28}$$

Function $h(\Delta t)$ in Equation 3.28 only depends on Δt, and not on the past time t. The only continuous probability distribution that satisfies Equation 3.28 is the exponential distribution. In the discrete time case, requirement in Equation 3.28 leads to the geometric distribution.

In a Markovian degradation structure, the transition between two states at time t depends only on the two states involved and is independent of the history of the process before time t (memoryless property). The fixed transition probabilities/rates and the geometric/exponential sojourn time distribution limit the use of a Markov chain to model the degradation process of real systems. For the degradation process of some systems, the probability of making the transition from one state to a more degraded state may increase with the age and the probability that it continuously stays at the current state will decrease. That is, $p_{ii}(t + \Delta t) \leq p_{ii}(t)$ and $\sum_{j=i+1}^{n} p_{ij}(t+\Delta t) \geq \sum_{j=i+1}^{n} p_{ij}(t)$. Therefore, the transition probabilities and transition rates are not constant during the time and an extension of the Markovian model, which is called aging Markovian deterioration model, is used to include this aging effect.

For the discrete-time aging Markovian model, $P(t)$ is one-step transition probability matrix at time t and $p_{ij}(t)$ represents the transition probability from state i to state j at time t. As shown in Chen and Wu (2007), each row of $P(t)$ represents a state probability distribution given the current state at i that will form a bell-shape distribution. Let N_i satisfy $p_{i,N_i}(t) = \max\left\{p_{i,j}(t), j = 0, 1, \ldots, M\right\}$, where N_i represents the peak transition probability in the bell-shape distribution. Then:

$$P_i^L(t) \equiv \sum_{j=1}^{N_i} p_{ij}(t) ; \; P_i^R(t) \equiv \sum_{j=N_i+1}^{M} p_{ij}(t) \tag{3.29}$$

$P_i^L(t)$ and $P_i^R(t)$ are left-hand side and right-hand side cumulated probabilities, respectively. Since $\sum_{j=1}^{M} p_{ij}(t) = 1$, then $P_i^R(t) = 1 - P_i^L(t)$. For $j \leq N_i$, $p_{ij}(t+1) \leq p_{ij}(t)$ and for $j > N_i$, $p_{ij}(t+1) \leq p_{ij}(t)$. When the system becomes older, P_i^L increases while P_i^R decreases, therefore:

$$P_i^L(t) \geq P_i^L(t+1) ; \; P_i^R(t) \leq P_i^R(t+1) \tag{3.30}$$

Then $P(t+1)$ can be modified as:

$$p_{ij}(t+1) \equiv p_{ij}(t) \frac{P_i^L(t+1)}{P_i^L(t)} \; \forall j \leq N_i; \; p_{ij}(t+1) \equiv p_{ij}(t) \frac{P_i^R(t+1)}{P_i^R(t)} \; \forall j > N_i \tag{3.31}$$

The aging factor δ $(0 \leq \delta < 1)$ is defined by Chen and Wu (2007) as $\delta = \frac{P_i^R(t+1)}{P_i^R(t)} - 1$ that can be estimated from historical data. Therefore, Equation 3.31 is represented as:

$$p_{ij}(t+1) \equiv p_{ij}(t) \cdot \left(1 - \frac{P_i^R(t+1)}{P_i^L(t)}\right) \forall j \leq N_i; \; p_{ij}(t+1) \equiv p_{ij}(t) \cdot (1+\delta) \; \forall j > N_i \tag{3.32}$$

Starting with the initial transition probability matrix $P(0)$, the values of the $P(t)$, which are changing during the time, can be calculated according to Equation 3.32.

For the continuous-time aging Markovian model, which is called the non-homogeneous continuous-time Markov process, the amount of time that the system spends in each state before proceeding to the degraded state does not follow the exponential distribution. Usually, the transition times are assumed to obey Weibull distribution because of its flexibility, which allows considering hazard functions both increasing and decreasing over time, at different speeds.

To get the state probabilities at each time t, we have to solve the Chapman-Kolmogorov equations as:

$$\frac{dp_j(t)}{dt} = \sum_{\substack{i=0 \\ i \neq j}}^{M} p_i(t)\lambda_{ij}(t) - p_j(t)\sum_{\substack{i=0 \\ i \neq j}}^{M} \lambda_{ji}(t), \quad j = 0,1,\ldots,M \tag{3.33}$$

Equation 3.33 can be written in the matrix form as:

$$\frac{d\,p(t)}{dt} = p(t)\lambda(t),$$

$$p(t) = \begin{bmatrix} p_0(t),\ldots,p_M(t) \end{bmatrix}, \quad \lambda(t) = \begin{pmatrix} \lambda_{00}(t) & \lambda_{01}(t) & \ldots & \lambda_{0M}(t) \\ \lambda_{10}(t) & \lambda_{11}(t) & \ldots & \lambda_{1M}(t) \\ \ldots & \ldots & \ldots & \ldots \\ \lambda_{M0}(t) & \lambda_{M1}(t) & \ldots & \lambda_{MM}(t) \end{pmatrix} \tag{3.34}$$

The transition rate matrix $\lambda(t)$ has the same properties as the transition matrix in Equation 3.27. To find the state probabilities at time t, many methods have been used to solve Equation 3.34 such as state–state integration method (Liu and Kapur 2007) and recursive approach (Sheu and Zhang 2013). Equation 3.34 can be recursively solved from state 0 to state M as follows:

$$p_0(t) = e^{\int_0^t \lambda_{00}(s)ds} \tag{3.35}$$

$$p_j(t) = \sum_{i=0}^{j-1} \int_0^t p_i(\tau_{i+1})\lambda_{ij}(\tau_{i+1})e^{\int_{\tau_{i+1}}^t \lambda_{jj}(s)ds}\,d\tau_{i+1}, \quad j = 1,\ldots, M-1 \tag{3.36}$$

$$p_M(t) = 1 - \sum_{j=0}^{M-1} p_j(t) \tag{3.37}$$

The initial conditions are assumed to be $p(t) = \begin{bmatrix} p_0(0) = 1, p_1(0) = 0, \ldots p_M(0) = 0 \end{bmatrix}$.

Example 3.3.1.2

(Sheu and Zhang 2013; Shu et al. 2015) Assume that a system degrades through five different possible states, $S = \{0,1,2,3,4\}$ and state 0 is the best state and state 4 is the worst state. The time T_{ij} spent in each state i before moving to the next state j follows the Weibull distribution $T_{ij} \sim Weibull(1/(i-0.5j),3)$ with scale parameter

$\alpha_{ij} = 1/(i - 0.5j)$ and shape parameter $\beta = 3$. The nonhomogeneous continuous time Markov process is used to model the degradation process. The transition rate from state i to state j at time t is $\lambda_{ij}(t) = 3t^2/(i - 0.5j)^3 \, \forall i, j \in S, i > j$. Based on the demand level, the states 3 and 4 are unacceptable states. The goal is to compute the system reliability at time $t(0 < t < 4)$.

Solution 3.3.1.2: The transient degradation rate matrix is:

$$\lambda(t) = \begin{pmatrix} \lambda_{00}(t) & \lambda_{01}(t) & \lambda_{02}(t) & \lambda_{03}(t) & \lambda_{04}(t) \\ 0 & \lambda_{11}(t) & \lambda_{12}(t) & \lambda_{13}(t) & \lambda_{14}(t) \\ 0 & 0 & \lambda_{22}(t) & \lambda_{23}(t) & \lambda_{24}(t) \\ 0 & 0 & 0 & \lambda_{33}(t) & \lambda_{34}(t) \\ 0 & 0 & 0 & 0 & 0 \end{pmatrix} =$$

$$\begin{pmatrix} -0.419945t^2 & 0.1920t^2 & 0.1111t^2 & 0.06997t^2 & 0.046875t^2 \\ 0 & -0.6781t^2 & 0.3750t^2 & 0.1920t^2 & 0.1111t^2 \\ 0 & 0 & -1.2639t^2 & 0.8889t^2 & 0.3750t^2 \\ 0 & 0 & 0 & -3t^2 & 3t^2 \\ 0 & 0 & 0 & 0 & 0 \end{pmatrix}$$

$p_0(0) = 1$, $p_j(0) = 0$ $j = 1, 2, \ldots, M$.
The state probabilities can be obtained using Equations 3.36 and 3.37 as:

$$p_0(t) = e^{-0.14t^3}$$

$$p_1(t) = -0.7439(e^{-0.226t^3} - e^{-0.14t^3})$$

$$p_2(t) = 0.0139e^{-0.4213t^3} + 0.4623e^{-0.14t^3} - 0.4761e^{-0.226t^3}$$

$$p_3(t) = -0.005109e^{-t^3} + 0.24178e^{-0.14t^3} - 0.24381e^{-0.226t^3} + 0.007117e^{-0.4213t^3}$$

$$p_4(t) = 1 - p_0(t) - p_1(t) - p_2(t) - p_3(t)$$

$$= 1 - 2.44798e^{-0.14t^3} + 1.46381e^{-0.226t^3} - 0.021017e^{-0.4213t^3} + 0.005109e^{-t^3}$$

The system state probabilities are shown in Figure 3.8.

As the states 3 and 4 are unacceptable states, the reliability of the system at time t is $R_s(t) = p_0(t) + p_1(t) + p_2(t)$. Figure 3.9 shows the system reliability as a function of time.

The aging Markovian models used to overcome the limitations of Markov chain structures can be framed as a semi-Markov process. Semi-Markovian structures consider the history of the degradation process and consider arbitrary sojourn time distributions at each state. Semi-Markovian models as an extension of Markovian-based models will be explained in the next section.

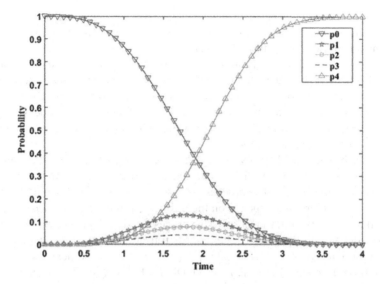

FIGURE 3.8   System state probabilities: Example 3.3.1.2.

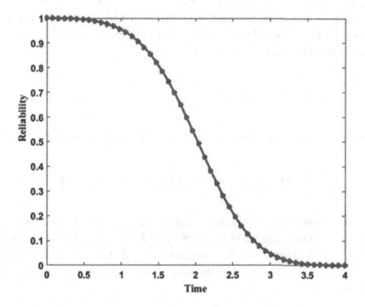

FIGURE 3.9   System reliability as a function of time.

3.3.2   SEMI-MARKOV PROCESS

The semi-Markov process can be applied to model the degradation process of some systems whose degradation process cannot be captured by a Markov process. For example, Ng and Moses (1998) used the semi-Markov process to model bridge degradation behavior. They described the semi-Markov process in terms of a transition matrix and a holding time or sojourn time matrix. A transition matrix has a set of transition probabilities between states that describe the embedded Markov chain. The holding time matrix has a set of probabilities obtained from the probability density function of the holding times between states.

For Markov models, the transition probability of going from one state to another does not depend on how the item arrived at the current state or how long it has been there. However, semi-Markov models relax this condition to allow the time spent in a state to follow an arbitrary probability distribution. Therefore, the process stays in a particular state for a random duration that depends on the current state and on the next state to be visited (Ross 1995).

To describe the semi-Markov process $X \equiv \{X(t) : t \geq 0\}$, consider the degradation process of a system with finite state space $S = \{0,1,2,...,M\}$ ($M + 1$: the total number of possible states). The process visits some state $i \in S$ and spends a random amount of time there that depends on the next state it will visit, $j \in S, i \neq j$. Let T_n denote the time of the nth transition of the process, and let $X(T_n)$ be the state of the process after the nth transition. The process transitions from state i to state $j \neq i$ with the probability $p_{ij} = P(X(T_{n+1}) = j | X(T_n) = i)$. Given the next state is j, the sojourn time from state i to state j has a CDF, F_{ij}. For a semi-Markov process, the sojourn times can follow any distribution, and p_{ij} is defined also as the transition probability of the embedded Markov chain.

The one-step transition probability of the semi-Markov process transiting to state j within a time interval less than or equal to t, provided starting from state, is expressed as (Cinlar 1975):

$$Q_{ij}(t) = \Pr\left(X(T_{n+1}) = j, T_{n+1} - T_n \leq t, \middle| X(T_n) = i \right) \quad t \geq 0 \tag{3.38}$$

The random time between every transition $(T_{n+1} - T_n)$, sojourn time, has a CDF as:

$$F_{ij}(t) = \Pr\left(T_{n+1} - T_n \leq t \middle| X(T_{n+1}) = j, X(T_n) = i \right) \tag{3.39}$$

If the sojourn time in a state depends only on the current visited state, then the unconditional sojourn time in state i is $F_{ij}(t) = F_i(t) = \sum_{j \in S} Q_{ij}(t)$. The transition probabilities of the semi-Markov process $(Q(t) = [Q_{ij}(t)], i, j \in S)$, which is called semi-Markov kernel, is the essential quantity of a semi-Markov process and satisfies the relation:

$$Q_{ij}(t) = p_{ij} F_{ij}(t) \tag{3.40}$$

Equation 3.40 indicates that the transition of the semi-Markov model has two steps. Figure 3.10 shows a sample degradation path of a system. The system is in the state i at the initial time instance and transits to the next worse state j with transition probability p_{ij}. As the process is a monotone non-increasing function without considering the maintenance, $j = i + 1$ with probability one. Before moving into the next state j, the process will wait for a random time with CDF $F_{ij}(t)$. This process continues until

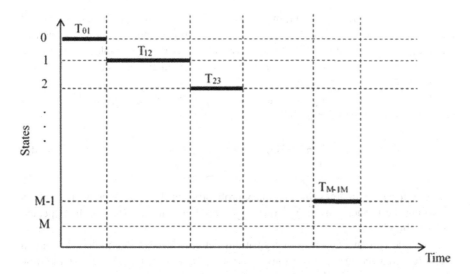

FIGURE 3.10   A sample degradation process.

the process enters the state M that is an absorbing state. For this example the transition probability matrix is given as:

$$P = \begin{pmatrix} 0 & 1 & 0 & \cdots & 0 \\ 0 & 0 & 1 & \cdots & 0 \\ \cdots & \cdots & \cdots & \cdots & \cdots \\ 0 & 0 & 0 & \cdots & 1 \end{pmatrix} \qquad (3.41)$$

When the semi-Markov process is used to model the degradation process, the initial state of the process, the transition probability matrix P, and matrix $F(t)$ must be known. Another way of defining the semi-Markov process is knowing the kernel matrix and the initial state probabilities.

   Like previous models, it is important to find the state probabilities of the semi-Markov process. The probability that a semi-Markov process will be in state j at time $t \geq 0$ given that it entered state i at time zero, $\pi_{ij}(t) \equiv \Pr\{X(t) = j \mid X(0) = i\}$, is found as follows (Howard 1960; Kulkarni 1995):

$$\pi_{ij}(t) = \delta_{ij}\left[1 - F_i(t)\right] + \sum_{k \in S}\int_0^t q_{ik}(\vartheta)\pi_{kj}(t - \vartheta)d\vartheta \qquad (3.42)$$

$$q_{ik}(\vartheta) = \frac{dQ_{ik}(\vartheta)}{d\vartheta} \qquad (3.43)$$

$$\delta_{ij} = \begin{cases} 1 & i = j \\ 0 & i \neq j \end{cases} \qquad (3.44)$$

In general, it is difficult to obtain the transition functions, even when the kernel matrix is known. Equation 3.42 can be solved using numerical methods such as quadrature method (Blasi et al. 2004; Corradi et al. 2004) and Laplace and inverse Laplace transforms (Dui et al. 2015) or simulation methods (Sánchez-Silva and Klutke 2016).

Moreover, the stationary distribution $\pi = (\pi_j; j \in S)$ of the semi-Markov process is defined, when it exists, as:

$$\pi_j := \lim_{t\to\infty} \pi_{ij}(t) = \frac{\upsilon_j w_j}{\sum_{i=0}^{M} \upsilon_i w_i} \tag{3.45}$$

where υ_j for $j \in S$ denotes the stationary probability of the embedded Markov chain satisfying the property: $\upsilon_j = \sum_{i=0}^{M} \upsilon_i p_{ij}$, $\sum_{i=0}^{M} \upsilon_i = 1$, and w_j for $j \in S$ is the expected sojourn time in state j.

For some systems, degradation transitions between two states and may depend on the states involved in the transitions, the time spent at the current state (t), the time that the system reached the current state (s), and/or the total age of the system ($t+s$). As another extension, a nonhomogeneous semi-Markov process is used for modeling the degradation of such systems in which degradation transition can follow an arbitrary distribution.

The associated non-homogeneous semi-Markov kernel is defined by:

$$Q_{ij}(s,t) = \Pr\Big(X(T_{n+1}) = j, T_{n+1} \le t, \big| X(T_n) = i, T_n = s \Big) \, t \ge 0 \tag{3.46}$$

In non-homogeneous semi-Markov, the state probabilities are defined and obtained using the following equation:

$$\pi_{ij}(t) = \Pr\{X(t) = j \mid X(0) = i\} = \delta_{ij}[1 - F_i(t,s)] + \sum_{k \in S} \int_{s}^{t} q_{ik}(s,\vartheta)\pi_{kj}(t-\vartheta)(d\vartheta) \tag{3.47}$$

The obtained state probabilities can be used to find different availability and reliability indexes.

Example 3.3.2

Consider a system (or a component) whose possible states during its evolution in time are $S = \{0, 1, 2\}$. Denote by $U = \{0, 1\}$ the subset of working states of the system and by $D = \{2\}$ the failure state. In this system, both minor and major failures are possible. The state transition diagram is shown in Figure 3.11.

The holding times are normally distributed, i.e., $F_{ij} \sim N(\mu_{ij}, \sigma_{ij})$. Therefore, the CDF of the holding time from state i to state j is:

$$F_{ij}(t) = \frac{1}{\sqrt{2\pi\sigma_{ij}^2}} \int_{0}^{t} e^{-\left[\frac{(u-\mu_{ij})}{2\sigma_{ij}}\right]} du \qquad \forall i, j \in S$$

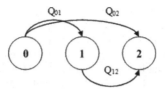

FIGURE 3.11 State transition diagram for semi-Markov model.

The goal is to find the system reliability at time t given the best state is the initial state of the system.

Solution 3.3.2: As the system is at state 0 at the beginning, the reliability of the system at time t is the probability of transition from state 0 to state 2 at time t, $\pi_{02}(t)$.

First, we find the kernel matrix of the semi-Markov process $Q(t) = [Q_{ij}(t)], i, j \in S$:

$$Q(t) = \begin{bmatrix} 0 & Q_{01}(t) & Q_{02}(t) \\ 0 & 0 & Q_{12}(t) \\ 0 & 0 & 0 \end{bmatrix}$$

$Q_{01}(t)$ is the probability that the process transitions from state 0 to 1 within a time interval less than or equal to t that can be determined as the probability that the time of transition from state 0 to 1 (T_{01}) is less than or equal to t and the time of transition from state 0 to 2 (T_{02}) is greater than t.

$$Q_{01}(t) = \Pr\left(T_{01} \le t \text{ and } T_{02} > t\right) = \int_0^t \left[1 - F_{02}(t)\right] dF_{01}(t)$$

Other values of the kernel matrix are obtained as:

$$Q_{02}(t) = \Pr\left(T_{02} \le t \text{ and } T_{01} > t\right) = \int_0^t \left[1 - F_{01}(t)\right] dF_{02}(t)$$

$$Q_{12}(t) = \Pr\left(T_{12} \le t\right) = F_{12}$$

According to Equation 3.42, the following system of equations has to be solved to obtain the system reliability $(\pi_{02}(t))$:

$$\begin{cases} \pi_{02}(t) = \int_0^t q_{01}(\vartheta)\pi_{12}(t - \vartheta)d\vartheta \\ \\ \pi_{12}(t) = \int_0^t q_{12}(\vartheta)\pi_{22}(t - \vartheta)d\vartheta \\ \\ \pi_{22}(t) = 1 \end{cases}$$

All these models presented are based on the assumption that the degradation process is directly observable. However, in many cases, the degradation level is not directly observable due to the complexity of the degradation process or the nature of the product type. Therefore, to deal with indirectly observed states, models such as hidden Markov models (HMM) and hidden semi-Markov models (HSMM) have been developed. The HMM deals with two different stochastic processes: the unobservable degradation process and measurable characteristics (which is dependent on the actual degradation process). In HHMs, finding a stochastic relationship between unobservable degradation process and the output signals of the observation process is a critical prerequisite for condition monitoring and reliability analysis. As discussed, the details of HMM are beyond the scope of this chapter, interested readers can refer to Shahraki et al. (2017 and Si et al. (2011) for more details.

3.4 SUMMARY AND CONCLUSIONS

This chapter presented the application of stochastic processes in degradation modeling to assess product/system performances. All the stochastic processes are categorized into continuous state and discrete state processes. Among the continuous state stochastic processes, the Wiener, Gamma, and IG processes are discussed and applied for degradation modeling of engineering systems using accelerated degradation data. The lifetime and reliability estimation approaches also are derived based on stochastic degradation models. For accurately assessing the product performances, appropriate selection of the stochastic process is crucial. The graphical and statistical methods are presented to assist in successful selection of the best-fitted degradation model for a case specific situation.

In addition, discrete state stochastic processes have been discussed and applied to model the degradation of systems when their degraded states take values from discrete space. The discrete- and continuous-time Markov chain models are used to model the degradation process when the state transitions will happen at a discrete or continuous time, respectively. In Markov chain models, the next state of the system only depends on the current and not the history of the system (memoryless property) that limits their application for some systems. As the extensions of the Markovian model, aging Markovian deterioration and semi-Markov models are applied to capture the influence of the age and the history on the future states. The system reliability is calculated for systems that are degrading with time after modeling their degradation process using proper models.

REFERENCES

Blasi, A., Janssen, J. and Manca, R., 2004. Numerical treatment of homogeneous and non-homogeneous semi-Markov reliability models. *Communications in Statistics, Theory and Methods* 33(3): 697–714.

Chaluvadi, V. N. H., 2008. Accelerated life testing of electronic revenue meters. PhD dissertation, Clemson, SC: Clemson University.

Chen, A. and Wu, G.S., 2007. Real-time health prognosis and dynamic preventive maintenance policy for equipment under aging Markovian deterioration. *International Journal of Production Research* 45(15): 3351–3379.

Cinlar E., 1975. *Introduction to Stochastic Processes*. Englewood Cliffs, NJ: Prentice-Hall.

Corradi, G., Janssen, J. and Manca, R., 2004. Numerical treatment of homogeneous semi-Markov processes in transient case—a straightforward approach. *Methodology and Computing in Applied Probability* 6(2): 233–246.

Dui, H., Si, S., Zuo, M. J. and Sun, S., 2015. Semi-Markov process-based integrated importance measure for multi-state systems. *IEEE Transactions on Reliability* 64(2): 754–765.

Howard R. 1960. *Dynamic Programming and Markov Processes*, Cambridge, MA: MIT press.

Kulkarni, V. G. 1995. *Modeling and Analysis of Stochastic Systems*, London, UK: Chapman and Hall.

Limon, S., Yadav, O. P. and Liao, H., 2017a. A literature review on planning and analysis of accelerated testing for reliability assessment. *Quality and Reliability Engineering International* 33(8): 2361–2383.

Limon, S., Yadav, O. P. and Nepal, B., 2017b. Estimation of product lifetime considering gamma degradation process with multi-stress accelerated test data. *IISE Annual Conference Proceedings*, pp. 1387–1392.

Limon, S., Yadav, O. P. and Nepal, B., 2018. Remaining useful life prediction using ADT data with Inverse Gaussian process model. *IISE Annual Conference Proceedings*, pp. 1–6.

Lisnianski, A., Frenkel, I. and Ding, Y., 2010. *Multi-state System Reliability Analysis and Optimization for Engineers and Industrial Managers*, Berlin, Germany: Springer Science & Business Media.

Lisnianski, A. and Levitin, G., 2003. *Multi-state System Reliability: Assessment, Optimization, and Applications*, Singapore: World scientific.

Liu, Y. W. and Kapur, K. K. C., 2007. Customer's cumulative experience measures for reliability of non-repairable aging multi-state systems. *Quality Technology & Quantitative Management* 4(2): 225–234.

Moghaddass, R. and Zuo, M. J., 2014. An integrated framework for online diagnostic and prognostic health monitoring using a multistate deterioration process. *Reliability Engineering & System Safety* 124: 92–104.

Narendran, N. and Gu, Y., 2005. Life of led-based white light sources. *Journal of Display Technology* 1: 167–171.

Nelson, W., 2004. *Accelerated Testing: Statistical Models, Test Plans and Data Analysis* (2nd ed.), New York: John Wiley & Sons.

Ng, S. K. and Moses, F., 1998. Bridge deterioration modeling using semi-Markov theory. *A. A. Balkema Uitgevers B. V, Structural Safety and Reliability* 1: 113–120.

O'Connor, P. D. D. T. and Kleyner, A., 2012. *Practical Reliability Engineering* (5th ed.), Chichester, UK: Wiley.

Park, C. and Padgett, W. J., 2005. Accelerated degradation models for failure based on geometric Brownian motion and gamma processes. *Lifetime Data Analysis* 11: 511–527.

Park, J. I. and Yum, B. J., 1997. Optimal design of accelerated degradation tests for estimating mean lifetime at the use condition. *Engineering Optimization* 28: 199–230.

Ross, S., 1995. *Stochastic Processes*, New York: Wiley.

Sánchez-Silva, M. and Klutke, G. A., 2016. *Reliability and Life-cycle Analysis of Deteriorating Systems* (Vol. 182). Cham, Switzerland: Springer International Publishing.

Shahraki, A. F. and Yadav, O. P., 2018. Selective maintenance optimization for multi-state systems operating in dynamic environments. In *2018 Annual Reliability and Maintainability Symposium (RAMS)*. IEEE: pp. 1–6.

Shahraki, A. F., Yadav, O. P. and Liao, H., 2017. A review on degradation modelling and its engineering applications. *International Journal of Performability Engineering* 13(3): 299.

Sheu, S. H. and Zhang, Z. G., 2013. An optimal age replacement policy for multi-state systems. *IEEE Transactions on Reliability* 62(3): 722–735.

Sheu, S. H., Chang, C. C., Chen, Y. L. and Zhang, Z. G., 2015. Optimal preventive mainte-
    nance and repair policies for multi-state systems. *Reliability Engineering & System
    Safety*, 140, 78–87.
Si, X. S., Wang, W., Hu, C. H. and Zhou, D. H., 2011. Remaining useful life estimation:
    A review on the statistical data driven approaches. *European Journal of Operational
    Research* 213(1): 1–14.
Trivedi, K, 2002. *Probability and Statistics with Reliability, Queuing and Computer Science
    Applications*, New York: Wiley.
Wang, X. and Xu, D., 2010. An inverse Gaussian process model for degradation data.
    Technometrics 52: 188–197.
Ye, Z. S. and Chen, N., 2014. The inverse Gaussian process as a degradation model.
    Technometrics 56: 302–311.
Ye, Z. S., Wang, Y., Tsui, K. L. and Pecht, M., 2013. Degradation data analysis using Wiener
    processes with measurement errors. *IEEE Transactions on Reliability* 62: 772–780.

4 Building a Semi-automatic Design for Reliability Survey with Semantic Pattern Recognition

Christian Spreafico and Davide Russo

CONTENTS

4.1 INTRODUCTION

Almost 70 years after its introduction, Failure Modes and Effects Analysis (FMEA) has been applied in a large series of cases from different sectors, such as automotive, electronics, construction and services, and has become a standard procedure in many companies for quality control and for the design of new products. FMEA has also a great following in the scientific community as testified by the vast multitude of related documents from scientific and patent literature; to date, more than 3,600 papers in Scopus DB and 146 patents in Espacenet DB come up by just searching for FMEA without synonyms, with a trend of constant growth over the years.

The majority of those contributions deals with FMEA modifications involving the procedure and the integrations with new methods and tools to enlarge the field of application and to improve the efficiency of the analysis, such as by reducing the required time and by finding more results.

To be able to orientate among the many contributions, the surveys proposed in the literature can play a fundamental role, which have been performed according to different criteria of data gathering and classification.

In [1] the authors analyzed scientific papers about the description and review of basic principles, the types, the improvements, the computer automation codes, the combination with other techniques, and specific applications of FMEA.

The literature survey in [2] analyzes the FMEA applications for enhancing service reliability by determining how FMEA is focused on profit and supply chain-oriented service business practices. The significant contribution consists in comparing what previously was mentioned about FMEA research opportunities and in observing how FMEA is related to enhancement in Risk Priority Number (RPN), reprioritization, versatility of its application in service supply chain framework and non-profit service sector, as well as in combination with other quality control tools, which are proposed for further investigations.

In [3], the authors studied 62 methodologies about risk analysis by separating them into three different phases (identification, evaluation, and hierarchization) and by studying their inputs (plan or diagram, process and reaction, products, probability and frequency, policy, environment, text, and historical knowledge), the implemented techniques to analyze risk (qualitative, quantitative, deterministic, and probabilistic), and their output (management, list, probabilistic, and hierarchization).

In [4], the authors analyzed the innovative proposed approaches to overcome the limitations of the conventional RPN method within 75 FMEA papers published between 1992 and 2012 by identifying which shortcomings attract the most attention, which approaches are the most popular, and the inadequacy of approaches.

Other authors focused on analyzing specific applications of the FMEA approach. In [5] the authors studied how 78 companies of motor industry in the United Kingdom apply FMEA by identifying some common difficulties such as time constraints, poor organizational understanding of the importance of FMEA, inadequate training, and lack of management commitment.

However, despite the results achieved by these surveys, no overview considers all the proposals presented, including patents, and analyzes at a higher level than "simple" document counting within the cataloging classes and tools used.

To fulfill this aim, a previous survey [6] considerably increased the number of analyzed documents, by including also patents. In addition, the analysis of the content was improved by carrying out the analysis on two related levels: followed strategies of intervention (e.g., reduce time of application) and integrated tools (e.g., fuzzy logic). Although the results achieved are remarkable, the main limitations of this analysis are the onerous amount of time required along with the number of correlations between different aspects (e.g., problems and solutions, methods and tools, etc.).

This chapter proposes a semi-automatic semantic analysis about documents related to FMEA modifications and the subsequent manual review for reassuming each of them through a simple sentence made by a causal chain including the declaration of the goals, the followed strategies (FMEA modifications), and integrations with methods/tools.

This chapter is organized as follows. Section 4.2 presents the proposed procedure of analysis, Section 4.3 proposes the results and the discussions, and Section 4.4 draws conclusions.

4.2 RESEARCH METHODOLOGY AND POOL DEFINITION

The first step of this work is the definition of the pool of documents to be analyzed: starting from the same pool of documents in [6] proposing FMEA modifications. This pool counts 286 documents, 177 scientific papers (165 from academia and 12 from industry), and 109 patents (23 from academia and 86 from industry). Figure 4.1 shows the time distribution for patents and for scientific publications. The number of patents is increasing, except for the last period that does not include all potential patents since they are not disclosed for the first 18 months.

4.2.1 DEFINITION OF THE ELECTRONIC POOL

In order to automatically process the collected documents through available tools for semantic analysis, for each document, an XML file was manually created, which was nominated with a unique ID and compiled according to a rigid structure where each part of the original document was inserted within specific text fields (e.g., Title, Abstract, Introduction, State_of_the_Art, Proposal).

The objective of this classification is to divide the original proposals from each document, within the field Proposal, from the previous ones, reported within the field State_of_the_Art, so as not distort the survey with redundant results, and to provide the possibility to separately process the different parts to achieve specific purposes (e.g., keywords investigation). In addition, the comparison with the ID allows referencing the content to the specific document.

4.2.2 DEFINITION OF THE FEATURES OF ANALYSIS

An additional preliminary activity deals with the definition of the features to be analyzed. Since one purpose of the proposed method is to perform a deeper analysis by relating different aspects, the features deliberately consider heterogeneous aspects (goal, strategies of interventions, and integrations) and they work at different levels of detail (e.g., goals and sub-goals, methods and tools).

Some features have been hypothesized a priori by considering previous FMEA surveys, while others iteratively emerged during the analysis.

In the following discussion, the features are presented in detail.

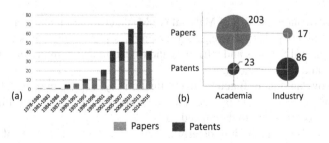

FIGURE 4.1 (a) Time distribution (priority date) of the collected documents and (b) composition of the final set of documents (papers vs. patents and academia vs. industry).

4.2.2.1 Goals

These features deal with targets that the authors who is proposing the analyzed FMEA modifications wants to achieve through them. All of them focus on improving the main aspects related to the applicability of the method (e.g., reducing the required input, improving expected output, ameliorating the approach of the involved actors):

- Reduce FMEA time/costs of application by applying the modified FMEA version to reduce: the number of participant (e.g., experts), the time required to gather the useful information and perform the analysis ([9], [30], [36], [37], [47], [52], [56], [64], [72], [78], [80], [90], [99], [132]).
- Reduce production time/costs of the considered product by using FMEA modifications for finding and preventing possible faults during production that can cause possible delays or extra costs, without modifying product design ([35], [43], [57], [63], [88], [89], [93], [109], [110], [119], [121], [126]).
- Improve design of the product by applying a modified FMEA during design process in order to specifically change the design of the product in order to make it: more robust (i.e., robust design), more able to meet the requirements, or to not dissatisfy them (i.e., product re-design), more easily to be manufactured (i.e., design for manufacturing) though a radically change of product's shape and components, more easily been repaired (i.e., design for maintenance) ([15], [19], [23], [24], [25], [27], [39], [40], [49], [58], [61], [62], [65], [69], [70], [76], [79], [87], [92], [94], [96], [100], [103], [104], [107], [114]).
- Analyze complex systems. If the modified version of FMEA has been specifically improved to manage products with a high number of component and functionalities ([26], [31], [32], [82], [98], [118], [117], [124], [128]).
- Ameliorate human approach. If the modified version of FMEA is able to improve the user interface, reduce its tediousness and better involve the user in a more pro-active approach ([10], [13], [16], [22], [28], [29], [33], [34], [41], [42], [46], [48], [50], [51], [53], [54], [55], [59], [60], [66], [68], [71], [73], [74], [77], [81], [83], [84], [85], [86], [105], [106], [108], [111], [112], [113], [115], [116], [122], [123], [125], [130], [131]).

4.2.2.2 Strategies (FMEA Interventions)

These features investigate the strategies of intervention on FMEA structure, or the parts/steps of the traditional procedure that are modified by the considered documents:

- Improve/automate Bill of Material (BoM) determination to provide criteria to (1) identify the parts (e.g., sub-assemblies and single components) and their useful features and attributes and (2) facilitate the management of the parts and their relations.
- Improve/automate function determination by suggesting modalities to identify and describe product requirements, functions and sub-functions, and associate them to the related parts.
- Improve/automate failure determination to increase the number of considered failure modes, effects and causes, identify their relations, and improve their representation by introducing supporting models.

- Improve/automatize Risk Analysis by overcoming the main limitations of traditional indexes by providing explanations about their uses or new complementary or alternative methods ([14], [18], [20], [21], [44], [45], [55], [75], [91], [95], [120]).
- Improve/automate problem solving by improving the decision making and solving phase.

4.2.2.3 Integrations

The following kinds of integrations have been collected:

- Templates (e.g., tables and matrices) to organize and manage the bill of material, the list of functions and faults, and the related risk.
- Database (DB) containing information about product parts, functions, historical failures, risk, and the related economic quantifications. They are used to automatically or manually gather the content for the analysis.
- Tools for fault analysis (Fault A.) including Fault Tree Analysis (FTA), Fishbone diagram and Root Cause Analysis (RCA) ([17], [38]).
- Interactive graphical interfaces or software that directly involve user interactions through graphical elements and representations (e.g., plant schemes and infographics) for data entry and visualization.
- Artificial Intelligence (AI) based tools involving Semantic Recognition and Bayesian Networks ([12], [67], [102], [125], [127], [129], [133]).

Other considered integrations are function analysis (FA), fuzzy logic, Monte Carlo method, quality function deployment (QFD), hazard and operability study (HAZOP), ontologies, theory of inventive problem solving (TRIZ), guidelines, automatic measurements (AM) methods, brainstorming techniques, and cognitive maps (C Map).

4.3 SEMI-AUTOMATIC ANALYSIS

At this point, the defined features have been semi-automatically investigated within the collected pool using a software for semantic analysis. The first step of the procedure deals with the manual translation of each considered feature into one or more search queries consisting of single keywords (e.g., name, verb, adjective).

For each keyword, the software provides its main linguistic relations with other term found within the specific sentences of the documents through semantic analysis.

The kinds of relations are different depending on the linguistic nature of the used keyword. If a substantive (e.g., FMEA) is used, then the following can be identified: the modifiers, or adjectives or substantives acting as adjectives (e.g., traditional FMEA, fuzzy FMEA, cost-based FMEA), nouns and verbs modified by the keyword (e.g., FMEA table, FMEA sheet), verbs with the keyword used as object (e.g., executing FMEA, evaluate FMEA), verbs with the keyword used as subject (e.g., FMEA is …, FMEA generates …), substantives linked to the keyword through AND/OR relations (e.g., FMEA and QFD, FMEA and risk), prepositional phrases

TABLE 4.1
Keywords Used to Explain the Features Through the Queries

	Generic terms		
Name	**Verbs**	**FMEA Terms**	**Methods/Tool**
FMEA, Human, Approach,	Improve, Anticipate,	Failures, Modes,	Fuzzy, TRIZ, Database,
Design, Production,	Ameliorate,	Effects, Cause,	Artificial Intelligence,
Maintenance, Time,	Automatize, Analyze,	Risk, Solving,	QFD, Function
Costs, Problem	Reduce, Eliminate,	Decision making	Analysis, etc.
	Solve		

(e.g., … of FMEA, … through FMEA). When a verb is used as keyword, the following can be identified: its modifiers (e.g., effectively improve), the objects (e.g., improve quality, improve design), the subjects (e.g., QFD improves), and other particles used before or after the verb (e.g., improve and evaluate).

In this way, by using the restricted number of keywords, reported in Table 4.1, all the features can be easily investigated.

Thus, the translation of a generic feature (e.g., ameliorate human approach) depends on the manual formulation of a keyword (e.g., ameliorate), the automatic processing, and the manual research of the more suitable relations to express the features itself (e.g., ameliorate + human approach).

However, since the features can be expressed in a variety of ways, by increasing the number of alternative keywords, the number of pertinent identified documents also increases (recall). What achieves this aim is the expansion of the synonyms (e.g., improve in addition to ameliorate) and the research of the alternative forms that can be used to express the feature (e.g., Reduce Tediousness and Reduce Subjectivity for Improve Human Approach).

The research of specific terms, such as the name of the integrated tools (e.g., fuzzy, TRIZ, QFD), can instead be carried out according to different strategies: (1) including them within the keywords, (2) using verbs (e.g., introduce, integrate), and searching the tools among the objects (e.g., Introduce fuzzy logic), (3) using the modifiers of FMEA (e.g., fuzzy FMEA), and (4) searching the relations between FMEA and linguistic particles (e.g., FMEA and TRIZ).

Then, for each interesting relation identified, the software provides the list of the related sentences for each document manually checked in order evaluate its adherence with the investigated feature.

At this point, each selected sentence is summarized through a triad consisting of subject + verb + object.

Table 4.2 shows the followed steps to define the triads in the paper proposed in [7].

All the identified triads are then collected within a table (as shown in Table 4.3), the data for each document (row) is organized according to the features (columns), where, in each cell, the subject of a triad is reported (e.g., The improved failure

TABLE 4.2

Example of the Strategy Used to Build the Triads

Considered document

Investigated Features	Used Keyword	Syntactic Parser	Related Sentence	Triad Subject + Verb + Object
Ameliorate Human Approach	Improve	Improve + Human Approach	The objective of this paper is to propose a new approach for simplifying FMEA by determining the failures in a more practical way by better involving the problem solver in a more pro-active and creative approach	The improved Failure Modes Determination ameliorates human approach
Improve Failure Modes determination	Improve	Improve + Failure Modes	Perturbed Functional Analysis is proposed in order to improve the capability of determine Failure Modes	Perturbed Function Analysis improves Failure Modes determination
Introduce TRIZ	TRIZ	TRIZ + Perturbed Function Analysis (Modifier)	Specifically, an inedited version of TRIZ function analysis, called "Perturbed Function Analysis" is proposed	The authors propose the Perturbed Function Analysis

Source: Spreafico, C. and Russo, D., Can TRIZ functional analysis improve FMEA? *Advances in Systematic Creativity Creating and Managing Innovations*, Palgrave Macmillan, Cham, Switzerland, pp. 87–100, 2019.

TABLE 4.3

An Extract from the Table of Comparison of the Documents and the Triads

	Features					
	Goal		**Strategy**		**Methods/Tools**	
Document	**Ameliorate Human Approach**	...	**Improve Failures Determination**	...	**Introduce Perturbed Function Analysis**	...
[7]	The improved failure modes	...	Perturbed Function Analysis	...	The authors	...
...	...	...	...	...	...	...

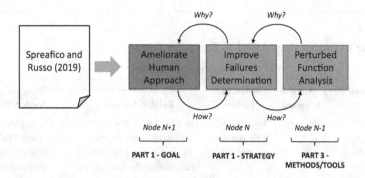

FIGURE 4.2   Example of a causal chain constituted by goal, strategy, and method/tool.

modes) related to a determined feature that has been redefined by using the verb and the object of the triad (e.g., ameliorates human approach).

Therefore, the identified subjects are used as links to build the causal chains, starting from the latter ones, related to the integrations with methods and tools. For example, the causal chain resulting from the previous example (Table 4.3) is the authors introduce the Perturbed Function Analysis (METHOD/TOOL) IN ORDER TO Improve the failure identification (STRATEGY) IN ORDER TO Ameliorate Human Approach (GOAL).

By reading the causal chain in this manner, the logic on its base is the following: each node provides the explanation of the existence of the previous one (WHY?) and it represents a way to obtain the next one (HOW?).

Figure 4.2 shows an example of the simpler causal chain that can be built, which is constituted by one goal (i.e., Ameliorate Human Approach), one strategy (e.g., Improve Failure Determination), and one integration with methods or tools (i.e., The Perturbed Function Analysis).

This example represents the simplest obtained causal chain, consisting of only three nodes arranged in sequence: one for the goals, one for the strategies, and one for the integrations with methods/tools.

However, the structure of the causal chain can be more complex because the number of nodes can increase and their reciprocal disposition can change from series to parallel and by a mix of both.

In the first case (nodes in series), each intermediate node is preceded (on the left) by another node expressing its motivation (WHY?—relation) and it is followed by another representing a way to realize it (HOW?—relation). More goals can be connected in the same way, through their hierarchization: e.g., the goal "reduce the number of experts" can be preceded by the more generic goal "reduce FMEA costs." The same reasoning is valid for the strategies and the integrations with methods/ tools. In particular, in this case, we stratified them into four hierarchical levels: (1) theories and logics (e.g., fuzzy logic), (2) methods (e.g., TRIZ), (3) tool, which can be included in the methods (e.g., FA is part of TRIZ), and (4) knowledge sources (e.g., costs DB).

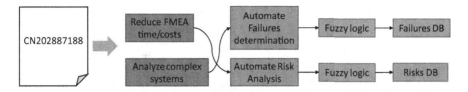

FIGURE 4.3   Example of a complex causal chain obtained from the patent. (From Ming, X. et al., System capable of achieving failure mode and effects analysis (FMEA) data multi-dimension processing, CN202887188, filed June 4, 2012, and issued April 17, 2013. Representation is courtesy of the authors.)

In the second case (nodes in parallel), two or more nodes can concurrently provide a motivation for a previous node or be two possibilities to realize the subsequent node.

As example of a more complex causal chain, consider the Chinese patent [8]. Table 4.4 represents an extract from the table of comparison relative to this document: as can be seen, the resulting relations between the included subjects and the features are more complex and interlaced in comparison to the example shown in Table 4.3.

Figure 4.3 represents the causal chain obtained for this document. In this case the two nodes reduce FMEA time/costs and analyze complex systems represent the two main independent goals pursued by this contribution. The two nodes Automate Failure Determination and Automate Risk Analysis are the two followed strategies both for reduce FMEA time/cost" and to analyze Complex Systems. Finally, the node fuzzy logic represents a high-level integration to realize the two strategies, while a failure DB and a risk DB have been used to provide the knowledge for a fuzzy logic-based reasoning in two different ways: the first one is used for Automate the Failure determination (through fuzzy logic) and the second one is to Automate Risk Analysis (through fuzzy logic).

4.4   RESULTS AND DISCUSSION

The proposed methodology has been tested during two distinct phases. During phase 1 (automatic semantic analysis), all the documents in the selected pool were processed because the algorithm of semantic parsing of the used tool is strictly influenced by number of analyzed sentences in terms of founded linguistic synonyms and relations. During phase 2 (manual review and causal chains building), instead a restricted set of documents was considered to test the methodology in a restricted time period under a temporal burden of required operations.

To obtain a significant sample, the documents were selected based on the typology (papers or patents), date of publication, kind of source (for papers—journal or proceedings), and nationality (for patents). The resulting sample counts 127 documents consisting of 80 papers and 47 patents.

After the sample was processed, the features were investigated, and the documents were classified, one causal chain was built for each document, which usually

TABLE 4.4
An Extract from the Table of Comparison of the Documents and the Triads, Line of the Document

Document	Goal		Strategy		Features	Methods/Tools	
	Reduce FMEA Time/Costs	Analyze Complex Systems	Automate Failure Determination	Automate Risk Analysis	Introduce Fuzzy Logic	Introduce Failure DB	Introduce Risk DB
[8]	Automate Failure Determination Automate Risk Analysis	Automate Failure Determination Automate Risk Analysis	Fuzzy logic	Fuzzy logic	Failure DB Risk DB	The authors	The authors

Source: Ming, X. et al., System capable of achieving failure mode and effects analysis (FMEA) data multi-dimension processing, CN202887188, filed June 4, 2012, and issued April 17, 2013.

consists of more than four nodes, including at least one for each part (goal, strategy, and integration). The total number of the causal chains is the same of the analyzed document (127), since their correspondence is biunivocal: for each document there was only one causal chain and vice versa.

In general, the more followed goals are Improve Design and Improve Human Approach, which together are contained within 61 percent of the triads, while the more considered strategies are related to the failure determination (automate and improve), followed by Automate Risk Analysis.

Among the integrations with methods and tools, fuzzy logic and databases are the most diffused, respectively, with 37 and 28 occurrences within the causal chains, followed by the interface with 23 occurrences.

More detailed considerations are possible by analyzing the relations between goals and strategies. In fact, the two more diffused strategies are considered differently: those for failure determination are implemented to realize all the goals, while those for Improving Risk Analysis are especially considered to Improve Human Approach but practically ignored for achieving other purposes (i.e., Improve Design and Analyze Complex Systems).

Other considerations can be done by comparing the couplings between multiple goals, strategies, and tools.

By comparing the combinations between goals, the most considered combinations found are: Improve Design—Improve Human Approach (8 occurrences) and Improve Design—Analyze Complex Systems (7 occurrences), and Improve Human Approach—Reduce Production Time/Costs (7 occurrences).

Among the combinations of the strategies that emerged, the most considered combinations are: Automate Failure Determination—Automate Risk Analysis (12 occurrences) and Automate Failure Determination—Improve Risk Analysis (7 occurrences).

Finally, the analysis of the multiple integrations revealed that the common coupling is between fuzzy logic and DBs with 6 occurrences.

A deeper analysis can be done by considering the causal chains. Among the different possibilities, the most significant deals with the comparison of the common triads, or the combinations of three nodes: goal, strategy, and integration. In this way, a synthetic but sufficiently significant indication is obtained to understand how the authors are working to improve FMEA.

Figure 4.4 shows the tree map of the common triads, where the five main areas are the goals, their internal subdivisions (colored) represent the strategies, in turn divided between the integrations, where are reported the documents index (please refer to the legend).

For example, analyzing the graph shows that the three documents [11,97,101] propose modified versions of FMEA based on the same common triad, or with the objective to Improve Design phase, by improving the determination of the failures through the introduction of databases (DB). Other goals, strategies, or integrations differentiate the three contributions.

Analysis of the common triad shows that the most diffused consider the goal Improve Human Approach: Improve Human Approach—Improve Risk Analysis—Fuzzy (8 documents), Improve Human Approach—Improve Function

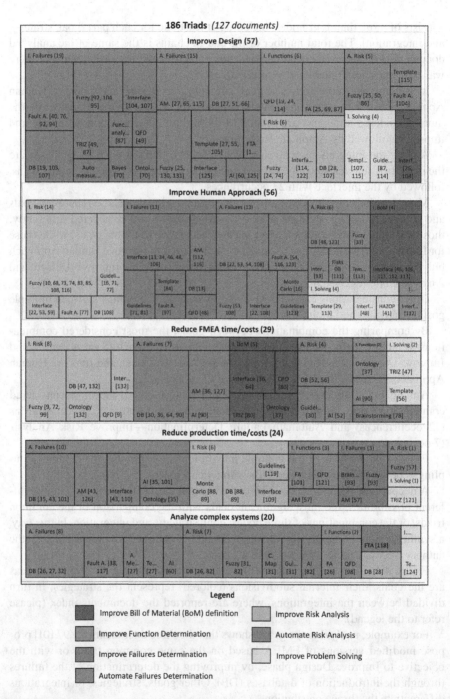

FIGURE 4.4 Main solutions proposed in papers and patents to improve FMEA, represented through triads (goal, strategy, and method/tool).

Determination—Interface (5 documents), and Improve Human Approach—Improve BoM Determination—Interface (5 documents).

By considering the triads, some more interesting observations can be made about the integrations. In general, their distribution is quite heterogeneous in relation to strategies and goals. In fact, fuzzy logic almost always has been introduced to Improve and Automate Risk Analysis to achieve all the goals, while has been used for improving Failure Determination or automate it, but only in order to Improve Design but not for other purposes.

Another case is represented by the interfaces, which have been introduced to improve almost the strategies and goals.

Other integrations instead are related almost exclusively to same strategy for achieving each goal. This is the case of the databases used to Automate Risk Analysis and secondly to Automate Failures Determination and guidelines that generally are used to Automate or to Improve Risk Analysis.

4.5 CONCLUSIONS

In this chapter a method for performing semi-automatic semantic analysis about FMEA documents has been presented and applied on a pool of 127 documents, consisting of paper and patents, selected from international journals, conference proceedings, and international patents.

As a result, each document has been summarized through a specific causal chain including its considered goals (i.e., Improve Design, Improve Human Approach, Reduce FMEA Time/Costs, Reduce Production Time/Costs, Analyze Complex Systems), its strategies of intervention (Improve/Automate BoM, Function, Failures Determination, Risk Analysis and Problem solving) and the integrated methods, tools, and knowledge sources.

The main output of this work is summarized in an infographic based on a Treemap diagram style comparing all the considered documents on the basis of the common elements in their causal chains, which highlights the more popular direction at different levels of detail (i.e., strategies, methods, and tools) of intervention in relation to the objective to pursue.

The consistent reduction of required time along with the number of considered analyzed sources and the level of deepening of the same, represented by the ability to determine the relationships between the different parameters of the analysis within the causal chain, are elements of novelty compared to previous surveys, which could positively impact scientific research in the sector.

The main limitations of the approach consist of the complexity of the manual operations required to define the electronic pool and to create part of the relations within the causal chains, which will be partly solved by automating the method for future developments.

REFERENCES

1. Bouti, A., and Kadi, D. A. 1994. A state-of-the-art review of FMEA/FMECA. *International Journal of Reliability Quality and Safety Engineering* 1(04): 515–543.
2. Sutrisno, A., and Lee, T. J. 2011. Service reliability assessment using failure mode and effect analysis (FMEA): Survey and opportunity roadmap. *International Journal of Engineering Science and Technology* 3(7): 25–38.
3. Tixier, J., Dusserre, G., Salvi, O., and Gaston, D. 2002. Review of 62 risk analysis methodologies of industrial plants. *Journal of Loss Prevention in the Process Industries* 15(4): 291–303.
4. Liu, H. C., Liu, L., and Liu, N. 2013. Risk evaluation approaches in failure mode and effects analysis: A literature review. *Expert Systems with Applications* 40(2): 828–838.
5. Dale, B. G., and Shaw, P. 1990. Failure mode and effects analysis in the UK motor industry: A state-of-the-art study. *Quality and Reliability Engineering International* 6(3): 179–188.
6. Spreafico, C., Russo, D., and Rizzi, C. 2017. A state-of-the-art review of FMEA/FMECA including patents. *Computer Science Review* 25: 19–28.
7. Spreafico, C., & Russo, D. (2019). Case: Can TRIZ Functional Analysis Improve FMEA? In *Advances in Systematic Creativity* (pp. 87–100). Palgrave Macmillan, Cham.
8. Ming, X., Zhu, B., Liang, Q., Wu, Z., Song, W., Xia R., and Kong, F. 2013. System capable of achieving failure mode and effects analysis (FMEA) data multi-dimension processing. CN202887188, filed June 4, 2012, and issued April 17, 2013.
9. Ahmadi, M., Behzadian, K., Ardeshir, A., and Kapelan, Z. 2017. Comprehensive risk management using fuzzy FMEA and MCDA techniques in highway construction projects. *Journal of Civil Engineering and Management* 23(2): 300–310.
10. Almannai, B., Greenough, R., and Kay, J. 2008. A decision support tool based on QFD and FMEA for the selection of manufacturing automation technologies. *Robotics and Computer-Integrated Manufacturing* 24(4): 501–507.
11. Arcidiacono, G., and Campatelli, G. 2004. Reliability improvement of a diesel engine using the FMETA approach. *Quality and Reliability Engineering International* 20(2): 143–154.
12. Augustine, M., Yadav, O. P., Jain, R., and Rathore, A. 2009. Modeling physical systems for failure analysis with rate cognitive maps. *Industrial Engineering and Engineering Management. IEEM 2009 IEEE International Conference* 1758–1762.
13. Lai, J., Zhang, H., & Huang, B. (2011, June). The object-FMA based test case generation approach for GUI software exception testing. In the Proceedings of 2011 9th International Conference on *Reliability, Maintainability and Safety* (pp. 717–723). IEEE.
14. Banghart, M., and Fuller, K. 2014. Utilizing confidence bounds in Failure Mode Effects Analysis (FMEA) hazard risk assessment. *Aerospace Conference, 2014 IEEE* 1–6.
15. Bertelli, C. R., and Loureiro, G. 2015. Quality problems in complex systems even considering the application of quality initiatives during product development. *ISPE CE* 40–51.
16. Bevilacqua, M., Braglia, M., and Gabbrielli, R. 2000. Monte Carlo simulation approach for a modified FMECA in a power plant. *Quality and Reliability Engineering International* 16(4): 313–324.
17. Bluvband, Z., Polak, R., and Grabov, P. 2005. Bouncing failure analysis (BFA): The unified FTA-FMEA methodology. *Reliability and Maintainability Symposium Proceedings Annual* 463–467.
18. Bowles, J. B., and Peláez, C. E. 1995. Fuzzy logic prioritization of failures in a system failure mode, effects and criticality analysis. *Reliability Engineering & System Safety* 50(2): 203–213.

19. Braglia, M., Fantoni, G., and Frosolini, M. 2007. The house of reliability. *International Journal of Quality & Reliability Management* 24(4): 420–440.
20. Braglia, M., Frosolini, M., and Montanari, R. 2003. Fuzzy TOPSIS approach for failure mode, effects and criticality analysis. *Quality and Reliability Engineering International* 19(5): 425–443.
21. Doskocil, D. C., and Offt, A. M. 1993. Method for fault diagnosis by assessment of confidence measure. CA2077772, filed September 9, 1992, and issued April 25, 1993.
22. Draber S. 2000. Method for determining the reliability of technical systems. CA2300546, filed March 7, 2000, and issued September 8, 2000.
23. Chang, K. H., and Wen, T. C. 2010. A novel efficient approach for DFMEA combining 2–tuple and the OWA operator. *Expert Systems with Applications* 37(3): 2362–2370.
24. Chen, L. H., and Ko, W. C. 2009. Fuzzy linear programming models for new product design using QFD with FMEA. *Applied Mathematical Modelling* 33(2): 633–647.
25. Chin, K. S., Chan, A., and Yang, J. B. 2008. Development of a fuzzy FMEA based product design system. *The International Journal of Advanced Manufacturing Technology* 36(7–8): 633–649.
26. Zhang, L., Liang, W., and Hu, J. 2011. Modeling method of early warning model of mixed failures and early warning model of mixed failures. CN102262690, filed June 7, 2011, and issued November 30, 2011.
27. Pan, L., Chin, X., Liu, X., Wang, W., Chen, C., Luo, J., Peng, X. et al., 2012. Intelligent integrated fault diagnosis method and device in industrial production process. CN102637019, filed February 10, 2011, and issued August 15, 2012.
28. Ming, X., Zhu, B., Liang, Q., Wu, Z., Song, W., Xia, R., and Kong, F. 2012. System for implementing multidimensional processing on failure mode and effect analysis (FMEA) data, and processing method of system. CN102810112, filed June 4, 2012, and issued December 5, 2012.
29. Li, G., Zhang, J., and Cui, C. 2012. FMEA (Failure Mode and Effects Analysis) process auxiliary and information management method based on template model and text matching. CN102831152, filed June 28, 2012, and issued December 19, 2012.
30. Li, R., Xu, P., and Xu, Y. 2012. Accidence safety analysis method for nuclear fuel reprocessing plant. CN102841600, filed August 24, 2012, and issued December 26, 2012.
31. Jia, Y., Shen, G., Jia, Z., Zhang, Y., Wang, Z., and Chen, B. 2013. Reliability comprehensive design method of three kinds of functional parts. CN103020378, filed December 26, 2012, and issued April 3, 2013.
32. Chen, Y., Zhang, X., Gao, L., and Kang, R. 2014. Newly-developed aviation electronic product hardware comprehensive FMECA method. CN103760886, filed December 2, 2013, and issued April 30, 2014.
33. Liu, Y., Deng, Z., Liu, S., Chen, X., Pang, B., Zhou, N., and Chen, Y. 2014. Method for evaluating risk of simulation system based on fuzzy FMEA. CN103902845, filed April 25, 2014, and issued July 2, 2014.
34. He, C., Zhao, H., Liu, X., Zong, Z., Li, L., Jiang, J., and Zhu, J. 2014. Data mining-based hardware circuit FMEA (Failure Mode and Effects Analysis) method. CN104198912, filed July 24, 2014, and issued December 10, 2014.
35. Xu, H., Wang, Z., Ren, Y., Yang D., and Liu, L. 2015. Failure knowledge storage and push method for FMEA (failure mode and effects analysis) process. CN104361026, filed October 22, 2014, and issued February 18, 2015.
36. Tang, Y., Sun, Q., and Lü, Z. 2015. Failure diagnosis modeling method based on designing data analysis. CN104504248, filed December 5, 2014, and issued April 8, 2015.
37. David, P., Idasiak, V., and Kratz, F. 2010. Reliability study of complex physical systems using SysML. *Reliability Engineering & System Safety* 95(4): 431–450.

38. Demichela, M., Piccinini, N., Ciarambino, I., and Contini, S. 2004. How to avoid the generation of logic loops in the construction of fault trees. *Reliability Engineering & System Safety* 84(2): 197–207.

39. Deshpande, V. S., and Modak, J. P. 2002. Application of RCM to a medium scale industry. *Reliability Engineering & System Safety* 77(1): 31–43.

40. Doble, M. 2005. Six Sigma and chemical process safety. *International Journal of Six Sigma and Competitive Advantage* 1(2): 229–244.

41. Van Bossuyt, D., Hoyle, C., Tumer, I. Y., and Dong, A. 2012. Risk attitudes in risk-based design: Considering risk attitude using utility theory in risk-based design. *AI EDAM* 26(4): 393–406.

42. Ebrahimipour, V., Rezaie, K., and Shokravi, S. 2010. An ontology approach to support FMEA studies. *Expert Systems with Applications* 37(1): 671–677.

43. Draber, C. D. 2000. Method for determining the reliability of technical systems. EP1035454, filed March 8, 1999, and issued September 8, 2000.

44. Eubanks, C. F., Kmenta, S., and Ishii, K. 1996. System behavior modeling as a basis for advanced failure modes and effects analysis. *ASME Computers in Engineering Conference,* Irvine, CA, pp. 1–8.

45. Eubanks, C. F., Kmenta, S., and Ishii, K. 1997. Advanced failure modes and effects analysis using behavior modeling. *ASME Design Engineering Technical Conferences,* Sacramento, CA, pp. 14–17.

46. Gandhi, O. P., and Agrawal, V. P. 1992. FMEA—A diagraph and matrix approach. *Reliability Engineering & System Safety* 35(2): 147–158.

47. Hartini, S., Nugroho, W. P., and Subekti, K. R. 2010. Design of Equipment Rack with TRIZ Method to Reduce Searching Time in Change Over Activity (Case Study: PT. Jans2en Indonesia). *Proceedings of the Apchi Ergo Future.*

48. Hassan, A., Siadat, A., Dantan, J. Y., and Martin, P. 2010. Conceptual process planning–an improvement approach using QFD, FMEA, and ABC methods. *Robotics and Computer-Integrated Manufacturing* 26(4): 392–401.

49. Hu, C. M., Lin, C. A., Chang, C. H., Cheng, Y. J., and Tseng, P. Y. 2014. Integration with QFDs, TRIZ and FMEA for control valve design. *Advanced Materials Research Trans Tech Publications* 1021: 167–180.

50. Jenab, K., Khoury, S., and Rodriguez, S. 2015. Effective FMEA analysis or not. *Strategic Management Quarterly* 3(2): 25–36.

51. Jong, C. H., Tay, K. M., and Lim, C. P. 2013. Application of the fuzzy failure mode and effect analysis methodology to edible bird nest processing. *Computers and Electronics in Agriculture* 96: 90–108.

52. Koizumi, A., Shimokawa K., and Isaki, Y. 2003. Fmea system. JP2003036278, filed July 25, 2001, and issued February 7, 2003.

53. Wada, T., Miyamoto, Y., Murakami, S., Sugaya, A., Ozaki Y., Sawai, T., Matsumoto, S. et al., 2003. Diagnosis rule structuring method based on failure mode analysis, diagnosis rule creating program, and failure diagnosis device. JP2003228485, filed February 6, 2002, and issued August 15, 2003

54. Yatake, H., Konishi, H., and Onishi T. 2009. Fmea sheet creation support system and creation support program. JP2011008355, filed June 23, 2009.

55. Suzuki, K., Hayata, A., and Yoshioka, M. 2009. Reliability analysis device and method. JP2011113217, issued November 25, 2009.

56. Kawai, M., Hirai, K., and Aryoshi, T. 1990. Fmea simulation method for analyzing circuit. JPH0216471, filed July 4, 1988, and issued January 19, 1990.

57. Sonoda, Y., and Kageyama., T. 1992. Plant diagnostic device. JPH086635, filed May 9, 1990, and issued January 23, 1992.

58. Kim, J. H., Kim, I. S., Lee, H. W., and Park, B. O. 2012. A Study on the Role of TRIZ in DFSS. *SAE International Journal of Passenger Cars-Mechanical Systems* 5(2012–01–0068): 22–29.
59. Kimura, F., Hata, T., and Kobayashi, N. 2002. Reliability-centered maintenance planning based on computer-aided FMEA. *Proceeding of the 35th CIRP-International Seminar on Manufacturing Systems* 506–511.
60. Kmenta, S., and Ishii, K. 2000. Scenario-based FMEA: A life cycle cost perspective. *Proceedings of ASME Design Engineering Technical Conference*, Baltimore, MD.
61. Kmenta, S., and Ishii, K. 2004. Scenario-based failure modes and effects analysis using expected cost. *Journal of Mechanical Design* 126(6): 1027–1035.
62. Kmenta, S., and Ishii, K. 1998. Advanced FMEA using meta behavior modeling for concurrent design of products and controls. *Proceedings of the 1998 ASME Design Engineering Technical Conferences*.
63. Kmenta, S., Cheldelin, B., and Ishii, K. 2003. Assembly FMEA: A simplified method for identifying assembly errors. *ASME 2003 International Mechanical Engineering Congress and Exposition* 315–323.
64. Lee, M. S., and Lee, S., H. 2013. Real-time collaborated enterprise asset management system based on condition-based maintenance and method thereof. KR20130065800, filed November 30, 2011, and issued June 24, 2013.
65. Choi, S. H., Kim, G. H., Cho, C. H., and Kim, Y., G. 2013. Reliability centered maintenance method for power generation facilities. KR20130118644, filed April 20, 2012, and issued December 12, 2013.
66. Lim, S. S., and Lee, J., Y. 2014. Intelligent failure asset management system for railway car. KR20140036375, filed September 12, 2012, and issued March 3, 2014.
67. Ku, C., Chen, Y. S., and Chung, Y. K. 2008. An intelligent FMEA system implemented with a hierarchy of back-propagation neural networks. *Cybernetics and Intelligent Systems IEEE Conference* 203–208.
68. Kutlu, A. C., and Ekmekçioğlu, M. 2012. Fuzzy failure modes and effects analysis by using fuzzy TOPSIS-based fuzzy AHP. *Expert Systems with Applications* 39(1): 61–67.
69. Laaroussi, A., Fiès, B., Vankeisbelckt, R., and Hans, J. 2007. Ontology-aided FMEA for construction products. Bringing ITC knowledge to work. *Proceedings of W78 Conference* 26(29): 6.
70. Lee, B. H. 2001. Using FMEA models and ontologies to build diagnostic models. *AI EDAM* 15(4): 281–293.
71. Lindahl, M. 1999. E-FMEA—a new promising tool for efficient design for environment. *Proceedings of Environmentally Conscious Design and Inverse Manufacturing* 734–739.
72. Liu, H. T. 2009. The extension of fuzzy QFD: From product planning to part deployment. *Expert Systems with Applications* 36(8): 11131–11144.
73. Liu, J., Martínez, L., Wang, H., Rodríguez, R. M., and Novozhilov, V. 2010. Computing with words in risk assessment. *International Journal of Computational Intelligence Systems* 3(4): 396–419.
74. Liu, H. C., Liu, L., Liu, N., and Mao, L. X. 2013. Risk evaluation in failure mode and effects analysis with extended VIKOR method under fuzzy environment. *Expert Systems with Applications* 40(2): 828–838.
75. Grantham, K. (2007). Detailed risk analysis for failure prevention in conceptual design: RED (Risk in early design) based probabilistic risk assessments.
76. Mader, R., Armengaud, E., Grießnig, G., Kreiner, C., Steger, C., and Weiß, R. 2013. OASIS: An automotive analysis and safety engineering instrument. *Reliability Engineering & System Safety* 120: 150–162.
77. Mandal, S., and Maiti, J. 2014. Risk analysis using FMEA: Fuzzy similarity value and possibility theory-based approach. *Expert Systems with Applications* 41(7): 3527–3537.

78. Montgomery, T. A., and Marko, K. A. 1997. Quantitative FMEA automation. *Proceedings of Reliability and Maintainability Symposium* 226–228.

79. Moratelli, L., Tannuri, E. A., and Morishita, H. M. 2008. Utilization of FMEA during the preliminary design of a dynamic positioning system for a Shuttle Tanker. ASME 27th International Conference on Offshore Mechanics and Arctic Engineering, Estoril, Portugal, pp. 787–796.

80. Ormsby, A. R. T., Hunt, J. E., and Lee, M. H. 1991. Towards an automated FMEA assistant. *Applications of Artificial Intelligence in Engineering VI*, Springer, the Netherlands 739–752.

81. Ozarin, N. 2008. What's wrong with bent pin analysis, and what to do about it. *Reliability and Maintainability Symposium,* Washington, DC: IEEE Computer Society, pp. 386–392.

82. Pelaez, C. E., and Bowles, J. B. 1995. Applying fuzzy cognitive-maps knowledge-representation to failure modes effects analysis. *Proceedings of Reliability and Maintainability Symposium*: 450–456.

83. Pang, L. M., Tay, K. M., and Lim, C. P. 2016. Monotone fuzzy rule relabeling for the zero-order TSK fuzzy inference system. *IEEE Transactions on Fuzzy Systems* 24(6): 1455–1463.

84. Kim, J. H., Jeong, H. Y., and Park, J. S. 2009. Development of the FMECA process and analysis methodology for railroad systems. *International Journal of Automotive Technology* 10(6): 753.

85. Petrović, D. V., Tanasijević, M., Milić, V., Lilić, N., Stojadinović, S., & Svrkota, I. 2014. Risk assessment model of mining equipment failure based on fuzzy logic. *Expert Systems with Applications* 41(18): 8157–8164.

86. Price, C. J. 1996. Effortless incremental design FMEA. *Reliability and Maintainability Symposium Proceedings. IEEE International Symposium on Product Quality and Integrity* 43–47.

87. Regazzoni, D., and Russo, D. 2011. TRIZ tools to enhance risk management. *Procedia Engineering* 9: 40–51.

88. Rhee, S. J., and Ishii, K. 2003. Using cost based FMEA to enhance reliability and serviceability. *Advanced Engineering Informatics* 17(3–4): 179–188.

89. Rhee, S. J., and Ishii, K. 2002. Life cost-based FMEA incorporating data uncertainty. *ASME International Design Engineering Technical Conferences and Computers and Information in Engineering Conference* 309–318.

90. Russomanno, D. J., Bonnell, R. D., and Bowles, J. B. 1994. Viewing computer-aided failure modes and effects analysis from an artificial intelligence perspective. *Integrated Computer-Aided Engineering* 1(3): 209–228.

91. Shahin, A. 2004. Integration of FMEA and the Kano model: An exploratory examination. *International Journal of Quality & Reliability Management* 21(7): 731–746.

92. Sharma, R. K., and Sharma, P. 2010. System failure behavior and maintenance decision making using, RCA, FMEA and FM. *Journal of Quality in Maintenance Engineering* 16(1): 64–88.

93. Sharma, R. K., Kumar, D., and Kumar, P. 2005. Systematic failure mode effect analysis (FMEA) using fuzzy linguistic modelling. *International Journal of Quality & Reliability Management* 22(9): 986–1004.

94. Sharma, R. K., Kumar, D., and Kumar, P. 2007. Modeling and analysing system failure behaviour using RCA, FMEA and NHPPP models. *International Journal of Quality & Reliability Management,* 24(5): 525–546.

95. Sharma, R. K., Kumar, D., and Kumar, P. 2008. Fuzzy modeling of system behavior for risk and reliability analysis. *International Journal of Systems Science* 39(6): 563–581.

96. Su, C. T., and Chou, C. J. 2008. A systematic methodology for the creation of Six Sigma projects: A case study of semiconductor foundry. *Expert Systems with Applications* 34(4): 2693–2703.

97. Suganthi, S., and Kumar, D. 2010. FMEA without fear AND tear. In *Management of Innovation and Technology (ICMIT)IEEE International Conference* 1118–1123.

98. Ming Tan, C. 2003. Customer-focused build-in reliability: A case study. *International Journal of Quality & Reliability Management* 20(3): 378–397.

99. Meng Tay, K., and Peng Lim, C. 2006. Fuzzy FMEA with a guided rules reduction system for prioritization of failures. *International Journal of Quality & Reliability Management* 23(8): 1047–1066.

100. Teng, S. H., and Ho, S. Y. 1996. Failure mode and effects analysis: An integrated approach for product design and process control. *International Journal of Quality & Reliability Management* 13(5): 8–26.

101. Teoh, P. C., and Case, K. 2004. Failure modes and effects analysis through knowledge modelling. *Journal of Materials Processing Technology*153: 253–260.

102. Throop, D. R., Malin, J. T., and Fleming, L. D. 2001. Automated incremental design FMEA. *IEEE Aerospace Conference. Proceedings* 7: 7–3458.

103. Johnson, T., Azzaro, S., and Cleary, D., 2004. Method, system and computer product for integrating case-based reasoning data and failure modes, effects and corrective action data. US2004103121, filed November 25, 2002, and issued May 27, 2004.

104. Johnson, T. L., Cuddihy, P. E., and Azzaro, S. H. 2004. Method, system and computer product for performing failure mode and effects analysis throughout the product life cycle. US2004225475, filed November 25, 2002, and issued November 11, 2004.

105. Chandler, F. T., Valentino, W. D., Philippart, M. F., Relvini, K. M., Bessette, C. I. and Shedd, N. P. 2004. Human factors process failure modes and effects analysis (hf pfmea) software tool. US2004256718, filed April 15, 2004, and issued December 23, 2004.

106. Liddy, R., Maeroff, B., Craig, D., Brockers, T., Oettershagen, U., and Davis, T. 2005. Method to facilitate failure modes and effects analysis. US2005138477, filed November 25, 2003, and issued June 23, 2005.

107. Lonh, K. J., Tyler, D. A., Simpson, T. A., and Jones, N. A. 2006. Method for predicting performance of a future product. US2006271346, filed May 31, 2005, and issued November 30, 2006.

108. Mosleh, A., Wang, C., and Groen, F. J. 2007. System and methods for assessing risk using hybrid causal logic. US2007011113, filed March 17, 2006, and issued July 11, 2007.

109. Coburn, J. A., and Weddle, G. B. 2009. Facility risk assessment systems and methods. US20090138306, filed September 25, 2008, and issued May 28, 2009.

110. Singh, S., Holland, S. W. and Bandyopadhyay, P. 2012. Graph matching system for comparing and merging fault models. US2012151290, filed December 9, 2010, and issued June 14, 2012.

111. Harsh, J. K., Walsh, D. E., and Miller, E., M. 2012. Risk reports for product quality planning and management. US2012254044, filed March 30, 2012, and issued October 4, 2012.

112. Abhulimen, K. E. 2012. Design of computer-based risk and safety management system of complex production and multifunctional process facilities-application to fpso's, US2012317058, filed June 13, 2011, and issued December 13, 2012.

113. Oh, K., P. 2013. Spreadsheet-based templates for supporting the systems engineering process. US2013013993, filed August 24, 2011, and issued January 10, 2013.

114. Chang, Y. 2014. Product quality improvement feedback method. US20140081442, filed September 18, 2012, and issued March 20, 2014.

115. Barnard, R. F., Dohanich, S. L., and Heinlein, P., D. 1996. System for failure mode and effects analysis. US5586252, filed May 24, 1994, and issued December 17, 1996.

116. Williams, E., and Rudoff, A. 2006. System and method for performing automated system management. US7120559, filed June 29, 2004, and issued October 10, 2006.

117. Williams, E., and Rudoff, A. 2008. System and method for automated problem diagnosis. US7379846, filed June 29, 2004, and issued May 27, 2008.

118. Williams, E., and Rudoff A., 2009. System and method for providing a data structure representative of a fault tree. US7516025, filed June 29, 2004, and issued April 7, 2009.

119. Dreimann, M., Ehlers, P., Goerisch, A., Maeckel, O., Sporer, R., and Sturm, A. 2007. Method for analyzing risks in a technical project. US8744893, filed April 11, 2006, and issued November 1, 2007.

120. Vahdani, B., Salimi, M., and Charkhchian, M. 2015. A new FMEA method by integrating fuzzy belief structure and TOPSIS to improve risk evaluation process. *The International Journal of Advanced Manufacturing Technology* 77(1–4): 357–368.

121. Wang, C. S., and Chang, T. R. 2010. Systematic strategies in design process for innovative product development. *Industrial Engineering and Engineering Management Proceedings*: 898–902.

122. Wang, M. H. 2011. A cost-based FMEA decision tool for product quality design and management. *IEEE Intelligence and Security Informatics Proceedings* 297–302.

123. Wirth, R., Berthold, B., Krämer, A., and Peter, G. 1996. Knowledge-based support of system analysis for the analysis of failure modes and effects. *Engineering Applications of Artificial Intelligence* 9(3): 219–229.

124. Selvage, C. 2007. Look-across system. WO2007016360, filed July 28, 2006, and issued February 28, 2007.

125. Bovey, R. L., and Senalp, E., T. 2010. Assisting with updating a model for diagnosing failures in a system, WO2010038063, filed September 30, 2009, and issued April 8, 2010.

126. Snooke, N. A. 2010. Assisting failure mode and effects analysis of a system, WO2010142977, filed June 4, 2010, and issued December 16, 2010.

127. Snooke, N. A. 2012. Automated method for generating symptoms data for diagnostic systems, WO2012146908, filed April 12, 2012, and issued November 1, 2012.

128. Xiao, N., Huang, H. Z., Li, Y., He, L., and Jin, T. 2011. Multiple failure modes analysis and weighted risk priority number evaluation in FMEA. *Engineering Failure Analysis* 18(4): 1162–1170.

129. Yang, C., Letourneau, S., Zaluski, M., and Scarlett, E. 2010. APU FMEA validation and its application to fault identification. *ASME International Design Engineering Technical Conferences and Computers and Information in Engineering Conference* 959–967.

130. Zafiropoulos, E. P., and Dialynas, E. N. 2005. Reliability prediction and failure mode effects and criticality analysis (FMECA) of electronic devices using fuzzy logic. *International Journal of Quality & Reliability Management* 22(2): 183–200.

131. Yang, Z., Bonsall, S., and Wang, J. 2008. Fuzzy rule-based Bayesian reasoning approach for prioritization of failures in FMEA. *IEEE Transactions on Reliability* 57(3), 517–528.

132. Zhao, X., and Zhu, Y. 2010. Research of FMEA knowledge sharing method based on ontology and the application in manufacturing process. *Database Technology and Applications (DBTA), 2nd International Workshop* 1–4.

133. Zhou, J., and Stalhaane, T. 2004. Using FMEA for early robustness analysis of Web-based systems. In *Computer Software and Applications Conference Proceedings* (2): 28–29.

5 Markov Chains and Stochastic Petri Nets for Availability and Reliability Modeling

Paulo Romero Martins Maciel, Jamilson Ramalho Dantas, and Rubens de Souza Matos Júnior

CONTENTS

5.1 INTRODUCTION

Due to the ubiquitous provision of services on the internet, dependability has become an attribute of prime concern in hardware/software development, deployment, and operation. Providing fault-tolerant services is related inherently to the adoption of redundancy. Redundancy can be exploited either in time or in space. Replication of services usually is provided through distributed hosts across the world so that whenever the service, the underlying host, or network fails another service is ready to take over. Dependability of a system can be understood as the ability to deliver a specified functionality that can be justifiably trusted. Functionality might be a set of roles or services (functions) observed by an outside agent (a human being, another system, etc.) that interacts with system at its interfaces; and the specified functionality of a system is what the system is intended.

Two fundamental dependability attributes are reliability and availability. The task of estimating reliability and availability metrics may be undertaken by adopting combinatorial models such as reliability block diagrams and fault trees. These models, however, lack the modeling capacity to represent dynamic redundancies. State-based models such as Markov chains and stochastic Petri nets have higher modeling power, but the computation cost for performing the evaluation is usually an issue to be considered. This chapter studies the reliability and availability modeling of a system through Markov chains and stochastic Petri nets.

This chapter is divided into four sections. After the introduction follows a glance on some key authors and papers of area. Section 5.3 brings out background concepts on Markov chains and Stochastic Petri Nets. Section 5.4 presents some availability and reliability models for computer systems. Section 5.5 closes the chapter.

5.2   A GLANCE AT HISTORY

This section provides a summary of early work related to dependability and briefly describes some seminal efforts as well as the respective relations with current prevalent methods. This effort is undoubtedly incomplete; nonetheless, the intent is that it provides key events, people, and noteworthy research related to what is now called dependability modeling [28].

Dependability is related to disciplines such as fault tolerance and reliability. The concept of dependable computing first appeared in the 1820s when Charles Babbage carried out the initiative to conceive and build a mechanical calculating engine to get rid of the risk of human errors [1,2]. In his book, *On the Economy of Machinery and Manufacture*, he remarks "The first objective of every person who attempts to make an article of consumption is, or ought to be, to produce it in perfect form" [3]. In the nineteenth century, reliability theory advanced from probability and statistics as a way to support estimating maritime and life insurance rates. In the early twentieth century, methods had been proposed to estimate survivorship of railroad equipment [4,5].

The first IEEE (formerly AIEE and IRE) public document to mention reliability is "Answers to Questions Relative to High Tension Transmission" that archives the meeting of the Board of Directors of the American Institute of Electrical Engineers held on September 26, 1902 [6]. In 1905, H. G. Stott and H. R. Stuart discuss "Time-Limit Relays and Duplication of Electrical Apparatus to Secure Reliability of Services" at New York [4] and Pittsburg [5]. In these works, the concept of reliability was chiefly qualitative. In 1907, A. A. Markov began the study of a notable sort of chance process. In this process, the outcome of a given experiment can modify the outcome of the next experiment. This sort of process is now called a Markov chain [7]. Markov's classic textbook, *Calculus of Probabilities*, was published four times in Russian and was translated into German [9]. In 1926, 20 years after Markov's initial discoveries, a paper by Russian mathematician S. N. Bernstein used the term "Markov chain" [8]. In the 1910s, A. K. Erlang studied telephone traffic planning for reliable service provisioning [10].

The first generation of electronic computers was entirely undependable; thence many techniques were investigated for improving their reliability. Among such techniques, many researchers investigated design strategies and evaluation methods. Many methods then were proposed for improving system dependability such as error control codes, replication of components, comparison monitoring, and diagnostic routines. The leading researchers during that period were Shannon [13], Von Neumann [14], and Moore [15], who proposed and developed theories for building reliable systems by using redundant and less reliable components. These theories were the forerunners of the statistical and probabilistic techniques that form the groundwork of modern dependability theory [17].

In the 1950s, reliability turns out to be a subject of great interest because of the cold war efforts, failures of American and Soviet rockets, and failures of the first commercial jet—the British de Havilland Comet [18,19]. Epstein and Sobel's 1953 paper on the exponential distribution was a landmark contribution [20]. In 1954, the first Symposium on Reliability and Quality Control (it is now the IEEE Transactions on Reliability) was held in the United States, and in 1958 the First All-Union Conference on Reliability was held in Moscow [7,21]. In 1957, S. J. Einhorn and F. B. Thiess applied Markov chains for modeling system intermittence [22], and in 1960 P. M. Anselone employed Markov chains for evaluating the availability of radar systems [23]. In 1961, Birnbaum, Esary, and Saunders published a pioneering paper introducing coherent structures [24].

The reliability models might be classified as combinatorial (non-state space model) and state-space models. Reliability Block Diagrams (RBD) and Fault Trees (FT) are combinatorial models and the most widely adopted models in reliability evaluation. RBD is probably the oldest combinatorial technique for reliability analysis. Fault Tree Analysis (FTA) was initially developed in 1962 at Bell Laboratories by H. A. Watson to analyze the Minuteman I Intercontinental Ballistic Missile Launch Control System. Afterward, in 1962, Boeing and AVCO expanded the use of FTA to the entire Minuteman II [25]. In 1965, W. H. Pierce unified the Shannon, Von Neumann, and Moore theories of masking and redundancy as the concept of failure tolerance [26]. In 1967, A. Avizienis combined masking methods with error detection, fault diagnosis, and recovery into the concept of fault-tolerant systems [27].

The formation of the IEEE Computer Society Technical Committee on Fault-Tolerant Computing (now Dependable Computing and Fault Tolerance TC) in 1970 and of IFIP Working Group 10.4 on Dependable Computing and Fault Tolerance in 1980 was an essential mean for defining a consistent set of concepts and terminology. In early 1980s, Laprie coined the term dependability for covering concepts such as reliability, availability, safety, confidentiality, maintainability, security, and integrity [1,29].

In late 1970s some works were proposed for mapping Petri nets to Markov chains [30,32,47]. These models have been extensively adopted as high-level Markov chain automatic generation models and for discrete event simulation. Natkin was the first to apply what is now generally called stochastic Petri nets (SPNs) to dependability evaluation of systems [33].

5.3   BACKGROUND

This section provides a very brief introduction to Continuous Time Markov Chains (CTMCs) and SPNs, which are the formalism adopted to model availability and reliability in this chapter.

5.3.1   MARKOV CHAINS

Markov chains have been applied in many areas of science and engineering. They have been widely adopted for performance and dependability evaluation in manufacturing, logistics, communication, computer systems, and so forth [34]. The name Markov chains came from the Russian mathematician Andrei Andreevich Markov. Markov was born on June, 14, 1856, in Ryazan, Russia, and died on July 20, 1922, in Saint Petersburg [35].

The References offers many books on Markov chains [36–40]. These books cover Markov chain theory and applications in different depth and styles.

A stochastic process is defined as a family of random variables ($\{X_i(t): t \in T\}$) indexed through some parameter (t). Each random variable ($X_i(t)$) is defined on some probability space. The parameter t usually represents time, so $X_i(t)$ denotes the value assumed by the random variable at time t. T is called the parameter space and is a subset of R (the set of real numbers).

If T is discrete, that is, $T = \{0,1,2,...\}$, the process is classified as discrete-time parameter stochastic process. On the other hand, if T is continuous, that is, $T = \{t: 0 \leq t < \infty\}$, the process is a continuous-time parameter stochastic process. In CTMC, a change of state may occur at any point in time. A CTMC is a continuous time, discrete state-space stochastic process, that is, the state values are discrete, but parameter t has a continuous range over [0,∞].

A CTMC can be represented by a state-transition diagram in which the vertices represent states and the arcs between vertices i and j are labeled with the respective transition rates, that is, λ_{ij}, i ≠ j. Consider a chain composed of three states, s_0, s_1, and s_2, and their transition rates, α, β, γ, and λ. The model transitions from s_0 to s_1 with rate α; from state s_1, the model transitions to state s_0 with rate β, and to state s_2 with rate γ. When in state s_2, the model transitions to state s_1 with rate λ. The rate matrix, Q is:

$$Q = \begin{pmatrix} -\alpha & \alpha & 0 \\ \beta & -(\beta+\gamma) & \gamma \\ 0 & \lambda & -\lambda \end{pmatrix}$$

For time homogeneous CTMCs:

$$\frac{d\Pi(t)}{dt} = \Pi(t) \cdot Q, \tag{5.1}$$

that has the following solution [12,16]:

$$\Pi(t) = \Pi(0)\, e^{Qt} = \Pi(0)\left(I + \sum_{k=1}^{\infty} \frac{Qt^k}{k!} \right). \tag{5.2}$$

In many cases, however, the instantaneous behavior, $\Pi(t)$, of the Markov chain is more than needed. In many cases, often it is satisfied already when computing the steady-state probabilities, that is, $\Pi = \lim_{t \to \infty} \Pi(t)$. Hence, consider the system of differential equations presented in Equation 5.1. If the steady-state distribution exists, then $d\Pi(t)$:

$$\frac{d\Pi(t)}{dt} = 0$$

Consequently, for calculating the steady-state probabilities, the only necessity is to solve the system:

$$\Pi \cdot Q = 0, \qquad \sum_{\forall i} \pi_i = 1. \tag{5.3}$$

5.3.2 Stochastic Petri Nets

The first SPN extensions were proposed independently by Symons, Natkin, and Molloy [30,31,32]. After, many other stochastic extensions were introduced, Marsan et al. extended the basic SPNs by considering stochastic timed transitions and immediate transitions [41]. This model was named Generalized Stochastic Petri Nets (GSPN) [43]. Later on, Marsan and Chiola proposed an extension that also supported deterministic timed transitions [42], which was named Deterministic Stochastic Petri Nets (DSPN) [46]. Many other extensions followed, among them extended Deterministic Stochastic Petri Nets (eDSPN) [44,45] and Stochastic Reward Nets (SRN) [48].

The SPN considered here is a very general stochastic extension of Place-Transition nets. Its modeling capacity is well beyond that presented by Symons, Natkin, and Molloy. The original SPN considered only exponential distributions. GSPNs adopted, besides exponential distributions, immediate transitions. These models shared the memoryless property also presented in untimed Petri nets since reachable marking is only dependent on the current Petri net marking.

Stochastic Petri Nets—Let $SPN = (P, T, I, O, H, M_0, Atts)$ be an SPN, where P, T, I, O, and M_0 are defined as for Place-Transition nets, that is, P is the set of places, T is the set of transitions, I in input matrix, O is the output matrix, and M_0 is the initial marking. The set of transition, T, is, however, divided into immediate transitions (T_{im}), timed exponentially distributed transitions (T_{exp}), deterministic timed transitions (T_{det}), and timed generically distributed transitions (T_g):

$$T = T_{im} \cup T_{exp} \cup T_{det} \cup T_g.$$

Immediate transitions are graphically represented by thin black rectangles, timed exponentially distributed are depicted by white rectangles, deterministic timed transitions are represented by thick black rectangles, and timed generically distributed gray rectangles denote transitions. The matrices I and O represent the input and output arcs of transitions. These matrices may be marking dependent, that is the arc weights may be dependent on current marking:

$$I = (i_{p,t})_{|P| \times |T|}, \quad i_{p,t} : MD \times RS_{SPN} \to \mathbb{N},$$

and

$$O = (o_{p,t})_{|P|\times|T|}, \quad o_{p,t} : MD \times RS_{SPN} \to \mathbb{N},$$

where $MD = \{true, false\}$ is a set that specify if the arc between p and t is marking dependent or not. If the arc is marking dependent, the arc weight is dependent on the current marking $M \in RS_{SPN}$, RS_{SPN} is the reachability set of the net SPN. Otherwise, it is constant.

$$H = (h_{p,t})_{|P|\times|T|}, \quad h_{p,t} : MD \times RS_{SPN} \to \mathbb{N}$$

is a matrix of inhibitor arcs. These arcs may also be marking dependent, that is the arc weight may be dependent on current marking. $h_{p,t}: MD \times RS_{SPN} \to \mathbb{N}$, where $MD = \{true, false\}$ is a set that specify if the arc between p and t is marking dependent or not. If the arc is marking dependent, the arc weight is dependent on the current marking $M \in RS_{SPN}$. Otherwise, it is constant.

- $Atts = (\Pi, Dist, MDF, W, G, Policy, Concurrency)$ is set of attributes assigned to transitions, where:
- $\Pi: T \to N$ is a function that assigns a firing priority on transitions. The larger the number the higher is the firing priority. Immediate transitions have higher priorities than timed transitions, and timed deterministic transitions have higher priorities than random timed transitions, that is, $\pi(t_i) > \pi(t_j) > \pi(t_k)$, $t_i \in T_{im}$, $t_j \in T_{det}$, and $t_k \in T_{exp} \cup T_g$.
- $Dist: T_{exp} \cup T_g \to F$ is a function that assigns non-negative probability distribution function to random delay transitions. F is the set of functions.
- $MDF: T \to MD$ is a function that defines if the probability distribution functions assigned to delays of transitions are marking dependent or not. $MD = \{true, false\}$.
- $W: T_{exp} \cup T_{det} \cup T_{im} \to R^+$ is a function that assigns a non-negative real number to exponential, deterministic, and immediate transitions. For exponential transitions, these values correspond to the parameter values of the exponential distributions (rates). In the case of deterministic transitions, they are the deterministic delays assigned to transitions. Moreover, in the case of immediate transitions, they denote the weights assigned to transitions.
- $G: T \to 7N^{|P|}$ is a partial operator that assigns to transitions a guard expression. The guards are evaluated by $GE: (T \to 7N^{|P|}) \to \{true, false\}$ that results in true or false. The guard expressions are Boolean formulas composed of predicates specified regarding marking of places. A transition may be enabled only if its guard function is evaluated as true. It is worth noting that not every transition may be guarded.
- $Policy: T \to \{prd, prs\}$, where prd denotes pre-emptive repeat different (restart), and prs is pre-emptive resume (continue). The timers of transitions

with *prd* are discarded and new values are generated in the new marking. The timers of transitions with *prs* hold the present values.

- *Concurrency*: $T - T_{im} \rightarrow \{sss, iss\}$ is a function that assigns to each timed transition a timing semantics, where *sss* denotes single server semantics and *iss* is infinite server semantics.

SPNs are usually evaluated through numerical methods. However, if the state space is too big, infinite or even if non-phase-type distributions should be represented, the evaluation option may fall into the simulation. With simulation, there are no fundamental restrictions on the models that can be evaluated. Nevertheless, the simulation does have pragmatical constraints, since the amount of computer time and memory running a simulation can be prohibitively large. Therefore, the general advice is to pursue an analytical model wherever possible, even if simplifications and or decomposition is required.

For a detailed introduction to SPNs, refer to [43,45].

5.4 AVAILABILITY AND RELIABILITY MODELS FOR COMPUTER SYSTEMS

Dependability aspects deserve great attention for assuring of the quality of service provided by a computer system. Dependability studies look for determining reliability, availability, security, and safety metrics for the infrastructure under analysis [50]. RBD [51], FT [53] and Petri nets are, as well as Markov chains, widely used to capture the system behavior and allow the description and prediction of dependability metrics.

The most basic dependability aspects of a system are the failure and repair events, which may bring the system to different configurations and operational modes. The steady-state availability is a common measure extracted from dependability models. Reliability, downtime, uptime, and mean time to system failure are other metrics usually obtained as output from a dependability analysis in computer systems.

The combined analysis of performance and dependability aspects, so-called performability analysis, is another frequent necessity when dealing with computer systems, since many of them may continue working after partial failures. Such gracefully degrading systems [54] require specific methods to achieve an accurate evaluation of their metrics. Markov reward models constitute an essential framework for performability analysis. In this context, the hierarchical modeling approach is also a useful alternative in which distinct models may be used to represent the dependability relationships of the system in the upper level and performance aspects in the lower level, or vice versa [49,55,58].

For all kinds of Markov chain or SPN analyses, an important assumption must be kept in mind: the exponential distribution of transition rates or firing delays, respectively. The behavior of events in many computer systems may fit better to other probability distributions, but in some of these situations, the exponential distribution is a fair approximation, enabling the use of Markovian models. In cases when the exponential distribution is not a reasonable approximation, SPN extensions may be used that enable non-exponential distributions. Such a deviation from Markovian assumptions requires the adoption of simulation for a model solution [57,59–61]. It is possible also to adapt transitions to represent other distributions employing phase approximation or moment matching as shown in [36,52]. The use of such techniques allows the modeling of events described by distributions such as Weibull, hypoexponential, hyperexponential, and Erlang and Cox [13,16].

5.4.1 COMMON STRUCTURES FOR COMPUTATIONAL SYSTEMS MODELING

Consider a single component repairable system. This system may be either operational or in failure. If the time to failure (TTF) and the time to repair (TTR) are exponentially distributed with rates λ and μ, respectively, the CTMC shown in Figure 5.1a is its availability model. The state U (Up) represents the operational state, and the state D (Down) denotes the faulty system. If the system is operational, it may fail. The system failure is represented by the transition from state U to state D. The faulty system may be restored to its operational state by a repair. The repair is represented by the transitions from state D to state U. The matrix rate, Q, is presented in Figure 5.1b.

The instantaneous availability is the instantaneous probability of being in state U and D is, respectively:

$$A(t) = \pi_U(t) = \frac{\mu}{\lambda + \mu} + \frac{\lambda}{\lambda + \mu} e^{-(\lambda + \mu)t} \tag{5.4}$$

and

$$UA(t) = \pi_D(t) = \frac{\lambda}{\lambda + \mu} - \frac{\lambda}{\lambda + \mu} e^{-(\lambda + \mu)t}, \tag{5.5}$$

such that $\pi_U(t) + \pi_D(t) = 1$.

If $t \to \infty$, then the steady-state availability and unavailability is obtained, respectively:

$$A = \pi_U = \frac{\mu}{\lambda + \mu} \tag{5.6}$$

and

$$UA = \pi_D = \frac{\lambda}{\lambda + \mu}, \tag{5.7}$$

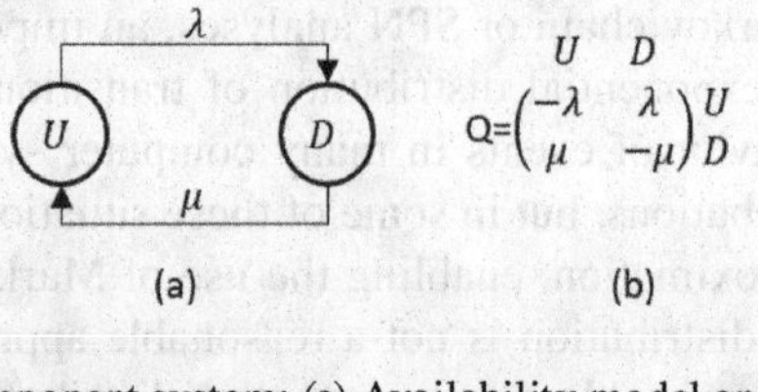

FIGURE 5.1 Single component system: (a) Availability model and (b) Matrix rate.

such that $\pi_U + \pi_D = 1$. The steady-state measures can be obtained also by solving:

$$\Pi \cdot Q = 0, \qquad \pi_U + \pi_D = 1,$$

where $\Pi = (\pi_U, \pi_D)$. The downtime in a period T is $DT = \pi_D \times T$. For a time period of 1 year (365 days), the number of hours T is 8760 h and 525,600 min. Now assume a CTMC that represents the system failure. This model has two states, U and D, and only one transition. This transition represents the system failure; that is, when the system is operational (U), it may fail, and this event is represented by the transition from the state U to state D, with failure rate (λ). Solving:

$$\frac{d\Pi(t)}{dt} = \Pi(t) \cdot Q,$$

where $\Pi(t) = (\pi_U(t), \pi_D(t))$ and $\pi_U(t) + \pi_D(t) = 1$, $\pi_U(t) = e^{-\lambda t}$ and $\pi_D(t) = 1 - e^{-\lambda t}$ are obtained. The system reliability is:

$$R(t) = \pi_U(t) = e^{-\lambda t} \tag{5.8}$$

and the unreliability is:

$$UR(t) = \pi_D(t) = 1 - e^{-\lambda t}. \tag{5.9}$$

It is worth mentioning $UR(t) = F(t)$, where $F(t)$ is cumulative distribution function of the time to failure. Consequently, as $MTTF = \int_0^\infty R(t)\, dt$, we have: $MTTF = \int_0^\infty e^{-\lambda t} dt = \frac{1}{\lambda}$. The mean time to failure (MTTF) also can be computed from the rate matrix Q [56,65].

5.4.1.1 Cold, Warm, and Hot Standby Redundancy

Systems with stringent dependability requirements demand methods for detecting, correcting, avoiding, and tolerating faults and failures. A failure in a large-scale system can mean catastrophic losses. Many techniques have been proposed and adopted to address dependability issues in computer systems in such a way that failures can be tolerated and circumvented. Many of those techniques are based on redundancy, i.e., the replication of components so that they work for a common purpose, ensuring data security and availability even in the event of some component failure. Three replication techniques deserve special attention due to its extensive use in clustered server infrastructures [28]:

- *Cold Standby*: The backup nodes, or modules, are turned off on standby and will only be activated if the primary node fails. One positive point for this technique is that the secondary node has low energy consumption. While in standby mode, the reliability of the unit is preserved, i.e., it will not fail or at least its mean time to failure is expected to be much higher than a fully active component. On the other hand, the secondary node needs significant time to be activated, and clients who were accessing information on the primary node lose all information with the failure of the primary node and must redo much of the work when the secondary node activates.

- *Hot Standby*: This type can be considered the most transparent of the replication modes. The replicated modules are synchronized with the operating module; thereby, the active and standby cluster participants are seen by the end user as a single resource. After a node fails, the secondary node is activated automatically and the users accessing the primary node will now access the secondary node without noticing the change of equipment.

- *Warm Standby*: This technique tries to balance the costs and the recovery time delay of cold and hot standby techniques. The secondary node is on standby, but not completely turned off, so it can be activated faster than in the cold standby technique, as soon as a monitor detects the failure of the primary node. The replicated node is synchronized partially with the operating node, so users who were accessing information on the operating node may lose some information that was being written close to the moment when the primary node failed. It is common to assume that in such a state the standby component has higher reliability than when receiving the workload (i.e., properly working).

Figure 5.2 depicts an example SPN for a cold-standby server system, comprising two servers (S1 and S2). There are two places (S1 _Up and S2 _Down) representing the operational status of the primary server, indicating when it is working or has failed, respectively. Three places (S1 Up, S2 Down, and S2 Waiting) represent the operational status of the spare server, indicating when it is working, failed, or waiting for activation in case of a primary server failure.

Notice that in the initial state of the cold-standby model, both places S1 _up and S2 Waiting have one token, denoting the primary server is up, and the spare server is in standby mode. The activation of the spare server occurs when the transition S1 Fail fires, consuming the token from S1 Up. Once the place S1 _Up is empty, the transition S2 Switch On becomes enabled, due to the inhibitor arc that connects it to S1 Up. Hence, S2 Switch On fires, removing the token from S2 Waiting, and putting one token in place S2 Up. This is the representation of the switchover process from the primary server to the secondary server, which takes an activation time specified in the S2 Switch On firing delay.

The repair of the primary server is represented by firing the S1 Repair transition. The places S1 Down and S2 Up become empty, and S1 _Up receives one token again. As previously mentioned, the time to failure of primary and secondary servers will be different after the spare server is preserved from the effects of wear and tear when it is on shut off or in standby mode. The availability can be numerically obtained from the expression:

$$A = P\left(\left(\#S1_{UP} = 1\right) \vee \left(\#S2_{UP} = 1\right)\right)$$

Figure 5.3 depicts an example CTMC for a warm-standby server system, originally shown in [49]. This model has many similarities to the SPN model for the

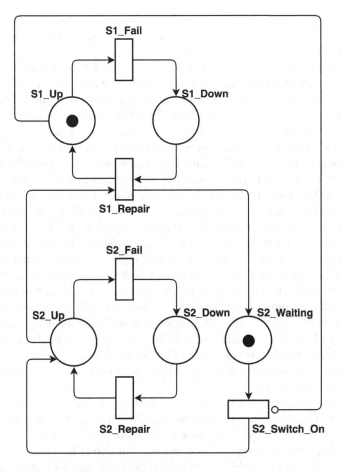

FIGURE 5.2   SPN for cold standby redundancy.

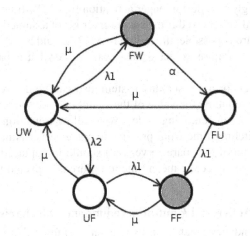

FIGURE 5.3   CTMC for warm standby redundancy.

cold-standby system, despite the distinct semantics and notation. It might be interesting to verify that both approaches can be used interchangeably, mainly when the state-space size is not a major concern.

The CTMC has five states: UW, UF, FF, FU, and FW, and considers one primary and one spare server. The first letter in each state indicates the primary server status, and the second letter indicates the secondary server status. The letter U stands for Up and active, F means Failed, and W indicates Waiting condition (i.e., the server is up but in standby waiting for activation). The shaded states represent that the system has failed (i.e., it is not operational anymore). The state UW represents the primary server (S1) is functional and secondary server (S2) in standby. When S1 fails, the system goes to state FW, where the secondary server has not yet detected the S1 failure. FU represents the state where S2 leaves the waiting condition and assumes the active role, whereas S1 is failed. If S2 fails before taking the active role, or before the repair of S1, the system goes to the state FF, when both servers have failed. For this model, we consider a setup where the primary server repair has priority over the secondary server repair. Therefore, when both servers have failed (state FF) there is only one possible outgoing transition: from FF to UF. If S2 fails when S1 is up, the system goes to state UF and returns to state UW when the S2 repair is accomplished. Otherwise, if S1 also fails, the system transitions to the state FF. The failure rates of S1 and S2, when they are active, are denoted by $\lambda 1$. The rate $\lambda 2$ denotes the failure rate of the secondary server when it is inactive. The repair rate assigned to both S1 and S2 is μ. The rate α represents the switchover rate (i.e., the reciprocal of the mean time to activate the secondary server after a failure of S1).

The warm standby system availability is computed from the CTMC model by summing up the steady-state probabilities for UW, UF, and FU states, which denote the cases where the system is operational. Therefore, $A = \pi_{UW} + \pi_{UF} + \pi_{FU}$. System unavailability might be computed as $U = 1 - A$, but also as $U = \pi_{FF} + \pi_{FW}$.

A CTMC model for a cold standby system can be created with little adjustments to the warm standby model, described as follows. The switchover rate (α) must be modified accordingly to reflect a longer activation time. The transition from UW to the UF state should be removed if the spare server is not assumed to fail while inactive. If such a failure is possible, the failure rate ($\lambda 2$) should be adjusted to match the longer mean time to failure expected for a spare server that is partially or entirely turned off.

A CTMC model for a hot standby system also can be derived from the warm standby model by reducing the value of the switchover rate (α) to reflect a smaller activation time or even removing state FW to allow transition from UW to FU directly if the switching time from primary to spare server is negligible. In every case, the failure rate of the spare server ($\lambda 2$) should be replaced by the same rate of the primary server since the mean time to failure is expected to be the same for both components.

5.4.1.2 Active-Active and k-out-of-n Redundancy Mechanisms

Active-active redundancy means that two operational units share the workload, but workload can be served with acceptable quality by a single unit.

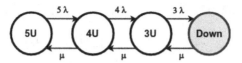

FIGURE 5.4   CTMC for 3-out-of-5 redundancy.

The concept of active-active redundancy can be generalized by assuming that a system may depend strictly only on a subset of its components. Consider a system composed of n identical and independent components that is operational if at least k out of its n components are working correctly. This sort of redundancy is named k-out-of-n.

Combinatorial models, such as RBD [62], are widely used for representing k-out-of-n arrangements, but they also might be modeled and analyzed with CTMC models with equivalent accuracy and even more flexibility [28,57]. Figure 5.4 depicts an example of CTMC model for a 3-out-of-5 redundant server system.

In such a CTMC, the 5U state represents that all five servers are operational. The failure rate of a single server is denoted by λ, whereas the repair rate is denoted by μ. The transition from state 5U to state 4U occurs with the rate 5λ, according to the properties of exponential distribution that is assumed in a Markov chain, considering that the failure of each unit is statistically independent of one to each other, which simply means they may fail concurrently. Similarly, the model goes from state 4U to state 3U with a rate of 4λ after there are only four operational servers remaining. If the model is in state 3U, another failure brings it to the Down state, which represents that the whole system is not operational anymore, and the other servers are turned off, and hence no other failure can occur. Only the repair of at least one server can bring the system to an operational state again. This model considers that only one server can be repaired at a time, which may be the case in many companies where the maintenance team has a limited number of members or equipment needed for the repair. For such a reason, the repair occurs with a μ rate for all transitions outgoing from Down, 3U, and 4U states.

The availability for such a system may be computed as:

$$A = P\{5U\} + P\{4U\} + P\{3U\} \tag{5.10}$$

$$A = 1 - \frac{60\lambda^3}{60\lambda^3 + 20\lambda^2\mu + 5\lambda\mu^2 + \mu^3}$$

The capacity-oriented availability (COA) allows to estimate how much service the system can deliver considering failure states [63,64]:

$$COA = \frac{5 \times P\{5U\} + 4 \times P\{4U\} + 3 \times P\{3U\}}{5} \tag{5.11}$$

$$COA = \frac{\lambda\left(60\lambda^2 + 16\lambda\mu + 3\mu^2\right)}{60\lambda^3 + 20\lambda^2\mu + 5\lambda\mu^2 + \mu^3}$$

The mean time to failure is:

$$MTTF = \frac{400\lambda^4 + 275\lambda^3\mu + 107\lambda^2\mu^2 + 13\lambda\mu^3 + \mu^4}{60\lambda^3\left(20\lambda^2 + 5\lambda\mu + \mu^2\right)}$$

5.4.2 EXAMPLES OF MODELS FOR COMPUTATIONAL SYSTEMS

To demonstrate how to analyze the availability and reliability of computing systems, an example of architecture that is presented in Figure 5.5 is used. The system is composed of a switch/router and server subsystem. The system fails if the switch/router fails or if the server subsystem fails. The server subsystem comprises two servers, S1 and S2. S1 is the main server, and S2 is the spare server. They are configured in cold standby mechanism, that is, S2 starts as soon as S1 fails. The startup time of S2 may be configured according to the adopted switching mechanism. If the start-up time of S2 is equal to zero, then it is perfect switching.

For computing the availability and reliability for such a system, a modeling strategy consisting of Markov chains and SPN models is used.

5.4.2.1 Markov Chains

5.4.2.1.1 Availability CTMC Model

The architecture described in Figure 5.5 enables availability analysis through a heterogeneous modeling approach. Many formalisms may be used to compute such metrics. However, the redundancy mechanism used in the systems requires the use of state-based models, such as Markov chains or SPNs. Therefore, this example depicts the use of CTMC model to compute availability and reliability measures.

Figure 5.6 represents the CTMC availability model. The CTMC represents the detailed behavior of the system which employs redundancy, the start-up time of S2 is zero. The CTMC has six states as a tuple: (D, S2,D), (S1,S2,D), (S1,S2,SR), (D, S2,SR), (D, D, D), and (D, D, SR), and considers one primary and one spare server, S1 and S2 respectively, and one switch/router.

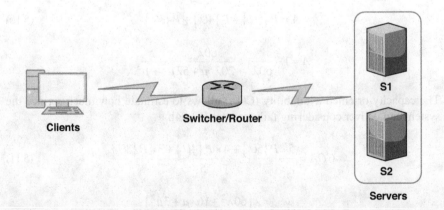

FIGURE 5.5 A simple example.

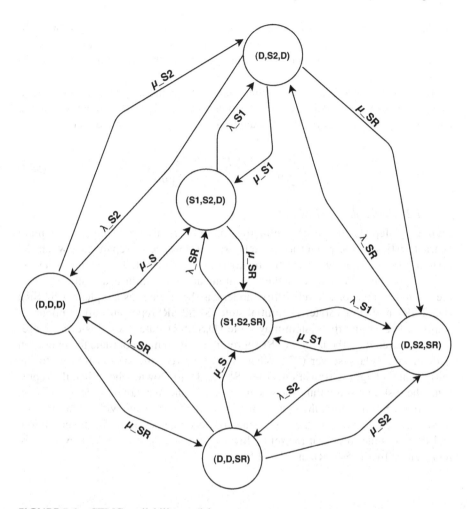

FIGURE 5.6 CTMC availability model.

Each state name comprises three parts. The first one represents the server one (S1), the second denotes the server two (S2), and the third letter describes the switch/router component (SR). The S1 denotes that S1 is running and operational, the S2 represents the S2 is running and operational, and SR represents the Switch/router is running and operational. The letter D represents the failure state. The initial state (S1,S2,SR) represents the primary server (S1) is running and operational, the secondary server (S2) is the spare server, and the switch/router (SR) is functional. When S1 fails, the system goes to the state (D,S2,SR), outgoing transition: from (S1,S2,SR) to (D,S2,SR), when S1 repair, the system returns to the initial state. Once in the state (D,S2,SR), the system may go to the state (D,S2,D) through the SR failure or, the system may go to state (D, SR) through the S2 failure. In both cases, the system may return to the previous state across the SR repair rate or S2 repair rate, respectively. As soon as the state (D,D,SR) is achieved, the system may go to the state (D,D,D) with the SR failure, or returns to the initial state (S1,S2,SR), when the repair is accomplished (i.e., the repair of the systems S1 and S2). The failure rates of S1, S2, and SR are denoted by λ_S1, λ_S2, and λ_SR, respectively, as well as the repair rates for each component μ_S1, μ_S2, and μ_SR. The μ_S denotes the repair rate when the two servers are in a failure state.

The CTMC that represents the architecture enables obtaining a closed-form equation to compute the availability (see Equation 5.12). It is important to stress that the parameters $\mu\_S1=\mu\_S2=\mu\_SR$ are equal to μ and $\lambda\_S1=\lambda\_S2$ are equal to λ.

$$A = \frac{\mu\big(\mu(\mu+\mu_s)+\lambda(\mu+2\mu_s)\big)}{(\lambda_{SR}+\mu)\big(\lambda^2+\mu(\mu+\mu_s)+\lambda(\mu+2\mu_s)\big)} \tag{5.12}$$

5.4.2.1.2 Reliability CTMC Model

Figure 5.7 depicts the CTMC reliability model for this architecture. The main characteristics of the reliability models are the absence of repair, i.e., when the system goes to the failure state the repair is not considered. This action is necessary to compute with more ease the system mean time to failure, and subsequently the reliability metric. The reliability model has three states as a tuple: (S1,S2,SR); (D,S2,SR); and Down state. The initial state (S1,S2,SR) represents all components running. If S1 fails, the system may go to (D,S2,SR) state, then this event represents that even with the failure of S1 server, the system may continue the operation with the secondary server (S2). When S1 is repaired, the system returns to the initial state. Outgoing transition: from (S1,S2,SR) to Down, when SR fails, represents the system failure; thus, the system is offline and may not provide the service. Once in (D,S2,SR) state, the system may go to the Down state with S2 failure rate or SR failure rate. Once in the Down state, the system goes to the failure condition, and it is possible to obtain the reliability metric. The up states of the system are represented by (S1,S2,SR) and (D,S2,SR).

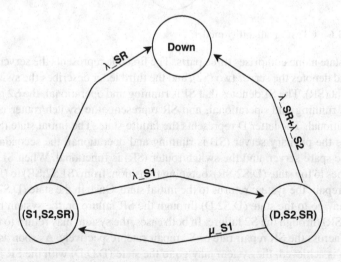

FIGURE 5.7 CTMC reliability model.

5.4.2.1.3 Results

Table 5.1 presents the values of failure and repair rates, which are the reciprocal of the MTTF and mean time to repair (MTTR) of each component represented in Figures 5.6 and 5.7. Those values were estimated and were used to compute the availability and reliability metrics.

It is important to stress that the μS represents twice the repair rate of $\mu S1$ considering just one maintenance team. The availability and reliability measures were computed herein for the architecture described in Figure 5.5, using the mentioned input parameters. The results are shown in Table 5.2, including steady-state availability, number of nines, annual downtime, reliability, and unreliability, considering 4,000 h of activity.

The downtime provides a view of how much time the system is unavailable for its users for 1 year. The downtime value of 10.514278 h indicates that the system can be improved; this downtime indicates that the system stands still for 10 hours of total outage through a year. At 4,000 h of activity, the system has a reliability a little over 80 percent.

5.4.2.2 SPN Models

5.4.2.2.1 Availability SPN Model

An SPN model may be used to represent the same system already analyzed with the CTMC model discussed in the previous section, and to obtain availability and reliability measures similarly.

TABLE 5.1
Input Parameters

Variable	Value (h^{-1})
λ_SR	1/20,000
$\lambda\_S1 = \lambda S2$	1/15,000
$\mu S1 = \mu S2 = \mu\_SR$	1/24
μS	1/48

TABLE 5.2
CTMC Results

Availability	0.9987997
Number of nines	2.9207247
Downtime (h/yr)	10.514278
Reliability (4,000 h)	0.8183847
Unreliability (4000 h)	0.1816152

The redundant mechanism is employed to represent switch/router component and two servers, S1 and S2. The servers are configured in cold Standby; that is, S2 starts as soon as S1 fails. The start-up time of S2 is denoted by *S2 Switching-On* transition. Figure 5.8 shows the SPN model adopted to estimate the availability and downtime of the servers with cold standby redundancy.

The markings of the places *SR OK* and *S1 OK* denote the operational states of the switch/router and S1 server. The marking of the *S2 _OFF* indicates the waiting state before the activation of S2 server. When the place *S2 OK* is marked, the server S2 is operational and in use. The places *SR F*, *S1 F*, and *S2 F* indicate the failure states of these components. When the main module fails (S1), the transition *S2 Switching-On* is enabled. Its firing represents the start of the spare in operational state (S2). This period is the Mean Time to Activate (MTA).

The following statement is adopted for estimating availability and unavailability:

$$A = \left(P \left(\left(\# SR\_OK = 1 \right) AND \left(\left(\# S1\_OK = 1 \right) OR \left(\# S2\_OK = 1 \right) \right) \right) \right)$$

$$UA = 1 - \left(P \left(\left(\# SR\_OK = 1 \right) AND \left(\left(\# S1\_OK = 1 \right) OR \left(\# S2\_OK = 1 \right) \right) \right) \right)$$

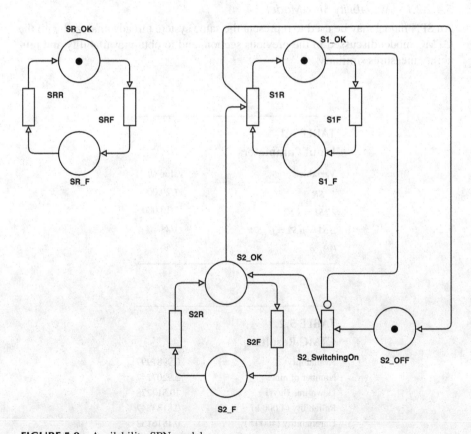

FIGURE 5.8   Availability SPN model.

5.4.2.2.2   Reliability SPN Model

Figure 5.9 shows the SPN reliability model for architecture presented in Figure 5.5. The main difference between models of Figures 5.8 and 5.9 is the repair time for the entire system, i.e., the system reliability considers the time until the first failure. The model represents an active/active redundancy, with the failure of S1 and S2 servers the immediate transition is enabled and may be fired, marking the

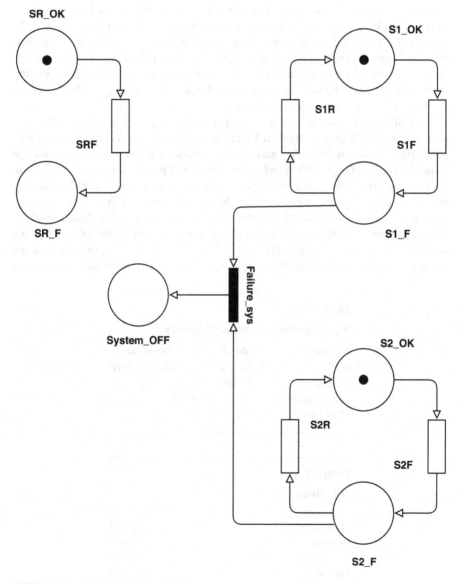

FIGURE 5.9   SPN Reliability model.

place System OFF with a token. The following expressions are adopted for estimating reliability and unreliability, respectively:

$$R(t) = 1 - P((\# SR\,F = 1)V\,(\# System\,OFF = 1))(t)$$

$$UR(t) = P((\# SR\,F = 1)V\,(\# System\,OFF = 1))(t)$$

5.4.2.2.3 Results

Table 5.3 presents the values of mean time to failure (MTTF) and mean time to repair (MTTR) used for computing availability and reliability metrics for the SPN models. We computed the availability and reliability measures using the mentioned input parameters. The results are shown in Table 5.4, including steady-state availability, number of nines, annual downtime, reliability, and unreliability, considering 4,000 h of activity. The switching time considered is 10 minutes, which are enough for the system startup and software loading.

This SPN model enables the computation of the reliability function of this system over time, which is plotted in Figure 5.10, considering the baseline setup of parameters shown in Table 5.3, and also a scenario with improved values for the switch/router MTTF (30,000 h) and both servers MTTR (8 h). It is noticeable that, in the baseline setup, the system reliability reaches 0.50 at around 15,000 h, and after 60,000 h (about 7 years), the system reliability is almost zero. When the improved version of the system is considered, the reliability has a smoother decay, reaching 0.50 just around 25,000 h, and approaching zero only near to 100,000 h. For the sake of comparison, the reliability at 4,000 h is 0.8840773, wherein the baseline setup is 0.818385. Such an analysis might be valuable for systems administrators to

TABLE 5.3

Input Parameters for SPN Models

Transition	Value (h)	Description
SRF	20,000	Switch/Router MTTF
S1F = S2F	15,000	Servers MTTF
SRR = S1R = S2R	24	MTTR
S2 Switching On	0.17	MTA

TABLE 5.4

SPN Results

Availability	0.998799
Number of nines	2.920421
Downtime (h/yr)	10.521636
Reliability (4,000 h)	0.818385
Unreliability (4,000 h)	0.181615

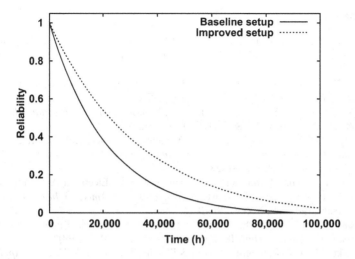

FIGURE 5.10 Reliability function for the example system.

make decisions regarding system maintenance and replacement of components to avoid failures that will cause significant damage for revenue, customer satisfaction, or other corporate goals.

5.5 FINAL COMMENTS

The process of analytical modeling for computational systems must consider a variety of strategies and characteristics of each available formalisms. The choice of one type of model may involve accuracy issues, expressiveness power, accessible software tools, and the complexity of the target system.

The concepts and examples presented in the chapter should be viewed as an introduction and motivation on possible methods to select when studying computing reliability and availability metrics. The conciseness and power of SPNs especially can be useful in many cases when complexity grows and many details must be represented. Nevertheless, CTMCs always will be kept as an option which provides enough resources for performing many kinds of analyses. Other modeling formalisms, such as FTs, RBDs, Reliability Graphs, and stochastic Automata networks, are also significantly important and enable different views for the same dependability concepts approached here.

The world is a place where information systems control almost every aspect of daily lives. The knowledge and framework exposed here may be increasingly required as regulatory agencies and big corporate customers demand the estimation of boundaries on how dependable their systems are.

ACKNOWLEDGMENT

This work was supported by a grant of contract number W911NF1810413 from the U.S. Army Research Office (ARO).

REFERENCES

1. Laprie, J.C. Dependable Computing and Fault Tolerance: Concepts and terminology. *Proceedings of the 15th IEEE International Symposium on Fault-Tolerant Computing.* 1985.
2. Schaffer, S. Babbage's Intelligence: Calculating Engines and the Factory System. *Critical Inquiry.* 1994, Vol. 21, No. 1, 203–227.
3. Blischke, W.R., Murthy, D.P. [ed.]. *Case Studies in Reliability and Maintenance.* Hoboken, NJ: John Wiley & Sons, 2003. p. 661.
4. Stott, H.G. Time-Limit Relays and Duplication of Electrical Apparatus to Secure Reliability of Services at New York. *Transactions of the American Institute of Electrical Engineers,* 1905, Vol. 24, 281–282.
5. Stuart, H.R. Time-Limit Relays and Duplication of Electrical Apparatus to Secure Reliability of Services at Pittsburg. *Transactions of the American Institute of Electrical Engineers,* 1905, vol. XXIV, 281–282.
6. Board of Directors of the American Institute of Electrical Engineers. Answers to Questions Relative to High Tension Transmission. s.l.: IEEE, 1902.
7. Ushakov, Igor. Is Reliability Theory Still Alive? *Reliability: Theory &Applications.* 2007, Vol. 2, No. 1, Mar. 2017, p. 10.
8. Bernstein, S. Sur l'extension du théorème limite du calcul des probabilités aux´ sommes de quantités dépendantes. *Mathematische Annalen.* 1927, Vol. 97, 1–59. http://eudml.org/doc/182666.
9. Basharin, G.P., Langville, A.N., Naumov, V.A. The Life and Work of A.A. Markov. *Linear Algebra and Its Applications.* 2004, Vol. 386, 3–26. doi:10.1016/j.laa.2003.12.041.
10. Principal Works of A. K. Erlang—The Theory of Probabilities and Telephone Conversations. First published in Nyt Tidsskrift for Matematik B. 1909, Vol. 20, 33–39.
11. Kotz, S., Nadarajah, S. (2000). *Extreme Value Distributions: Theory and Applications.* Imperial College Press. doi:10.1142/9781860944024.
12. Kolmogoroff, A. Über die analytischen Methoden in der Wahrscheinlichkeitsrechnung [in German] [Springer-Verlag]. *Mathematische Annalen.* 1931, Vol. 104, 415–458. doi:10.1007/BF01457949.
13. Shannon, C.E. A Mathematical Theory of Communication. *The Bell System Technical Journal.* 1948, Vol. 27, 379–423, 623–656.
14. Neumann, J.V. Probabilistic Logics and the Synthesis of Reliable Organisms from Unreliable Components. *Automata studies,* 1956, Vol. 34, 43–98.
15. Moore, E.F. Gedanken-Experiments on Sequential Machines. *The Journal of Symbolic Logic.* 1958, Vol. 23, No. 1, 60.
16. Cox, D. A Use of Complex Probabilities in the Theory of Stochastic Processes. *Mathematical Proceedings of the Cambridge Philosophical Society.* 1955, Vol. 51, No. 2, 313319. doi:10.1017/S0305004100030231.
17. Avizienis, A. Toward Systematic Design of Fault-Tolerant Systems. *IEEE Computer.* 1997, Vol. 30, No. 4, 51–58.
18. Barlow, R.E. *Mathematical Theory of Reliability.* New York: John Wiley & Sons, 1967. SIAM series in applied mathematics.
19. Barlow, R.E., Mathematical Reliability Theory: From the Beginning to the Present Time. *Proceedings of the Third International Conference on Mathematical Methods In Reliability, Methodology and Practice.* Trondheim, Norway, 2002.
20. Epstein, B., Sobel, M. Life Testing. *Journal of the American Statistical Association.* 1953, Vol. 48, No. 263, 486–502.
21. Gnedenko, B., Ushakov, I. A., & Ushakov, I. (1995). *Probabilistic reliability engineering.* John Wiley & Sons.

22. Thiess, S. J. Einhorn and F. B. Intermittence as a stochastic process. S. J. Einhorn and F. B. Thiess, *Intermittence as a stNYU-RCA Working Conference on Theory of Reliability.* New York: Ardsley-on-Hudson, 1957.
23. Anselone, P.M. Persistence of an Effect of a Success in a Bernoulli Sequence. *Journal of the Society for Industrial and Applied Mathematics.* 1960, Vol. 8, No. 2, 272–279.
24. Birnbaum, Z.W., Esary, J.D., Saunders, S.C. Multi-component Systems and Structures and Their Reliability. *Technometrics.* 1961, Vol. 3, No. 1, 55–77.
25. Ericson, C. Fault Tree Analysis—A History. *Proceedings of the 17th International Systems Safety Conference.* Orlando, FL, 1999.
26. Pierce, W.H. *Failure-Tolerant Computer Design.* New York: Academic Press, 1965, 65–69.
27. Avizienis A., Laprie J.C., Randell, B. Fundamental Concepts of Computer System Dependability. IARP/IEEE-RAS Workshop on Robot Dependability: Technological Challenge of Dependable Robots in Human Environments—Seoul, Korea, 2001
28. Maciel, P., Trivedi, K., Matias, R., Kim, D. Dependability Modeling. *Performance and Dependability in Service Computing: Concepts, Techniques and Research Directions ed.* Hershey, PA: IGI Global, 2011.
29. Laprie, J.C. *Dependability: Basic Concepts and Terminology.* s.l. New York: SpringerVerlag, 1992.
30. Natkin, S.O. (1980). *Les reseaux de Petri stochastiques et leur application a l'evaluation des systemes informatiques.* Conservatoire National des Arts et Metiers. PhD thesis. CNAM. Paris, France.
31. Molloy, M.K. (1982). On The Integration of Delay and Throughput Measures in Distributed Processing Models. PhD thesis. UCLA. Los Angeles, CA.
32. Symons, F.J.W. Modelling and Analysis of Communication Protocols using Numerical Petri Nets. PhD Thesis, University of Essex, also Dept of Elec. Eng. Science Telecommunications Systems Group Report No. 152, 1978.
33. Chiola, G., Franceschinis, G., Gaeta, R., Ribaudo, M. GreatSPN 1.7: Graphical Editor and Analyzer for Timed and Stochastic Petri Nets. *Performance Evaluation.* Vol. 25, No. 1–2, 47–68, 1995.
34. Haverkort, B.R. Markovian Models for Performance and Dependability Evaluation. *Lectures on Formal Methods and Performance Analysis.* Berlin, Germany: Springer, 2001.
35. Seneta, E. Markov and the Creation of the Markov Chains. *School of Mathematics and Statistics,* University of Sydney, NSW, Australia, 2006.
36. Trivedi, K.S. *Probability and Statistics with Reliability, Queuing, and Computer Science Applications,* 2nd ed. Hoboken, NJ: John Wiley & Sons, 2001.
37. Parzen, E. *Stochastic Processes.* Dover Publications. San Francisco, CA, 1962.
38. Stewart, W.J. *Probability, Markov Chains, Queues and Simulation.* Princeton, NJ: Princeton University Press, 2009.
39. Ash, R.B. *Basic Probability Theory.* New York: John Wiley & Sons, 1970.
40. Feller, W. *An Introduction to Probability Theory and Its Applications.* Vols. I, II. New York: John Wiley & Sons, 1968.
41. Marsan, M.A., Conte, G., Balbo, G. A Class of Generalized Stochastic Petri Nets for the Performance Evaluation of Multiprocessor Systems. *ACM Transactions on Computer System.* 1984, Vol. 2, No. 2, 93–122.
42. Ajmone Marsan, M., Chiola, G. On Petri Nets with deterministic and exponentially distributed firing times. In G. Rozenberg, editor, *Advances in Petri Nets 1987, Lecture Notes in Computer Science* 266, pp. 132–145. Springer-Verlag, 1987.
43. Marsan, M.A., Balbo, G., Conte, G., Donatelli, S., Franceschinis, G. *Modelling with Generalized Stochastic Petri Nets.* Wiley, 1995.

44. German, R., Lindemann, C. Analysis of Stochastic Petri Nets by the Method of Supplementary Variables. *Performance Evaluation.* 1994, Vol. 20, No. 1, 317–335.
45. German, R. *Performance Analysis of Communication Systems with NonMarkovian Stochastic Petri Nets.* New York: John Wiley & Sons, 2000.
46. Lindemann, C. (1998). Performance modelling with deterministic and stochastic Petri nets. *ACM sigmetrics performance evaluation review,* 26(2), 3.
47. Molloy, M.K. Performance Analysis Using Stochastic Petri Nets. *IEEE Transactions on Computers.* 1982, Vol. 9, 913–917.
48. Muppala, J., Ciardo, G., Trivedi, K.S. Stochastic Reward Nets for Reliability Prediction. *Communications in Reliability, Maintainability and Serviceability.* 1994, Vol. 1, 9–20.
49. Matos, R., Dantas, J., Araujo, J., Trivedi, K.S., Maciel, P. Redundant Eucalyptus Private Clouds: Availability Modeling and Sensitivity Analysis. *Journal Grid Computing.* 2017, Vol. 15, No. 1, 1–23.
50. Malhotra, M., Trivedi, K.S. Power-hierarchy of Dependability-Model Types. *IEEE Transactions on Reliability.* 1994, Vol. 43, No. 3, 493–502.
51. Shooman, M.L. The Equivalence of Reliability Diagrams and Fault-Tree Analysis. *IEEE Transactions on Reliability.* 1970, Vol. 19, No. 2, 74–75.
52. Watson, J.R., Desrochers, A.A. Applying Generalized Stochastic Petri Nets to Manufacturing Systems Containing Nonexponential Transition Functions. *IEEE Transactions on Systems, Man, and Cybernetics.* 1991, Vol. 21, No. 5, 1008–1017.
53. O'Connor P, Kleyner A. *Practical Reliability Engineering.* John Wiley & Sons; 2012 Jan 30.
54. Beaudry, M.D. Performance-Related Reliability Measures for Computing Systems. *IEEE Transactions on Computers.* 1978, Vol. 6, 540–547.
55. Dantas, J., Matos, R., Araujo, J., Maciel, P. Eucalyptus-based Private Clouds: Availability Modeling and Comparison to the Cost of a Public Cloud. *Computing.* 2015, Vol. 97, No. 11, 1121–1140.
56. Buzacott, J.A. Markov Approach to Finding Failure Times of Repairable Systems. *IEEE Transactions on Reliability.* 1970, Vol. 19, No. 4, 128–134.
57. Maciel, P., Matos, R., Silva, B., Figueiredo, J., Oliveira, D., Fe, I., Maciel, R., Dantas, J. Mercury: Performance and Dependability Evaluation of Systems with Exponential, Expolynomial, and General Distributions. In: *The 22nd IEEE Pacific Rim International Symposium on Dependable Computing (PRDC 2017).* January 22–25, 2017. Christchurch, New Zealand.
58. Guedes, E., Endo, P., Maciel, P. An Availability Model for Service Function Chains with VM Live Migration and Rejuvenation. *Journal of Convergence Information Technology.* Volume 14 Issue 2, April, 2019. Pages 42–53.
59. Silva, B., Matos, R., Callou, G., Figueiredo, J., Oliveira, D., Ferreira, J., Dantas, J., Junior, A.L., Alves, V., Maciel, P. Mercury: An Integrated Environment for Performance and Dependability Evaluation of General Systems. *Proceedings of Industrial Track at 45th Dependable Systems and Networks Conference (DSN-2015).* 2015. Rio de Janeiro, Brazil.
60. Silva, B., Maciel, P., Tavares, E., Araujo, C., Callou, G., Souza, E., Rosa, N. et al. ASTRO: A Tool for Dependability Evaluation of Data Center Infrastructures. *IEEE International Conference on Systems, Man, and Cybernetics,* 2010, Istanbul, Turkey. IEEE Proceeding of SMC, 2010.
61. Silva, B., Callou, G., Tavares, E., Maciel, P., Figueiredo, J., Sousa, E., Araujo, C., Magnani, F., Neves, F. *Astro: An Integrated Environment for Dependability and Sustainability Evaluation. Sustainable Computing: Informatics and Systems.* 2013 Mar 1;3(1):1–7.
62. Kuo, W., Zuo, M.J. *Optimal Reliability Modeling—Principles and Applications.* Wiley, 2003. p. 544.

63. Heimann, D., Mittal, N., Trivedi, K. Dependability Modeling for Computer systems. *Proceedings Annual Reliability and Maintainability Symposium*, 1991. IEEE, Orlando, FL, pp. 120–128.

64. Matos, R., Maciel, P.R.M., Machida, F., Kim, D.S., Trivedi, K.S. Sensitivity Analysis of Server Virtualized System Availability. *IEEE Transactions on Reliability.* 2012, Vol. 61, No. 4, 994–1006.

65. Reinecke, P., Bodrog, L., Danilkina, A. *Phase-Type Distributions. Resilience Assessment and Evaluation of Computing Systems*, Berlin, Germany: Springer, 2012.

6 An Overview of Fault Tree Analysis and Its Application in Dual Purposed Cask Reliability in an Accident Scenario

Maritza Rodriguez Gual, Rogerio Pimenta Morão, Luiz Leite da Silva, Edson Ribeiro, Claudio Cunha Lopes, and Vagner de Oliveira

CONTENTS

6.1 INTRODUCTION

Spent nuclear fuel is generated from the operation of nuclear reactors and must be safely managed following its removal from reactor cores. The Nuclear Technology Development Center (Centro de Desenvolvimento da Tecnologia Nuclear–CDTN), Belo Horizonte, Brazil constructed a dual-purpose metal cask in scale 1:2 for the transport and dry storage of spent nuclear fuel (SNF) that will be generated by research reactors, both plate-type material testing reactor (MTR) and TRIGA fuel rods. The CDTN is connected to the Brazilian National Nuclear Energy Commission (Comissão Nacional de Energia Nuclear—CNEN).

The dual purpose cask (DPC) development was supported by International Atomic Energy Agency (IAEA) Projects RLA4018, RLA4020, and RLA3008. The project began in 2001 and finished in 2006. Five Latin American countries participated—Argentina, Brazil, Chile, Peru, and Mexico. The cask is classified as a Type B package according to IAEA Regulations for the Safe Transport of Radioactive

Materials (IAEA, TS-R-1, 2009). The RLA/4018 cask was designed and constructed in compliance with IAEA Transport Regulations. The IAEA established the standards for the packages used in the transport of radioactive materials under both normal and accident conditions.

The general safety requirement concerns, among other issues, are package tiedown, lifting, decontamination, secure and closing devices, and material resistance to radiation, thermal, and pressure conditions likely to be found during transportation.

The regulations establish requirements that guarantee that fissile material is packaged and shipped in such a manner that they remain subcritical under the conditions prevailing during routine transport and in accidents.

6.2 OVERVIEW OF FAULT TREE ANALYSIS

Different techniques are applied widely in risk analyses of industrial process and equipment operating such as Failure Modes and Effects Analysis (FMEA) and its extension Failure Mode, Effects, and Criticality Analysis (FMECA) (Gual et al., 2014; Perdomo and Salomon, 2016), fault tree analysis (FTA) (Vesely, 1981), What-if (Gual, 2017), Layer of Protection Analysis (LOPA) (Troncoso, 2018a), and Hazards and Operability Study (HAZOP) (Troncoso, 2018b). Different methods of solving a fault tree as an advanced combinatorial method (Rivero et al., 2018) also has been applied. All techniques have advantages and limitations. The selection techniques chosen will depend on the documentation available and the objectives to be achieved.

The FTA is one of the most popular and visual techniques to identify risk for design operation, reliability, and safety.

The traditional FTA technique was selected to evaluate reliability and risk assessment of a DPC constructed at CDTN for the transport and storage of spent nuclear fuel that will be generated by research reactors.

FTA techniques were first developed at Bell Telephone Laboratories in the early 1960s. Since this time, FTA techniques have been adopted readily by a wide range of industries, such as power, rail transport, oils, nuclear, chemistry, and medicine, as one of the primary methods of performing reliability and safety analysis. FTA is a top down, deductive failure analysis in which an undesired state of a system is analyzed using Boolean logic to combine a series of lower-level events. This analysis method is used mainly in the field of safety engineering and reliability engineering to determine the probability of a safety accident or a specific system level (functional) failure. In 1981, the U.S. Nuclear Regulatory Commission (NRC) issued the Fault Tree Handbook, NUREG-0492 (Vesely, 1981). In 2002, the National Aeronautics and Space Administration (NASA) published the Fault Tree Handbook with Aerospace Applications (Stamatelatos et al., 2002). Today, FTA is used widely in all major fields of engineering.

FTA defined in NUREG-0492 is "An analytical technique, whereby an undesired state of the system is specified (usually a state that is critical from a safety standpoint), and the system is then analyzed in the context of its environment and operation to find all credible ways in which the undesired event can occur."

A Fault Tree always can be translated into entirely equivalent minimal cut sets (MCS), which can be considered the root causes for these fall fatalities

TABLE 6.1

Fundamental Laws of Boolean Algebra

Law	AND Form Representation	OR Form Representation
Commutative	$x + y = y + x$	$x \cdot y = y \cdot x$
Associative	$x + (y + z) = (x + y) + z$	$x \cdot (y \cdot z) = (x \cdot y) \cdot z$
Distributive	$x \cdot (y + z) = x \cdot y + x \cdot z$	$x \cdot y + x \cdot z$
Idempotent	$x \cdot x = x$	$x + x = x$
Absorption	$x \cdot (x + y) = x$	$x + x \cdot y = x$

(Vesely et al., 1981). The FTA begins by identifying multiple-cause combinations for each fatality. These multiple-cause combinations can be connected by an AND-gate (the output occurs only if all inputs occur), indicating that these two or three events contributed simultaneously to these fatal falls and OR gate (the output occurs if any input occurs). Fundamental laws of Boolean algebra (Whitesitt, 1995) (see Table 6.1) were applied to reduce all possible cause combinations to the smallest cut set (Vesely et al., 1981) that could cause the top event to occur. Eventually, all case combinations associated with each basic event can be simplified and presented in a fault tree diagram.

The fault tree gates are systematically substituted by their entries, applying the Boolean algebra laws in several stages until the top event Boolean expression contains only basic events. The final form of the Boolean equation is an irreducible logical union of minimum sets of events necessary and enough to cause of the top event, denominated MCSs. Then, the original fault tree is mathematically transformed into an equivalent MCS fault tree. The transformation process also ensures that any single event that appears repeatedly in various branches of the fault tree is not counted twice.

Fault trees graphically represent the interaction of failures and other events within a system. Basic events at the bottom of the fault tree are linked via logic symbols (known as gates) to one or more top events. These top events represent identified hazards or system failure modes.

A fault tree diagram (FTD) is a logic block diagram that displays the state of a system (top event) in terms of the states of its components or subsystems (basic events). The basic events are determined through a top-down approach. The relationship between causes and the top event (failure or fatalities) are represented by standard logic symbols (AND, OR, etc.).

FTA involves five steps to obtain an understanding of the system:

- Define the top event to study
- Obtain an understanding of the system (with functional diagram, design, for example)
- Construct the fault tree
- Analyze the fault tree qualitatively or quantitatively
- Evaluate the results and propose recommendations.

FTA is a simple, clear, and direct-vision method for effectively analyzing and estimating possible accidents or incidents and causes. FTA is useful to prioritize the preventive maintenance of the equipment that is contributing the most to the failure. Also, it is a quality assurance (QA) tool. The overall success of the FTA depends on the skill and experience of the analyst.

Qualitative analysis by FTA is an alternative for reliability assessment when historical data of undesirable event (fatalities or failure) are incomplete or unavailable for probabilistic calculation (quantitative).

FTA can be used for quantitative assessment of reliability if probability values are available.

For a large or very complex system that includes a large number of equipment and components, FTA can be time consuming. The complex FTA must be analyzed with a specialized computer program. However, there are still several practical cases in which fault trees are convenient as it is for the case study solved here.

This methodology (Vesely et al., 1981) is applicable to all fault trees, regardless of size of complexity, that satisfy the following conditions:

- All failures are binary in nature (either success or failure; ON/OFF). The partial failures do not exist.
- Transition between working and failed states occurs instantaneously (no time a delay is considered).
- All component failures are statistically independent.
- The failure rate of reach equipment item is constant.
- The repair rate for each equipment item is constant.
- After repair, the system will be as good as the old, not as good as new (i.e., the repaired component is returned to the same state, with the same failure characteristics that it would have had if the failure had not occurred; repair is not considered to be a renewal process).
- The fault tree for system failure is the same as the repair tree (i.e., repair of the failed component results in the immediate return to their normal state of all higher intermediate events that failed as a result of the failed component).

But, the biggest advantage of using FTA is that it starts from a top event that is selected by the user for a specific interest, and the tree developed will identify the root cause.

6.3 MINIMAL CUT SETS

There are several methods for determining MCS. In this case study, the classic Boolean reduction is used as was described previously.

The adjective minimal means that they are all essential. If just one of the single events is recovered, the system recuperates the success state, and when it fails, it causes the system failure.

Cut sets with fewer events are generally more likely to fail since only a few events must fail simultaneously. Therefore, these MCSs have a higher importance.

The MCSs can be ordered by number and the order (i.e., cut set size).

A cut set order is the number of elements in cut sets. The first-order MCS can be directly obtained, and the second-order MCS is obtained by the logical operation "OR." When the gate is "AND," it increases the order of the MCS and when it is "OR," the quantity of MCS is increased.

The lower the order, the more critical is the cut, which is only acceptable if this failure is of very low probability.

6.4 DESCRIPTION OF DUAL PURPOSE CASK

The difference between the conventional transport and storage packages of radioactive material and the DPC is that in addition to the cask body it has primary and secondary lids, an internal basket, and external shock absorbers (See Figure 6.1).

Figure 6.2 shows the photography of a DPC constructed in CDTN.

The RLA4018 cask consists of a robust cylindrical body provided with an internal cavity to accommodate a basket holding the spent fuel elements, a double lid (primary and secondary), and two external impact limiters (top and bottom). The impact limiters are structures made of an external stainless-steel skin and an energy-absorbing filling material.

The body and the primary lid are sandwich-like structures with stainless steel in the outside and lead in the inside for shielding purposes. The primary lid is provided with a double metallic sealing system. Bolts are used to fix the primary lid to the cask body.

The main function of the secondary lid is to protect the primary lid against impacts.

The internal basket is a square array stiffened by spacers and provided with a bottom plate and feet. It is designed typically to hold 21 MTR fuel elements. Each MTR fuel type element has 21 plates, with oxide fuel of U_3O_8 and clad in aluminum. The fuel elements are transferring into a basket of stainless steel and transported dry. Fuel elements are stored interim in dry conditions.

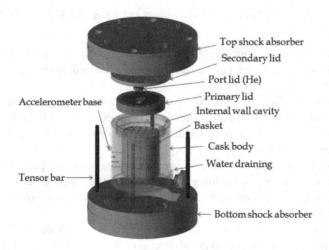

FIGURE 6.1 Spent fuel transport and storage RLA4018 design by CDTN.

FIGURE 6.2 Photography of spent fuel transport and storage RLA4018 design by CDTN.

The cask is provided with four lifting trunnions; two in the top half and two in its bottom half so that the cask can be easily rotated. The cask is vertically held down by four bottom screwed trunnions.

The process of loading spent fuel consists of submerging the transport cask in the reactor pool while spent fuel is transferred into the basket. The water is drained and the cask is dried to eliminate residual amounts of water in the cavity to ensure sub-criticality conditions.

The cask has one draining port for vacuum drying, while its primary lid is provided with another one for helium gas filling. After the water draining, a cask primary lid is installed on the cask body. Next, a vacuum drying system is connected to a cask to remove the moisture from the cask.

The shock absorbers provide protection to the whole cask during the 9 meters drop test prescribed for this type of package. They consist of a thin external stainless-steel shell encasing an energy absorbing material. Different materials have been used by the cask designers for this purpose, the most common being polyurethane foam, solid wood and wood composites, aluminum honeycomb, and aluminum foam. The currently selected cushioning material is high density rigid polyurethane foam.

It is important to note that the accelerometer base is not in the final cask. It is used only to measure the acceleration range during the impact tests.

Type B packages are designed to withstand very severe accidents in all the modes of transport without any unacceptable loss of containment or shielding.

The transport regulations and storage safety requirements to consider in the DPC package design (IAEA, 2014), under routine conditions of transport (RCT), normal conditions of transport (NCT), and accident conditions of transport (ACT) are:

- Containment of radioactive materials
- Shielding (control of external radiation levels)
- Prevention of nuclear criticality (a self-sustaining nuclear chain reaction)
- Prevention of damage caused by heat dissipation
- Structural integrity
- Stored spent fuel retrievability
- Aging

Aging effects in DPCs is considering because they are expected to be used for spent fuel interim storage for up to 20 years.

The objective of the regulations is to protect people and the environment from the effects of radiation during the transport of radioactive material.

Normal conditions that a spent fuel transport package must be able to resist include hot and cold environments, changes in pressure, vibration, water spray, impact, puncture, and compression.

To show that it can resist accident conditions, a package must pass impact, fire, and water immersion tests.

Reports from the United States (Nuclear Monitor 773, 2013) and the United Kingdom (Jones and Harvey, 2014) include descriptions of various accidents and incidents involving the transport of radioactive materials, which occurred until 2014, but none resulted in a release of radioactive material or a fatality due to radiation exposure. For this reason, this study is important.

6.5 CONSTRUCT THE FAULT TREE OF DUAL PURPOSE CASK

The fault tree is a directed acyclic graph consisting of two types of nodes: events (represented as circles) and gates.

An event is an occurrence within the system, typically the failure of a component or sub-system.

Events can be divided into:

- Basic events (BEs), which occur on their own
- Intermediate events (IEs), which are caused by other events

The root, called the top event (TE), is the undesired event of a tree.

Rectangle represents top event and middle events.

Circle represents basic events.

Logic OR gate, which is equivalent to the Boolean symbol +, represents a situation in which one of the events alone (input gate) is enough to contribute to the system fault (output event). OR gates increase the number of cut sets, but often lead to single component sets.

Logic AND gate, which is equivalent to the Boolean symbol, represents a situation in which all the events shown below the gate (input gate) are required for a system fault shown above the gate (output event). AND gates of the fault tree increase the number of components (order) in a cut set.

The analysis was performed according to the following steps:

- Definition of the system failure event of interest, known as the top event, as environmental contamination.
- Identification of contributing events (basic or intermediate), which might directly cause the top event to occur.

6.6 RESULTS

The specific case study analyzed to apply FTA is titled Environmental contamination (Top Event).

The fault tree was constructing within multidisciplinary teams working together, such as nuclear engineers, electrical engineers, and mechanical engineers. Working within multidisciplinary teams makes it possible to analyze the design of weak points.

The fault tree diagram is shown in Figure 6.3.

The basic events that led to the top event, Environmental contamination, are shown in Table 6.2 with the symbols given.

Table 6.3 describes the symbols for the Intermediary Events on the FTD.

The Boolean algebra analysis of the fault tree is shown in Table 6.4.

The MCSs are listed in Table 6.5.

Events B1, B2, B3, B4, B5, B7, B8, B9, and B10 are associated with human errors. Hence, B6 is susceptible to human error.

Boolean algebra laws reduced the amount of cause combinations and the redundancy of basic events.

MCS can be used to understand the structural vulnerability of a system. If the order of MCS is high, then the system will less vulnerable (or top event in fault trees) to these events combinations. In addition, if there are numerous MCSs, it indicates that the system has higher vulnerability. Cut sets can be used to detect single point failures (one independent element of a system that causes an immediate hazard to occur and/or causes the whole system to fail).

Two first-order and five second-order MCS were found.

- *1st order*: The occurrence of a BE implies the occurrence of the top or undesired event.
- *2nd order*: The simultaneous occurrences of BEs result in the loss of continuity of operation of the system.

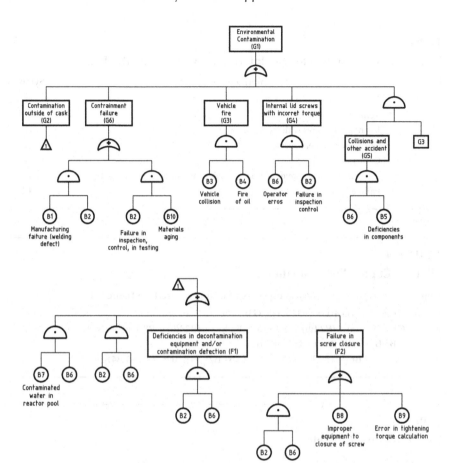

FIGURE 6.3 FTD of RLA4018 DPC before the MCS analysis.

TABLE 6.2
Description of Symbols for the Basic Events on the Fault Tree Diagram

Number	Basic Events	Symbols
1	Containment failure	B1
2	Failure in inspection, control, in testing program	B2
3	Vehicle collision	B3
4	Fire of oil	B4
5	Deficiencies in component	B5
6	Operator errors	B6
7	Contaminated water in reactor pool	B7
8	Improper equipment to closure of screw	B8
9	Error in tightening torque calculation	B9
10	Material aging	B10

TABLE 6.3
Description of Symbols for the Intermediary Events on the FTD

Number	Intermediary Events	Symbols
1	Contamination outside of cask	G2
2	Vehicle fire	G3
3	Internal lid screws with incorrect torque	G4
	Collision and other accidents	G5
4	Containment failure	G6
5	Deficiencies in decontamination equipment and/or contamination detection	F1
6	Failure in screw closure	F2

TABLE 6.4
Minimal Cut Set Determination Steps

Step	Boolean Expression for Top Event (G1) of Figure 6.3
1	$G1 = G2 + G6 + G3 + G4 + (G5 \cdot G3)$
2	$G1 = (B7 \cdot B6) + (B2 \cdot B6) + (B2 \cdot B6) + (B2 \cdot B6) + (B2 \cdot B6) + B8 + B9 + (B1 \cdot B2) + (B10 \cdot B2) ++ (B3 \cdot B4) + (B2 \cdot B6) + (B5 \cdot B6 \cdot B3 \cdot B4)$
3	$G1 = (B7 \cdot B6) + (B2 \cdot B6) + B8 + B9 + (B1 \cdot B2) + (B10 \cdot B2) + (B3 \cdot B4)$

TABLE 6.5
List of Minimal Cut Sets

Number	Minimal Cut Sets	Cause
1	B8	Improper equipment to closure of screw
2	B9	Error in tightening torque calculation
3	(B6,B7)	Operator errors and contaminated water in reactor pool
4	(B3,B4)	Vehicle collision and fire of oil
5	(B2,B6)	Failure in inspection, control, in testing program and operator errors
6	(B1,B2)	Containment failure and failure in inspection, control, in testing program
7	(B10,B2)	Failure in inspection, control, in testing program and material aging

Based on this, it is necessary early on to prevent the occurrence of the top event and take care more quickly with the most critical causes (i.e., those that represent the first or lowest order MCSs (B8 and B9). It shows the system is relatively safe because the first order MCSs are few. The system is relative dangerous however.

For this tree, seven root causes were found and, according to the MCSs, two of these causes are critical; they can happen independent of the others and cause the top event.

Human error in inspection, control, in testing program, decontamination, in contamination detection, manufacturing, in tightening torque calculation, in use of improper equipment to closure of the screws and preparation of operators for transportation as well as during loading and unloading of spent nuclear fuel must be considered.

Corrective actions are required to minimize the probability of fault occurrence, such as:

- Make sure the operator is well trained and qualified
- Create a preventive/predictive maintenance planning and scheduling
- Build a QA program

The diagrams created in the fault tree methods, in general, are more easily understood by non-probabilistic safety analysis (PSA) specialists, and therefore they can greatly assist in the documentation of the event model (IAEA-TECDOC-1267, 2002).

A PSA fault tree is a powerful tool that can be used to confirm assumptions that are commonly made in the deterministic calculation about the availability of systems, for example, to determine the potential for common cause failures or the minimum system requirements, to identify important single failures, and to determine the adequacy of technical specifications (IAEA-SSG-2, 2002).

The risk assessment has been seriously addressed within the IAEA staff in the Safety Analysis Report (SAR) and an assessment of PSA (IAEA-GSR-4, 2016) is included in the SAR.

The risk assessment for spent nuclear fuel transportation and storage are part of SAR of the CDTN. The constructed DPC is not yet licensed in Brazil. The SAR is an important document for the entire licensing process.

This study will form part of a future SAR of the CDTN and a safety operation manual for the DPC because it provides pertinent information.

6.7 CONCLUSION

The FTA of the DPC was established on the basis of the environmental contamination scenario of the DPC in this chapter.

Some main causes include the use of improper equipment for closure of screws and errors in calculation of the tightening torque. Appropriate precautions measures can be taken to decrease the probability of this occurrence.

The results revealed that a large proportion of undesired events were the result of human errors. Proposed corrective actions have been implemented to minimize the incident.

This evaluation system predicted the weak points existing in the DPC, as well as provided theoretical support to avoid the loss of DPC integrity.

Despite all the advantages previously discussed, it is important to note that this study is an initial work that must continue because other possible undesired events must be studied.

This is the first work in CDTN about FTA for DPCs that will contribute to many future studies in this system, and will involve quantitative derivation of probabilities.

This study provides an organized record of basic events that contribute to an environmental contamination of a DPC. Also is provided information pertinent to future SARs of nuclear installations of CDTN (in Portuguese, RASIN) and an operation manual for DPCs.

REFERENCES

Gual MR, Rival RR, Oliveira V, Cunha CL. 2017. Prevention of human factors and Reliability analysis in operating of sipping device on IPR-R1 TRIGA reactor, a case study. In. *Human Factors and Reliability Engineering for Safety and Security in Critical Infrastructures*, eds. Felice F. and Petrillo A., pp. 155–170, Springer Series in Reliability Engineering, Piscataway, NJ.

Gual MR, Perdomo OM, Salomon J, Wellesley J, Lora A. 2014. ASeC software application based on FMEA in a mechanical sample positioning system on a radial channel for irradiations in a nuclear research reactor with continuous full-power operation. *International Journal of Ecosystems and Ecology Science* 4(1):81–88.

International Atomic Energy Agency—IAEA. 2009. *Regulations for the Safe Transport of Radioactive Material*. Safety Requirements No. TS-R-1, Vienna, Austria.

IAEA-TECDOC-1267. 2002. *Procedures for Conducting Probabilistic Safety Assessment for Non-reactor Nuclear Facilities*. International Atomic Energy Agency Vienna International Centre, Vienna, Austria.

IAEA Specific Safety Guide No. SSG-2. 2009. *Deterministic Safety Analysis for Nuclear Power Plants Safety*. International Atomic Energy Agency Vienna International Centre, Vienna, Austria.

IAEA General Safety Requirements GSR-4. 2016. *Safety Assessment for Facilities and Activities*, IAEA-GSR-4, Vienna, Austria.

Jones AL, Harvey MP. 2014. *Radiological Consequences Resulting from Accidents and Incidents Involving the Transport of Radioactive Materials in the UK: 2012 Review, PHE-CRCE-014*. Education Public Health England, August, HPA-RPD-034.

Nuclear Monitor 773. 2013. Nuclear transport accidents and incidents. https://www.wiseinternational.org/nuclear-monitor/773/nuclear-transport-accidents-and-incidents (accessed May 30, 2018).

Perdomo OM, Salomon LJ. 2016. Expanded failure mode and effects analysis: Advanced approach for reliability assessments. *Revista Cubana de Ingeniería* VII (2):5–14.

Rivero JJ, Salomón LJ, Perdomo OM, Torres VA. 2018. Advanced combinatorial method for solving complex fault trees. *Annals of Nuclear Energy* 120:666–681.

Stamatelatos M, Vesely WE, Dugan J, Fragola J. 2002. *Fault Tree Handbook with Aerospace Applications*. NASA Office of Safety and Mission Assurance. NASA Headquarters. Washington, DC. August.

Troncoso M. 2018a. Estudio LOPA para la Terminal de Petrolíferos Veracruz (In Spanish) Internal Task: 9642-18-2017 (408).

Troncoso M. 2018b. Estudio HAZOP para la Terminal de Petrolíferos Puebla. (In Spanish) Internal Task: 9642-18-2015 (408).

Vesely W, Goldberg F, Roberts N, Haasl D. 1981. Fault tree handbook. Technical Report NUREG-0492, Office of Nuclear Regulatory Research U.S. Nuclear Regulatory Commission (NRC). Washington, DC.

Whitesitt J. Eldon. 1995. *Boolean Algebra and Its Applications*. Courier Corporation, 182 p., Dover Publications, Inc., New York.

7 An Overview on Failure Rates in Maintenance Policies

Xufeng Zhao and Toshio Nakagawa

CONTENTS

7.1 INTRODUCTION

Aging describes how an operating unit improves or deteriorates with its age and is usually measured by the term of the failure rate function [1,2]. The failure rate is the most important quantity in reliability theory and these properties were investigated in [2–4]. For an age replacement model, it has been supposed that an operating unit is replaced preventively at time T ($0 < T \leq \infty$) or correctively at failure time X ($X > 0$), whichever occurs first, in which the random variable X has a general distribution $F(t) \equiv \Pr\{X \leq t\}$ for $t \geq 0$ with finite mean $\mu \equiv \int_0^\infty \overline{F}(t)\mathrm{d}t$. The expected cost rate for the age replacement policy was given [2,4]:

$$C(T) = \frac{c_T + (c_F - c_T)F(T)}{\displaystyle\int_0^T \overline{F}(t)\mathrm{d}t},$$

(7.1)

where:

c_T = preventive replacement cost at time T,
c_F = corrective cost at failure time X,
$c_F > c_T$.

To obtain an optimum T^* to minimize Equation (7.1), it was assumed $F(t)$ has a density function $f(t) \equiv dF(t)/dt$, i.e., $F(t) = \int_0^t f(u)du$, where $\bar{\varphi} = 1 - \varphi(t)$ for any function $\varphi(t)$. Then, for $F(t) < 1$, the *failure rate* $h(t)$ is defined as [2]:

$$h(t) \equiv \frac{f(t)}{F(t)},$$ (7.2)

where $h(t)\Delta t \equiv \Pr\{t < X \le t + \Delta t\}$ for small $\Delta t > 0$ represents the probability that an operating unit with age t will fail during interval $(t, t + \Delta t)$. Therefore, optimum T^* to minimize $C(T)$ is a solution of:

$$\int_0^T \overline{F}(t)[h(T) - h(t)]dt = \frac{c_T}{c_F - c_T}.$$ (7.3)

It has been shown [4] that if $h(t)$ increases strictly to $h(\infty) \equiv \lim_{t\to\infty} h(t)$ and $h(\infty) > c_F / [\mu(c_F - c_T)]$, then a finite and unique T^* exists, and the optimum cost rate is given by the failure rate $h(t)$ as:

$$C(T^*) = (c_F - c_T)h(T^*).$$ (7.4)

Equations (7.3) and (7.4) indicate the optimum time T^* decreases while the expected cost rate $C(T^*)$ increases with the failure rate $h(t)$. This result means if we know more about the properties of the failure rate, we can make better replacement decisions for an operating unit in an economical way.

We recently proposed several new replacement models such as random replacement, replacement first, replacement last, replacement overtime, and replacement middle [5–8]. These models showed that the extended types of failure rates appeared, which played important roles in obtaining optimum replacement times in analytical ways. So it would be of interest to survey the reliability properties of the failure rates and their further applications for the recent maintenance models.

The standard failure rate $h(t)$ has been defined in Equation (7.2). We will formulate several extended failure rates in inequality forms by integrating $h(t)$ with replacement policy at time T. We show the examples of these failure rates appeared in replacement models. In Section 7.3, when the replacement time T and work number N become the decision variables, we introduce the failure rates that are found in age and random replacement models. In Section 7.4, the failure rates in periodic replacement models with minimal repairs are given and shown in periodic and random replacement models. In Section 7.5, the failure rates and their inequalities are shown for the model where replacement is done at failure number K.

The recent models of replacement first, replacement last, and replacement overtime are surveyed for these failure rates in the following sections. In addition, we give an appendix for the proofs of these extended failure rates.

7.2 INEQUALITIES OF FAILURE RATES

We give the *cumulative hazard function* $H(t)$, i.e., $H(t) \equiv \int_0^t h(u)du$, and obtain:

$$\overline{F}(t) = \exp\left(-\int_0^t h(u)du\right) = e^{-H(t)}, \quad \text{i.e.,} \quad H(t) = -\log \overline{F}(t). \tag{7.5}$$

This equation means the functions of $F(t)$, $h(t)$, and $H(t)$ can determine each other.

Suppose the failure distribution $F(t)$ has an increasing failure rate (IFR) property, and its failure rate $h(t)$ increases with t from $h(0)$ to $h(\infty) \equiv \lim_{t \to \infty} h(t)$, which might be infinity. Then we have the following inequalities [2,9]: For $0 < T < \infty$:

Inequality I:

$$\frac{\int_0^T F(t)dt}{\int_0^T \left[\int_0^t \overline{F}(u)du\right]dt} \le \frac{F(T)}{\int_0^T \overline{F}(t)dt} \le \frac{H(T)}{T} \le h(T)$$

$$\le \frac{\overline{F}(T)}{\int_T^\infty \overline{F}(t)dt} \le \frac{\int_T^\infty \overline{F}(t)dt}{\int_T^\infty \left[\int_t^\infty \overline{F}(u)du\right]dt}. \tag{7.6}$$

Inequality II:

$$\frac{F(T)}{\int_0^T \overline{F}(t)dt} \le \frac{\int_T^\infty F(t)dt}{\int_T^\infty \left[\int_0^t \overline{F}(u)du\right]dt} \le \frac{1}{\mu} \le \frac{\int_0^T \overline{F}(t)dt}{\int_0^T \left[\int_t^\infty \overline{F}(u)du\right]dt} \le \frac{\overline{F}(T)}{\int_T^\infty \overline{F}(t)dt}. \tag{7.7}$$

Generally, repeating the procedures of integral calculations in Equation (7.6), we obtain:

Inequality III:

$$\frac{\int_0^T \left\{\int_0^{t_n}[\cdots\int_0^{t_1} f(u)du\cdots]dt_{n-1}\right\}dt_n}{\int_0^T \left\{\int_0^{t_n}[\cdots\int_0^{t_1} \overline{F}(u)du\cdots]dt_{n-1}\right\}dt_n} \le \cdots \le \frac{\int_0^T f(t)dt}{\int_0^T \overline{F}(t)dt} \le h(T)$$

$$\le \frac{\int_T^\infty f(t)dt}{\int_T^\infty \overline{F}(t)dt} \le \cdots \le \frac{\int_T^\infty \left\{\int_{t_n}^\infty[\cdots\int_{t_1}^\infty f(u)du\cdots]dt_{n-1}\right\}dt_n}{\int_T^\infty \left\{\int_{t_n}^\infty[\cdots\int_{t_1}^\infty \overline{F}(u)du\cdots]dt_{n-1}\right\}dt_n} \quad (n = 1, 2, \cdots). \tag{7.8}$$

All of the above functions increase with T and become $h(T) = \lambda$ for $T \geq 0$ when $F(t) = 1 - e^{-\lambda t}$.

We next give some other applications of the failure rates in Equations (7.2), (7.6), and (7.7) to replacement policies planned at time T when $h(t)$ increases with t from $h(0)$ to $h(\infty)$.

Example 7.1

[4, p. 8] Suppose the unit only produces profit per unit of time when it is operating without failure, and it is replaced preventively at time T $(0 < T < \infty)$. Then the average time for operating profit during $[0, T]$ is:

$$l_0(T) = T \times \bar{F}(T) + 0 \times F(T) = T\bar{F}(T).$$

Optimum time T_0 to maximize $l_0(T)$ satisfies:

$$h(T) = \frac{1}{T}. \tag{7.9}$$

When $F(t) = 1 - e^{-\lambda t}$, optimum $T_0 = 1/\lambda$ means replacement should be made at the mean failure time.

Example 7.2

[4, p. 8] Suppose there is one spare unit available for replacement, the operating unit is replaced preventively with the spare one at time T $(0 < T \leq \infty)$, and the spare unit should operate until failure. When both units have an identical failure distribution $F(t)$ with mean μ, the mean time to failure of either unit is:

$$l_1(T) = \int_0^T t\,dF(t) + \bar{F}(T)(T + \mu) = \int_0^T \bar{F}(t)\,dt + \mu\bar{F}(T).$$

Optimum time T_1 to maximize $l_1(T)$ satisfies:

$$h(T) = \frac{1}{\mu}. \tag{7.10}$$

Next, suppose there are unlimited spare units available for replacement and each unit has an identical failure distribution $F(t)$. When preventive replacement is planned at time T $(0 < T \leq \infty)$, the mean time to failure of any unit is:

$$l(T) = \int_0^T t\,dF(t) + \bar{F}(T)[T + l(T)], \quad \text{i.e.,} \quad \frac{1}{l(T)} = \frac{F(T)}{\int_0^T \bar{F}(t)\,dt}, \tag{7.11}$$

which increases strictly with T from $h(0)$ to $1/\mu$.

Example 7.3

[4, p. 8] The failure distribution of an operating unit with age T $(0 \le T < \infty)$ is:

$$F(t;T) \equiv \Pr\{T < X \le T + t \mid X > T\} = \frac{F(t+T) - F(T)}{\overline{F}(T)}, \qquad (7.12)$$

which is also called failure rate. The mean time to failure is:

$$\frac{1}{\overline{F}(T)} \int_0^\infty [\overline{F}(T) - \overline{F}(t+T)] dt = \frac{1}{\overline{F}(T)} \int_T^\infty \overline{F}(t) dt, \qquad (7.13)$$

which decreases with T from μ to $1/h(\infty)$.

7.3   AGE AND RANDOM REPLACEMENT POLICIES

Suppose the unit operates for jobs with successive working times Y_j $(j = 1, 2, \cdots)$, where random variables Y_j are independent and have an identical distribution $G(t) \equiv \Pr\{Y_j \le t\} = 1 - e^{-\theta t}$ $(0 < 1/\theta < \infty)$. When the unit is replaced preventively at time T or at working number N, we give the following inequalities of the extended failure rates: For $0 < T < \infty$ and $N = 0, 1, 2, \cdots$:

Inequality IV:

$$\frac{\int_0^T (\theta t)^N e^{-\theta t} dF(t)}{\int_0^T (\theta t)^N e^{-\theta t} \overline{F}(t) dt} \le \frac{\int_0^T t^N dF(t)}{\int_0^T t^N \overline{F}(t) dt} \le h(T)$$

$$\le \frac{\int_T^\infty (\theta t)^N e^{-\theta t} dF(t)}{\int_T^\infty (\theta t)^N e^{-\theta t} \overline{F}(t) dt} \le \frac{\int_T^\infty t^N dF(t)}{\int_T^\infty t^N \overline{F}(t) dt}. \qquad (7.14)$$

Inequality V:

$$\frac{\int_0^T e^{-\theta t} dF(t)}{\int_0^T e^{-\theta t} \overline{F}(t) dt} \le \frac{F(T)}{\int_0^T \overline{F}(t) dt} \le \frac{\int_0^T t^N dF(t)}{\int_0^T t^N \overline{F}(t) dt} \le h(T) \le \frac{\overline{F}(T)}{\int_T^\infty \overline{F}(t) dt}$$

$$\le \frac{\int_T^\infty t^N dF(t)}{\int_T^\infty t^N \overline{F}(t) dt} \le \frac{\int_T^\infty e^{-\theta t} dF(t)}{\int_T^\infty e^{-\theta t} \overline{F}(t) dt}. \qquad (7.15)$$

Inequality VI:

$$\frac{\int_0^T (\theta t)^N [\int_0^t e^{-\theta u} dF(u)] dt}{\int_0^T (\theta t)^N [\int_0^t e^{-\theta u} \overline{F}(u) du] dt} \le \frac{\int_0^T e^{-\theta t} dF(t)}{\int_0^T e^{-\theta t} \overline{F}(t) dt} \le \frac{\int_T^\infty (\theta t)^N [\int_0^t e^{-\theta u} dF(u)] dt}{\int_0^T (\theta t)^N [\int_t^\infty e^{-\theta u} \overline{F}(u) du] dt}$$

$$\le \frac{\int_0^T (\theta t)^N [\int_t^\infty e^{-\theta u} dF(u)] dt}{\int_0^T (\theta t)^N [\int_t^\infty e^{-\theta u} \overline{F}(u) du] dt} \le \frac{\int_T^\infty e^{-\theta t} dF(t)}{\int_T^\infty e^{-\theta t} \overline{F}(t) dt} \qquad (7.16)$$

$$\le \frac{\int_T^\infty (\theta t)^N [\int_t^\infty e^{-\theta u} dF(u)] dt}{\int_T^\infty (\theta t)^N [\int_t^\infty e^{-\theta u} \overline{F}(u) du] dt}.$$

Note that all these functions increase with T and N.

Furthermore, we obtain:

Inequality VII:

$$\frac{\int_0^T e^{-\theta t} F(t) dt}{\int_0^T e^{-\theta t} [\int_0^t \overline{F}(u) du] dt} \le \frac{\int_0^T [\int_0^t e^{-\theta u} dF(u)] dt}{\int_0^T [\int_0^t e^{-\theta u} \overline{F}(u) du] dt} \le \frac{\int_0^T e^{-\theta t} dF(t)}{\int_0^T e^{-\theta t} \overline{F}(t) dt}$$

$$\le \frac{F(T)}{\int_0^T \overline{F}(t) dt} \le h(T) \le \frac{\int_T^\infty e^{-\theta t} dF(t)}{\int_T^\infty e^{-\theta t} \overline{F}(t) dt} \le \frac{F(T)}{\int_T^\infty \overline{F}(t) dt}$$

$$\le \frac{\int_T^\infty [\int_t^\infty e^{-\theta u} dF(u)] dt}{\int_T^\infty [\int_t^\infty e^{-\theta u} \overline{F}(u) du] dt} \le \frac{\int_T^\infty e^{-\theta t} \overline{F}(t) dt}{\int_T^\infty e^{-\theta t} [\int_t^\infty \overline{F}(u) du] dt}, \qquad (7.17)$$

where all these functions increase with T.

Example 7.4

[5, p. 30] Suppose when the random time Y has an exponential distribution $\Pr\{Y \le t\} = 1 - e^{-\theta t}$, the unit is replaced preventively at time T ($0 < T \le \infty$) or at time Y, whichever occurs first. Then the expected cost rate is:

$$C(T) = \frac{c_T + (c_F - c_T) \int_0^T e^{-\theta t} dF(t)}{\int_0^T e^{-\theta t} \overline{F}(t) dt}, \qquad (7.18)$$

where:

c_T = replacement cost at time T or at time Y,

c_F = replacement cost at failure with $c_F > c_T$.

Optimum time T to minimize $C(T)$ satisfies:

$$\int_0^T e^{-\theta t}\bar{F}(t)[h(T) - h(t)]dt = \frac{c_T}{c_F - c_T}. \qquad (7.19)$$

Example 7.5

[5, p. 44] Suppose the unit is replaced preventively at time T $(0 < T \le \infty)$ or at working number N $(N = 1, 2, \cdots)$, i.e., at $Y_1 + Y_2 + \cdots + Y_N$, whichever occurs first. Denoting that $G^{(j)}(t) \equiv \Pr\{Y_1 + Y_2 + \cdots + Y_j \le t\}$ $(j = 1, 2, \cdots)$ and $G^{(0)}(t) \equiv 1$ for $t > 0$, the expected cost rate is:

$$C(T, N) = \frac{c_T + (c_F - c_T)\int_0^T \left[1 - G^{(N)}(t)\right]dF(t)}{\int_0^T \left[1 - G^{(N)}(t)\right]\bar{F}(t)dt}. \qquad (7.20)$$

When $G(t) = 1 - e^{-\theta t}$, $G^{(N)}(t) = \sum_{j=N}^{\infty}[(\theta t)^j/j!]e^{-\theta t}$ $(N = 0, 1, 2, \cdots)$. Optimum time T to minimize $C(T, N)$ satisfies:

$$\sum_{j=0}^{N-1}\int_0^T \frac{(\theta t)^j}{j!}e^{-\theta t}\bar{F}(t)[h(T) - h(t)]dt = \frac{c_T}{c_F - c_T}, \qquad (7.21)$$

and optimum number N to minimize $C(T, N)$ satisfies:

$$\frac{\int_0^T (\theta t)^N e^{-\theta t}dF(t)}{\int_0^T (\theta t)^N e^{-\theta t}\bar{F}(t)dt}\sum_{j=0}^{N-1}\int_0^T \frac{(\theta t)^j}{j!}e^{-\theta t}\bar{F}(t)dt - \sum_{j=0}^{N-1}\int_0^T \frac{(\theta t)^j}{j!}e^{-\theta t}dF(t)$$

$$\ge \frac{c_T}{c_F - c_T}. \qquad (7.22)$$

Example 7.6

[5, p. 46] Suppose the unit is replaced preventively at time T $(0 < T \le \infty)$ or at working number N $(N = 1, 2, \cdots)$, whichever occurs last. Then the expected cost rate is:

$$C(T, N) = \frac{c_T + (c_F - c_T)\{F(T) + \int_T^{\infty}[1 - G^{(N)}(t)]dF(t)\}}{\int_0^T \bar{F}(t)dt + \int_T^{\infty}[1 - G^{(N)}(t)]\bar{F}(t)dt}. \qquad (7.23)$$

When $G(t) = 1 - e^{-\theta t}$, optimum time T to minimize $C(T, N)$ satisfies:

$$\int_0^T \overline{F}(t)[h(T) - h(t)]dt - \sum_{j=0}^{N-1} \int_T^\infty \frac{(\theta t)^j}{j!} e^{-\theta t} \overline{F}(t)[h(t) - h(T)]dt = \frac{c_T}{c_F - c_T}, \tag{7.24}$$

and optimum number N to minimize $C(T, N)$ satisfies:

$$\frac{\int_T^\infty (\theta t)^N e^{-\theta t} dF(t)}{\int_T^\infty (\theta t)^N e^{-\theta t} \overline{F}(t)dt} \left\{ \int_0^T \overline{F}(t)dt + \sum_{j=0}^{N-1} \int_T^\infty \frac{(\theta t)^j}{j!} e^{-\theta t} \overline{F}(t)dt \right\}$$

$$- F(T) - \sum_{j=0}^{N-1} \int_T^\infty \frac{(\theta t)^j}{j!} e^{-\theta t} dF(t) \geq \frac{c_T}{c_F - c_T}. \tag{7.25}$$

Example 7.7

[5, p. 34] Suppose the unit is replaced preventively at the next working time over time T, e.g., at time Y_{j+1} for $Y_j < T \leq Y_{j+1}$. When $G(t) = 1 - e^{-\theta t}$, the expected cost rate is:

$$C(T) = \frac{c_F - (c_F - c_T) \int_T^\infty \theta e^{-\theta(t-T)} \overline{F}(t)dt}{\int_0^T \overline{F}(t)dt + \int_T^\infty e^{-\theta(t-T)} \overline{F}(t)dt}, \tag{7.26}$$

where:
  c_T = replacement cost over time T,
  $c_T < c_F$.

Optimum time T to minimize $C(T)$ satisfies:

$$\frac{\int_T^\infty e^{-\theta t} dF(t)}{\int_T^\infty e^{-\theta t} \overline{F}(t)dt} \int_0^T \overline{F}(t)dt - F(T) = \frac{c_T}{c_F - c_T}. \tag{7.27}$$

Example 7.8

[5, p. 9] Suppose the unit is replaced preventively at the end of the next working number over time T or at working number N ($N = 1, 2, \cdots$), whichever occurs first. Then the expected cost rate is:

$C(T, N) =$

$$\frac{c_F - (c_F - c_T) \left\{ -\sum_{j=0}^{N-1} [(\theta T)^j / j!] e^{-\theta T} \int_T^\infty \theta e^{-\theta t} \overline{F}(t)dt - \sum_{j=0}^{N-1} [(\theta T)^j / j!] e^{-\theta T} \int_T^\infty \theta e^{-\theta t} \overline{F}(t)dt \right\}}{\sum_{j=0}^{N-1} \left\{ \int_0^T [(\theta t)^j / j!] e^{-\theta t} \overline{F}(t)dt + [(\theta T)^j / j!] e^{-\theta T} \int_T^\infty e^{-\theta t} \overline{F}(t)dt \right\}}.$$

$$\tag{7.28}$$

Optimum time T to minimize $C(T,N)$ satisfies:

$$\frac{\int_T^\infty e^{-\theta t}dF(t)}{\int_T^\infty e^{-\theta t}\overline{F}(t)dt}\sum_{j=0}^{N-1}\int_0^T \frac{(\theta t)^j}{j!}e^{-\theta t}\overline{F}(t)dt - \sum_{j=0}^{N-1}\int_0^T \frac{(\theta t)^j}{j!}e^{-\theta t}dF(t) = \frac{c_T}{c_F - c_T}, \qquad (7.29)$$

and optimum number N to minimize $C(T,N)$ satisfies:

$$\frac{\int_0^T (\theta t)^{N-1}\left[\int_t^\infty e^{-\theta u}dF(u)\right]dt}{\int_0^T (\theta t)^{N-1}\left[\int_t^\infty e^{-\theta u}\overline{F}(u)du\right]dt}$$

$$\times \sum_{j=0}^{N-1}\left[\frac{(\theta T)^j}{j!}\int_T^\infty e^{-\theta t}\overline{F}(t)dt + \int_0^T \frac{(\theta t)^j}{j!}e^{-\theta t}\overline{F}(t)dt\right]$$

$$(7.30)$$

$$- \sum_{j=0}^{N-1}\left[\frac{(\theta T)^j}{j!}\int_T^\infty e^{-\theta t}dF(t) + \int_0^T \frac{(\theta t)^j}{j!}e^{-\theta t}dF(t)\right] \geq \frac{c_T}{c_F - c_T}.$$

Example 7.9

[6, p. 13] Suppose the unit is replaced preventively at the end of the next working number over time T or at working number N ($N = 1,2,\cdots$), whichever occurs last. Then the expected cost rate is:

$$C(T,N) =$$

$$\frac{c_F - (c_F - c_T)\left\{\int_T^\infty \theta[(\theta t)^{N-1}/(N-1)!]e^{-\theta t}\overline{F}(t)dt - \sum_{j=N}^\infty [(\theta T)^j/j!]e^{-\theta T}\int_T^\infty \theta e^{-\theta t}\overline{F}(t)dt\right\}}{\int_0^T \overline{F}(t)dt + \sum_{j=0}^{N-1}\int_T^\infty [(\theta t)^j/j!]e^{-\theta t}\overline{F}(t)dt + \sum_{j=N}^\infty [(\theta T)^j/j!]e^{-\theta T}\int_T^\infty e^{-\theta t}\overline{F}(t)dt}.$$

$$(7.31)$$

Optimum time T to minimize $C(T,N)$ satisfies:

$$\frac{\int_T^\infty e^{-\theta t}dF(t)}{\int_T^\infty e^{-\theta t}\overline{F}(t)dt}\left\{\int_0^T \overline{F}(t)dt + \sum_{j=0}^{N-1}\int_T^\infty \frac{(\theta t)^j}{j!}e^{-\theta t}\overline{F}(t)dt\right\}$$

$$(7.32)$$

$$- F(T) - \sum_{j=0}^{N-1}\int_T^\infty \frac{(\theta t)^j}{j!}e^{-\theta t}dF(t) = \frac{c_T}{c_F - c_T},$$

and optimum number N to minimize $C(T, N)$ satisfies:

$$
\frac{\int_T^\infty (\theta t)^N [\int_t^\infty e^{-\theta u} dF(u)] dt}{\int_T^\infty (\theta t)^N [\int_t^\infty e^{-\theta u} \overline{F}(u) du] dt} \left[\int_0^T \overline{F}(t) dt + \sum_{j=0}^{N-1} \int_T^\infty \frac{(\theta t)^j}{j!} e^{-\theta t} \overline{F}(t) dt \right.
$$

$$
+ \sum_{j=N}^\infty \frac{(\theta T)^j}{j!} e^{-\theta T} \int_T^\infty e^{-\theta t} \overline{F}(t) dt \tag{7.33}
$$

$$
\left. - \sum_{j=N}^\infty \left[\frac{(\theta T)^j}{j!} e^{-\theta T} \int_T^\infty e^{-\theta t} dF(t) - \int_T^\infty \frac{(\theta t)^j}{j!} e^{-\theta t} dF(t) \right] \ge \frac{c_T}{c_F - c_T}.
$$

7.4   PERIODIC AND RANDOM REPLACEMENT POLICIES

Suppose the unit operates for jobs with working times Y_j defined in Section 7.3 and undergoes minimal repairs at failures. When the unit is replaced at time T or at working number N, we give the following inequalities of the extended failure rates: For $0 < T < \infty$ and $N = 0, 1, 2, \cdots$:

Inequality VIII:

$$
\frac{\int_0^T e^{-\theta t} h(t) dt}{\int_0^T e^{-\theta t} dt} \le \frac{\int_0^T (\theta t)^N e^{-\theta t} h(t) dt}{\int_0^T (\theta t)^N e^{-\theta t} dt} \le \frac{\int_0^T t^N h(t) dt}{\int_0^T t^N dt} \le h(T)
$$

$$
\le \int_0^\infty \theta e^{-\theta t} h(t+T) dt \le \frac{\int_T^\infty (\theta t)^N e^{-\theta t} h(t) dt}{\int_T^\infty (\theta t)^N e^{-\theta t} dt}. \tag{7.34}
$$

Inequality IX:

$$
\frac{\int_0^T (\theta t)^N \left[\int_0^t e^{-\theta u} h(u) du \right] dt}{\int_0^T (\theta t)^N \left[\int_0^t e^{-\theta u} du \right] dt} \le \frac{\int_0^T e^{-\theta t} h(t) dt}{\int_0^T e^{-\theta t} dt} \le \frac{\int_T^\infty (\theta t)^N \left[\int_0^t e^{-\theta u} h(u) du \right] dt}{\int_T^\infty (\theta t)^N \left[\int_0^t e^{-\theta u} du \right] dt}
$$

$$
\le \frac{\int_0^T (\theta t)^N \left[\int_t^\infty e^{-\theta u} h(u) du \right] dt}{\int_0^T (\theta t)^N \left[\int_t^\infty e^{-\theta u} du \right] dt} \le \int_0^\infty e^{-\theta t} h(t+T) dt \tag{7.35}
$$

$$
\le \frac{\int_T^\infty (\theta t)^N \left[\int_t^\infty e^{-\theta u} h(u) du \right] dt}{\int_T^\infty (\theta t)^N \left[\int_t^\infty e^{-\theta u} du \right] dt}.
$$

Note that all these functions increase with T and N.

Example 7.10

[5, p. 65] Suppose the unit is replaced at time T $(0 < T \leq \infty)$ or at time Y, whichever occurs first. Then the expected cost rate is:

$$C(T) = \frac{c_T + c_M \displaystyle\int_0^T e^{-\theta t} h(t) dt}{\displaystyle\int_0^T e^{-\theta t} dt}, \tag{7.36}$$

where:
   c_T = replacement cost at time T or at time Y,
   c_M = cost of minimal repair at each failure.

Optimum time T to minimize $C(T)$ satisfies:

$$\int_0^T e^{-\theta t} [h(T) - h(t)] dt = \frac{c_T}{c_M}. \tag{7.37}$$

Example 7.11

[5, p. 77] Suppose the unit is replaced at time T $(0 < T \leq \infty)$ or at working number N $(N = 1, 2, \cdots)$, whichever occurs first. Then, the expected cost rate is:

$$C(T, N) = \frac{c_T + c_M \displaystyle\int_0^T [1 - G^{(N)}(t)] h(t) dt}{\displaystyle\int_0^T [1 - G^{(N)}(t)] dt}. \tag{7.38}$$

When $G(t) = 1 - e^{-\theta t}$, optimum time T to minimize $C(T, N)$ satisfies:

$$\sum_{j=0}^{N-1} \int_0^T \frac{(\theta t)^j}{j!} e^{-\theta t} [h(T) - h(t)] dt = \frac{c_T}{c_M}, \tag{7.39}$$

and optimum number N to minimize $C(T, N)$ satisfies:

$$\frac{\displaystyle\int_0^T (\theta t)^N e^{-\theta t} h(t) dt}{\displaystyle\int_0^T (\theta t)^N e^{-\theta t} dt} \sum_{j=0}^{N-1} \int_0^T \frac{(\theta t)^j}{j!} e^{-\theta t} dt - \sum_{j=0}^{N-1} \int_0^T \frac{(\theta t)^j}{j!} e^{-\theta t} h(t) dt \geq \frac{c_T}{c_M}. \tag{7.40}$$

Example 7.12

[5, p. 79] Suppose the unit is replaced at time T $(0 \leq T \leq \infty)$ or at working number N $(N = 0, 1, 2, \cdots)$, whichever occurs last. Then, the expected cost rate is:

$$C(T, N) = \frac{c_T + c_M \{H(T) + \int_T^\infty [1 - G^{(N)}(t)]h(t)dt\}}{T + \int_T^\infty [1 - G^{(N)}(t)]dt}. \tag{7.41}$$

When $G(t) = 1 - e^{-\theta t}$, optimum time T to minimize $C(T, N)$ satisfies:

$$\int_0^T [h(T) - h(t)]dt - \sum_{j=0}^{N-1} \int_T^\infty \frac{(\theta t)^j}{j!} e^{-\theta t} [h(t) - h(T)]dt = \frac{c_T}{c_M}, \tag{7.42}$$

and optimum number N to minimize $C(T, N)$ satisfies:

$$\frac{\int_T^\infty (\theta t)^N e^{-\theta t} h(t)dt}{\int_T^\infty (\theta t)^N e^{-\theta t} dt} \left[T + \sum_{j=0}^{N-1} \int_T^\infty \frac{(\theta t)^j}{j!} e^{-\theta t} dt \right] - H(T)$$

$$- \sum_{j=0}^{N-1} \int_T^\infty \frac{(\theta t)^j}{j!} e^{-\theta t} h(t)dt \geq \frac{c_T}{c_M}. \tag{7.43}$$

Example 7.13

[6, p. 39] Suppose the unit is replaced at the end of the next working number over time T. When $G(t) = 1 - e^{-\theta t}$, the expected cost rate is:

$$C(T) = \frac{c_T + c_M [H(T) + \int_0^\infty e^{-\theta t} h(t + T)dt]}{T + 1/\theta}. \tag{7.44}$$

Optimum time T to minimize $C(T)$ satisfies:

$$T \int_0^T \theta e^{-\theta t} [h(t + T) - h(t)]dt = \frac{c_T}{c_M}. \tag{7.45}$$

Example 7.14

[6, p. 41] Suppose the unit is replaced at the next working number over time T $(0 < T \le \infty)$ or at working number N $(N = 1, 2, \cdots)$, whichever occurs first. Then, the expected cost rate is:

$$C(T,N) = \frac{c_T + c_M \left\{ \int_0^T [1 - G^{(N)}(t)]h(t)dt + \sum_{j=0}^{N-1} \int_0^T [\int_T^\infty \overline{G}(u-t)h(u)du]dG^{(j)}(t) \right\}}{\int_0^T [1 - G^{(N)}(t)]dt + (1/\theta)[1 - G^{(N)}(T)]}. \quad (7.46)$$

When $G(t) = 1 - e^{-\theta t}$, optimum time T to minimize $C(T,N)$ satisfies:

$$\sum_{j=0}^{N-1} \int_0^T \frac{(\theta t)^j}{j!} e^{-\theta t} \left\{ \int_0^\infty \theta e^{-\theta u}[h(u+T) - h(t)]du \right\} dt = \frac{c_T}{c_M}, \quad (7.47)$$

and optimum number N to minimize $C(T,N)$ satisfies:

$$\frac{\int_0^T (\theta t)^{N-1}\left[\int_t^\infty e^{-\theta u}h(u)du\right]dt}{\int_0^T (\theta t)^{N-1}e^{-\theta t}dt} \sum_{j=0}^{N-1}\left[\int_0^T \frac{\theta(\theta t)^j}{j!}e^{-\theta t}dt + \frac{(\theta T)^j}{j!}e^{-\theta T}\right]$$

$$- \sum_{j=0}^{N-1}\left\{ \int_0^T \frac{(\theta t)^j}{j!}e^{-\theta t}h(t)dt + \frac{(\theta T)^j}{j!}\int_T^\infty e^{-\theta t}h(t)dt \right\} \ge \frac{c_T}{c_M}. \quad (7.48)$$

Example 7.15

[6, p. 44] Suppose the unit is replaced at the next working number over time T $(0 \le T \le \infty)$ or at working number N $(N = 0, 1, 2, \cdots)$, whichever occurs last. Then, the expected cost rate is:

$$C(T,N) = \frac{c_T + c_M \left\{ H(T) + \int_T^\infty [1 - G^{(N)}(t)]h(t)dt + \sum_{j=N}^\infty \int_0^T [\int_T^\infty \overline{G}(u-t)h(u)du]dG^{(j)}(t) \right\}}{T + \int_T^\infty [1 - G^{(N)}(t)]dt + (1/\theta)G^{(N)}(T)}. \quad (7.49)$$

When $G(t) = 1 - e^{-\theta t}$, optimum time T to minimize $C(T,N)$ satisfies:

$$\int_0^T \left\{ \int_0^\infty \theta e^{-\theta u} [h(u+T) - h(t)] du \right\} dt$$

$$+ \sum_{j=0}^{N-1} \int_T^\infty \frac{(\theta t)^j}{j!} e^{-\theta t} \left\{ \int_0^\infty \theta e^{-\theta u} [h(u+T) - h(t)] du \right\} dt = \frac{c_T}{c_M}, \tag{7.50}$$

and optimum number N to minimize $C(T, N)$ satisfies:

$$\frac{\int_T^\infty (\theta t)^{N-1} [\int_t^\infty e^{-\theta u} h(u) du] dt}{\int_T^\infty (\theta t)^{N-1} e^{-\theta t} dt} \left[1 + \theta T + \sum_{j=0}^{N-1} \frac{(\theta T)^j}{j!} e^{-\theta T} \right] - H(T)$$

$$- \int_0^\infty e^{-\theta t} h(t+T) dt - \sum_{j=0}^{N-1} \left[\int_T^\infty \frac{(\theta t)^j}{j!} e^{-\theta t} h(t) dt - \frac{(\theta T)^j}{j!} \int_0^\infty e^{-\theta t} h(t+T) dt \right] \geq \frac{c_T}{c_M}. \tag{7.51}$$

7.5 PERIODIC REPLACEMENT POLICIES WITH FAILURE NUMBERS

It is assumed that failures occur at a non-homogeneous Poisson process with mean value function $H(t) \equiv \int_0^t h(u) du$, then the probability that k failures occur in $[0, t]$ is [9, p. 27]:

$$p_k(t) \equiv \frac{H(t)^k}{k!} e^{-H(t)} \quad (k = 0, 1, 2 \cdots),$$

and the probability that more than k failures occur in $[0, t]$ is $P_k(t) = \sum_{j=k}^\infty p_j(t)$. Note that $\overline{P}_k(t) \equiv 1 - P_k(t) = \sum_{j=0}^{k-1} p_j(t)$. Suppose the unit undergoes minimal repair at failures and is replaced at time T or at failure number K, we give the following inequalities of the extended failure rates: For $0 < T < \infty$ and $K = 0, 1, 2, \cdots$:

Inequality X:

$$\frac{F(T)}{\int_0^T \overline{F}(t) dt} \leq \frac{\int_0^T p_K(t) h(t) dt}{\int_0^T p_K(t) dt} \leq h(T) \leq \frac{\overline{F}(T)}{\int_T^\infty \overline{F}(t) dt} \leq \frac{\int_T^\infty p_K(t) h(t) dt}{\int_T^\infty p_K(t) dt}. \tag{7.52}$$

Inequality XI:

$$\frac{\int_0^T p_K(t) h(t) dt}{\int_0^T p_K(t) dt} \leq \frac{\int_0^T p_K(t) h(t) dt}{\int_0^T p_K(t) [\int_t^\infty \overline{F}(u) du / \overline{F}(t)] dt} \leq \frac{\overline{F}(T)}{\int_T^\infty \overline{F}(t) dt}$$

$$\leq \frac{\int_T^\infty p_K(t) h(t) dt}{\int_T^\infty p_K(t) dt} \leq \frac{\int_T^\infty p_K(t) h(t) dt}{\int_T^\infty p_K(t) [\int_t^\infty \overline{F}(u) du / \overline{F}(t)] dt}. \tag{7.53}$$

Inequality XII:

$$\frac{\int_0^T p_K(t)h(t)dt}{\int_0^T p_K(t)dt} \leq \frac{1}{\int_0^\infty p_K(t)dt} \leq \frac{\int_T^\infty p_K(t)h(t)dt}{\int_T^\infty p_K(t)dt}, \tag{7.54}$$

$$\frac{1}{\mu} \leq \frac{\int_0^T p_K(t)h(t)dt}{\int_0^T p_K(t)\left[\int_t^\infty \overline{F}(u)du/\overline{F}(t)\right]dt} \leq \frac{1}{\int_0^\infty p_{K+1}(t)dt}$$

$$\leq \frac{\int_T^\infty p_K(t)h(t)dt}{\int_T^\infty p_K(t)\left[\int_t^\infty \overline{F}(u)du/\overline{F}(t)\right]dt}. \tag{7.55}$$

Example 7.16

[2, p. 104] Suppose the unit is replaced at failure number K ($K = 1, 2, \cdots$). Then, the expected cost rate is:

$$C(K) = \frac{c_K + c_M K}{\int_0^\infty \overline{P}_K(t)dt}, \tag{7.56}$$

where c_K = replacement cost at failure number K. Optimum K to minimize $C(K)$ satisfies:

$$\frac{1}{\int_0^\infty p_K(t)dt}\int_0^\infty \overline{P}_K(t)dt - K \geq \frac{c_K}{c_M}. \tag{7.57}$$

Example 7.17

[10] Suppose the unit is replaced at time T ($0 < T \leq \infty$) or at failure number K ($K = 1, 2, \cdots$), whichever occurs first. Then, the expected cost rate is:

$$C(T, K) = \frac{c_T + c_M \int_0^{T^-} \overline{P}_K(t)(t)h(t)dt}{\int_0^{T^-} \overline{P}_K(t)dt}, \tag{7.58}$$

where c_T = replacement cost at time T and at failure number K. Optimum T to minimize $C(T, K)$ satisfies:

$$h(T)\int_0^T \overline{P}_K(t)dt - \int_0^T \overline{P}_K(t)h(t)dt = \frac{c_T}{c_M}, \tag{7.59}$$

and optimum K to minimize $C(T,K)$ satisfies:

$$\frac{\int_0^T p_K(t)h(t)dt}{\int_0^T p_K(t)dt} \int_0^T \overline{P}_K(t)dt - \int_0^T \overline{P}_K(t)h(t)dt \geq \frac{c_T}{c_M}. \qquad (7.60)$$

Example 7.18

[10] Suppose the unit is replaced at time T $(0 \leq T \leq \infty)$ or at failure number K $(K = 0,1,2,\cdots)$, whichever occurs last. Then, the expected cost rate is:

$$C(T,K) = \frac{c_T + c_M[H(T) + \int_T^\infty \overline{P}_K(t)h(t)dt]}{T + \int_T^\infty \overline{P}_K(t)dt}. \qquad (7.61)$$

Optimum T to minimize $C(T,K)$ satisfies:

$$h(T)\left[T + \int_T^\infty \overline{P}_K(t)dt\right] - H(T) - \int_T^\infty \overline{P}_K(t)h(t)dt = \frac{c_T}{c_M}, \qquad (7.62)$$

and optimum K to minimize $C(T,K)$ satisfies:

$$\frac{\int_T^\infty p_K(t)h(t)dt}{\int_T^\infty p_K(t)dt}\left[T + \int_T^\infty \overline{P}_K(t)dt\right] - H(T) - \int_T^\infty \overline{P}_K(t)h(t)dt \geq \frac{c_T}{c_M}. \qquad (7.63)$$

Example 7.19

[6, p. 47] Suppose the unit is replaced at the first failure over time T $(0 \leq T < \infty)$. Then, the expected cost rate is:

$$C(T) = \frac{c_T + c_M[H(T) + 1]}{T + \int_T^\infty e^{-[H(t)-H(T)]}dt}, \qquad (7.64)$$

where c_T = replacement cost over time T. Optimum T to minimize $C(T)$ satisfies:

$$TQ(T) - H(T) = \frac{c_T}{c_M}, \qquad (7.65)$$

where:

$$Q(T) \equiv \frac{\overline{F}(T)}{\int_T^\infty \overline{F}(t)dt}.$$

Example 7.20

[6, p. 47] Suppose the unit is replaced at failure number K ($K = 1, 2, \cdots$) or at the first failure over time T ($0 \le T < \infty$), whichever occurs first. Then, the expected cost rate is:

$$C(T,K) = \frac{c_T + c_M[\int_0^T \bar{P}_K(t)h(t)dt + \bar{P}_K(T)]}{\int_0^T \bar{P}_K(t)dt + \bar{P}_K(T)\int_T^\infty e^{-[H(t)-H(T)]}dt}. \tag{7.66}$$

Optimum T to minimize $C(T,K)$ satisfies:

$$Q(T)\int_0^T \bar{P}_K(t)dt - \int_0^T \bar{P}_K(t)h(t)dt = \frac{c_T}{c_M}, \tag{7.67}$$

and optimum K to minimize $C(T,K)$ satisfies:

$$\frac{\int_0^T p_{K-1}(t)h(t)dt}{\int_0^T p_{K-1}(t)[h(t)/Q(t)]dt}\left[\int_0^T \bar{P}_K(t)dt + \bar{P}_K(T)\int_T^\infty e^{-[H(t)-H(T)]}dt\right]$$
$$- \int_0^T \bar{P}_K(t)h(t)dt - \bar{P}_K(T) \ge \frac{c_T}{c_M}. \tag{7.68}$$

Example 7.21

[6, p. 50] Suppose the unit is replaced at failure number K ($K = 0, 1, 2, \cdots$) or at the first failure over time T ($0 \le T < \infty$), whichever occurs last. Then, the expected cost rate is:

$$C(T,K) = \frac{c_T + c_M[H(T) + \int_T^\infty \bar{P}_K(t)h(t)dt + P_K(T)]}{T + \int_T^\infty \bar{P}_K(t)dt + P_K(T)\int_T^\infty e^{-[H(t)-H(T)]dt}}. \tag{7.69}$$

Optimum T to minimize $C(T,K)$ satisfies:

$$Q(T)\left[T + \int_T^\infty \bar{P}_K(t)dt\right] - \left[H(T) + \int_T^\infty \bar{P}_K(t)h(t)dt\right] = \frac{c_T}{c_M}, \tag{7.70}$$

and optimum K to minimize $C(T,K)$ satisfies:

$$\frac{\int_T^\infty p_{K-1}(t)h(t)dt}{\int_T^\infty p_{K-1}(t)[h(t)/Q(t)]dt}\left[T + \int_T^\infty \bar{P}_K(t)dt + P_K(T)\int_T^\infty e^{-[H(t)-H(T)]dt}\right]$$
$$- \left[H(T) + \int_T^\infty \bar{P}_K(t)h(t)dt + P_K(T)\right] \ge \frac{c_T}{c_M}. \tag{7.71}$$

7.6 CONCLUSIONS

We surveyed several extended failure rates that appeared in the recent age, random, and periodic replacement models. The reliability properties of these extended failure rates would be helpful in obtaining optimum maintenance times for complex systems. We also gave the inequalities of the failure rates, which would help greatly to compare their optimum replacement policies.

There are some examples for which we cannot give inequalities. For example:

1. $\dfrac{F(T)}{\displaystyle\int_0^T \overline{F}(t)\mathrm{d}t}$ and $\dfrac{\displaystyle\int_0^T (\theta t)^N \mathrm{e}^{-\theta t}\mathrm{d}F(t)}{\displaystyle\int_0^T (\theta t)^N \mathrm{e}^{-\theta t}\overline{F}(t)\mathrm{d}t}$.

2. $\dfrac{\overline{F}(T)}{\displaystyle\int_T^\infty \overline{F}(t)\mathrm{d}t}$ and $\dfrac{\displaystyle\int_T^\infty (\theta t)^N \mathrm{e}^{-\theta t}\mathrm{d}F(t)}{\displaystyle\int_T^\infty (\theta t)^N \mathrm{e}^{-\theta t}\overline{F}(t)\mathrm{d}t}$.

3. $h(T)$ and $\dfrac{1}{\displaystyle\int_0^\infty p_K(t)\mathrm{d}t}$.

However, it can be shown that:

1. $\displaystyle\int_0^T (\theta t)^N \mathrm{e}^{-\theta t}\mathrm{d}F(t) / \int_0^T (\theta t)^N \mathrm{e}^{-\theta t}\overline{F}(t)\mathrm{d}t$ increases with N from:

$$\frac{\displaystyle\int_0^T \mathrm{e}^{-\theta t}\mathrm{d}F(t)}{\displaystyle\int_0^T \mathrm{e}^{-\theta t}\overline{F}(t)\mathrm{d}t} \le \frac{F(T)}{\displaystyle\int_0^T \overline{F}(t)\mathrm{d}t} \quad \text{to} \quad h(T) \ge \frac{F(T)}{\displaystyle\int_0^T \overline{F}(t)\mathrm{d}t}.$$

2. $\displaystyle\int_T^\infty (\theta t)^N \mathrm{e}^{-\theta t}\mathrm{d}F(t) / \int_T^\infty (\theta t)^N \mathrm{e}^{-\theta t}\overline{F}(t)\mathrm{d}t$ increases with N from:

$$\frac{\displaystyle\int_T^\infty \mathrm{e}^{-\theta t}\mathrm{d}F(t)}{\displaystyle\int_T^\infty \mathrm{e}^{-\theta t}\overline{F}(t)\mathrm{d}t} \le \frac{\overline{F}(T)}{\displaystyle\int_T^\infty \overline{F}(t)\mathrm{d}t} \quad \text{to} \quad h(T) \ge \frac{\overline{F}(T)}{\displaystyle\int_T^\infty \overline{F}(t)\mathrm{d}t}.$$

3. $1/\displaystyle\int_0^\infty p_K(t)\mathrm{d}t$ increases with K from $1/\mu \ge h(0)$ to $h(\infty)$.

APPENDICES

Assuming that the failure rate $h(t)$ increases with t from $h(0)$ to $h(\infty)$, we complete the following proofs.

APPENDIX 7.1

Prove that for $0 < T < \infty$:

$$\frac{H(T)}{T} = \frac{1}{T} \int_0^T h(t) dt$$

increases with T from $h(0)$ to $h(\infty)$ and:

$$\frac{F(T)}{\int_0^T \overline{F}(t) dt} \leq \frac{H(T)}{T} \leq h(T). \tag{A7.1}$$

Proof. Note that:

$$\lim_{T \to 0} \frac{H(T)}{T} = \lim_{T \to 0} \frac{1}{T} \int_0^T h(t) dt = h(0),$$

$$\lim_{T \to \infty} \frac{H(T)}{T} = \lim_{T \to \infty} \frac{1}{T} \int_0^T h(t) dt = h(\infty),$$

and

$$\frac{d[H(T)/T]}{dT} = \frac{1}{T^2} [Th(T) - H(T)] = \frac{1}{T^2} \int_0^T [h(T) - h(t)] dt \geq 0.$$

This follows that $H(T)/T$ increases with T from $h(0)$ to $h(\infty)$.
    From Equation (A7.1), letting:

$$L(T) \equiv H(T) \int_0^T \overline{F}(t) dt - TF(T),$$

we have $L(0) = 0$ and:

$$\frac{dL(T)}{dT} = h(T) \int_0^T \overline{F}(t) dt + H(T)\overline{F}(T) - F(T) - Tf(T)$$

$$= \int_0^T [h(T) - h(t)][F(T) - F(t)] dt \geq 0,$$

which completes the proof of Equation (A7.1).

APPENDIX 7.2

Prove that for $0 < T < \infty$:

$$\frac{\int_T^\infty F(t)dt}{\int_T^\infty \left[\int_0^t \overline{F}(u)du\right]dt} \tag{A7.2}$$

increases with T to $1/\mu$:

$$\frac{\int_0^T \overline{F}(t)dt}{\int_0^T \left[\int_t^\infty \overline{F}(u)du\right]dt} \tag{A7.3}$$

increases with T from $1/\mu$, and:

$$\frac{\int_T^\infty F(t)dt}{\int_T^\infty \left[\int_0^t \overline{F}(u)du\right]dt} \leq \frac{\int_0^T \overline{F}(t)dt}{\int_0^T \left[\int_t^\infty \overline{F}(u)du\right]dt}. \tag{A7.4}$$

Proof. Note that from Equation (A7.1):

$$\lim_{T \to \infty} \frac{\int_T^\infty F(t)dt}{\int_T^\infty [\int_0^t \overline{F}(u)du]dt} = \lim_{T \to \infty} \frac{F(T)}{\int_0^T \overline{F}(t)dt} = \frac{1}{\mu},$$

$$\lim_{T \to 0} \frac{\int_0^T \overline{F}(t)dt}{\int_0^T [\int_t^\infty \overline{F}(u)du]dt} = \lim_{T \to 0} \frac{\overline{F}(T)}{\int_T^\infty \overline{F}(t)dt} = \frac{1}{\mu}.$$

Furthermore, because $F(T)/\int_0^T \overline{F}(t)dt$ increases with T:

$$\frac{\int_T^\infty F(t)dt}{\int_T^\infty [\int_0^t \overline{F}(u)du]dt} \geq \frac{F(T)}{\int_0^T \overline{F}(t)dt}.$$

Similarly, because $\overline{F}(T)/\int_T^\infty \overline{F}(t)dt$ increases with T:

$$\frac{\int_0^T \overline{F}(t)dt}{\int_0^T [\int_t^\infty \overline{F}(u)du]dt} \leq \frac{\overline{F}(T)}{\int_T^\infty \overline{F}(t)dt}.$$

Differentiating $\int_T^\infty F(t)dt / \int_T^\infty [\int_0^t \overline{F}(u)du]dt$ with respect to T:

$$-F(T)\int_T^\infty \left[\int_0^t \overline{F}(u)du\right]dt + \int_T^\infty F(t)dt\int_0^T \overline{F}(t)dt$$

$$= \int_0^T \overline{F}(t)dt\int_T^\infty \left[\int_0^t \overline{F}(u)du\right]dt \left[\frac{\int_T^\infty F(t)dt}{\int_T^\infty [\int_0^t \overline{F}(u)du]dt} - \frac{F(T)}{\int_0^T \overline{F}(t)dt}\right] \geq 0,$$

which proves that Equation (A7.2) increases with T to $1/\mu$.

Similarly, we can prove Equation (A7.3) increases with T from $1/\mu$ and then complete the proof of Equation (A7.4).

APPENDIX 7.3

For $0 < T < \infty$ and $N = 0,1,2,\cdots$:

$$\frac{\int_0^T t^N dF(t)}{\int_0^T t^N \overline{F}(t)dt} \tag{A7.5}$$

increases with T from $h(0)$ and increases with N from $F(T)/\int_0^T \overline{F}(t)dt$ to $h(T)$:

$$\frac{\int_0^T (\theta t)^N e^{-\theta t} dF(t)}{\int_0^T (\theta t)^N e^{-\theta t} \overline{F}(t)dt} \tag{A7.6}$$

increases with T from $h(0)$ and increases with N from $\int_0^T e^{-\theta t} dF(t)/\int_0^T e^{-\theta t} \overline{F}(t)dt$ to $h(T)$:

$$\frac{\int_0^T t^N dF(t)}{\int_0^T t^N \overline{F}(t)dt} \geq \frac{\int_0^T (\theta t)^N e^{-\theta t} dF(t)}{\int_0^T (\theta t)^N e^{-\theta t} \overline{F}(t)dt}. \tag{A7.7}$$

Proof. Note that:

$$\lim_{T\to 0} \frac{\int_0^T t^N dF(t)}{\int_0^T t^N \overline{F}(t)dt} = \lim_{T\to 0} \frac{\int_0^T (\theta t)^N e^{-\theta t} dF(t)}{\int_0^T (\theta t)^N e^{-\theta t} \overline{F}(t)dt} = h(0),$$

$$\lim_{N \to \infty} \frac{\int_0^T t^N \mathrm{d}F(t)}{\int_0^T t^N \overline{F}(t)\mathrm{d}t} = \lim_{N \to \infty} \frac{\int_0^T (\theta t)^N \mathrm{e}^{-\theta t}\mathrm{d}F(t)}{\int_0^T (\theta t)^N \mathrm{e}^{-\theta t}\overline{F}(t)\mathrm{d}t} = h(T).$$

Differentiating $\int_0^T t^N \mathrm{d}F(t) / \int_0^T t^N \overline{F}(t)\mathrm{d}t$ with respect to T:

$$T^N f(T) \int_0^T t^N \overline{F}(t)\mathrm{d}t - T^N \overline{F}(T) \int_0^T t^N \mathrm{d}F(t)$$

$$= T^N \overline{F}(T) \int_0^T t^N \overline{F}(t)[h(T) - h(t)]\mathrm{d}t \geq 0,$$

which follows that Equation (A7.5) increases with T from $h(0)$, forming:

$$\frac{\int_0^T t^{N+1}\mathrm{d}F(t)}{\int_0^T t^{N+1}\overline{F}(t)\mathrm{d}t} - \frac{\int_0^T t^N \mathrm{d}F(t)}{\int_0^T t^N \overline{F}(t)\mathrm{d}t},$$

and denoting:

$$L(T) \equiv \int_0^T t^{N+1}\mathrm{d}F(t) \int_0^T t^N \overline{F}(t)\mathrm{d}t - \int_0^T t^N \mathrm{d}F(t) \int_0^T t^{N+1}\overline{F}(t)\mathrm{d}t,$$

we have $L(0) = 0$ and

$$\frac{\mathrm{d}L(T)}{\mathrm{d}T} = T^{N+1}f(T) \int_0^T t^N \overline{F}(t)\mathrm{d}t + T^N \overline{F}(T) \int_0^T t^{N+1}\mathrm{d}F(t)$$

$$- T^N f(T) \int_0^T t^{N+1}\overline{F}(t)\mathrm{d}t - T^{N+1}\overline{F}(T) \int_0^T t^N \mathrm{d}F(t)$$

$$= T^N \overline{F}(T) \int_0^T t^N \overline{F}(t)(T-t)[h(T) - h(t)]\mathrm{d}t \geq 0,$$

which follows that Equation (A7.5) increases with N to $h(T)$.

Similarly, we can prove Equation (A7.6) increases with T from $h(0)$ and increases with N to $h(T)$.

From Equation (A7.7), letting:

$$L(T) \equiv \int_0^T t^N \mathrm{d}F(t) \int_0^T (\theta t)^N \mathrm{e}^{-\theta t}\overline{F}(t)\mathrm{d}t$$

$$- \int_0^T t^N \overline{F}(t)\mathrm{d}t \int_0^T (\theta t)^N \mathrm{e}^{-\theta t}\mathrm{d}F(t),$$

we have $L(0) = 0$ and

$$\frac{dL(T)}{dT} = T^N f(T) \int_0^T (\theta t)^N e^{-\theta t} \overline{F}(t) dt + (\theta T)^N e^{-\theta T} \overline{F}(T) \int_0^T t^N dF(t)$$

$$-T^N \overline{F}(T) \int_0^T (\theta t)^N e^{-\theta t} dF(t) - (\theta T)^N e^{-\theta T} f(T) \int_0^T t^N \overline{F}(t) dt$$

$$= T^N \overline{F}(T) \int_0^T (\theta t)^N \overline{F}(t) (e^{-\theta t} - e^{-\theta T})[h(T) - h(t)] dt \geq 0,$$

which completes the proof of Equation (A7.7).

APPENDIX 7.4

For $0 < T < \infty$ and $N = 0,1,2,\cdots$:

$$\frac{\displaystyle\int_T^\infty t^N dF(t)}{\displaystyle\int_T^\infty t^N \overline{F}(t) dt}$$

increases with T to $h(\infty)$ and increases with N from $\overline{F}(T) / \int_T^\infty \overline{F}(t) dt$ to $h(\infty)$:

$$\frac{\displaystyle\int_T^\infty (\theta t)^N e^{-\theta t} dF(t)}{\displaystyle\int_T^\infty (\theta t)^N e^{-\theta t} \overline{F}(t) dt}$$

increases with T to $h(\infty)$ and increases with N from $\int_T^\infty e^{-\theta t} dF(t) / \int_T^\infty e^{-\theta t} \overline{F}(t) dt$ to $h(\infty)$, and:

$$\frac{\displaystyle\int_T^\infty t^N dF(t)}{\displaystyle\int_T^\infty t^N \overline{F}(t) dt} \geq \frac{\displaystyle\int_T^\infty (\theta t)^N e^{-\theta t} dF(t)}{\displaystyle\int_T^\infty (\theta t)^N e^{-\theta t} \overline{F}(t) dt}.$$

Proof. Appendix 7.4 can be proved by using the similar discussions of Appendix 7.3.

APPENDIX 7.5

Prove that:

$$\frac{\displaystyle\int_0^T (\theta t)^N [\int_t^\infty e^{-\theta u} dF(u)] dt}{\displaystyle\int_0^T (\theta t)^N [\int_t^\infty e^{-\theta u} \overline{F}(u) du] dt} \tag{A7.8}$$

increases with T from $\int_0^\infty e^{-\theta t} dF(t) / \int_0^\infty e^{-\theta t} \overline{F}(t) dt$ and increases with N to: $\int_T^\infty e^{-\theta t} dF(t) / \int_T^\infty e^{-\theta t} \overline{F}(t) dt$,

$$\frac{\int_T^\infty (\theta t)^N [\int_0^t e^{-\theta u} dF(u)] dt}{\int_T^\infty (\theta t)^N [\int_0^t e^{-\theta u} \overline{F}(u) du] dt} \tag{A7.9}$$

increases with T to $\int_0^\infty e^{-\theta t} dF(t) / \int_0^\infty e^{-\theta t} \overline{F}(t) dt$ and increases with N to $\int_0^\infty e^{-\theta t} dF(t) / \int_0^\infty e^{-\theta t} \overline{F}(t) dt$, and:

$$\frac{\int_0^T (\theta t)^N [\int_t^\infty e^{-\theta u} dF(u)] dt}{\int_0^T (\theta t)^N [\int_t^\infty e^{-\theta u} \overline{F}(u) du] dt} \geq \frac{\int_T^\infty (\theta t)^N [\int_0^t e^{-\theta u} dF(u)] dt}{\int_T^\infty (\theta t)^N [\int_0^t e^{-\theta u} \overline{F}(u) du] dt}. \tag{A7.10}$$

Proof. Note that:

$$\lim_{T \to 0} \frac{\int_0^T (\theta t)^N [\int_t^\infty e^{-\theta u} dF(u)] dt}{\int_0^T (\theta t)^N [\int_t^\infty e^{-\theta u} \overline{F}(u) du] dt} = \frac{\int_0^\infty e^{-\theta t} dF(t)}{\int_0^\infty e^{-\theta t} \overline{F}(t) dt},$$

$$\lim_{N \to \infty} \frac{\int_0^T (\theta t)^N [\int_t^\infty e^{-\theta u} dF(u)] dt}{\int_0^T (\theta t)^N [\int_t^\infty e^{-\theta u} \overline{F}(u) du] dt} = \frac{\int_T^\infty e^{-\theta t} dF(t)}{\int_T^\infty e^{-\theta t} \overline{F}(t) dt},$$

and:

$$\lim_{T \to \infty} \frac{\int_T^\infty (\theta t)^N \left[\int_0^t e^{-\theta u} dF(u)\right] dt}{\int_T^\infty (\theta t)^N \left[\int_0^t e^{-\theta u} \overline{F}(u) du\right] dt} = \lim_{N \to \infty} \frac{\int_T^\infty (\theta t)^N \left[\int_0^t e^{-\theta u} dF(u)\right] dt}{\int_T^\infty (\theta t)^N \left[\int_0^t e^{-\theta u} \overline{F}(u) du\right] dt}$$

$$= \frac{\int_0^\infty e^{-\theta t} dF(t)}{\int_0^\infty e^{-\theta t} \overline{F}(t) dt}.$$

Furthermore, because $\int_T^\infty e^{-\theta t} dF(t) / \int_T^\infty e^{-\theta t} \overline{F}(t) dt$ increases with T:

$$\frac{\int_0^T (\theta t)^N [\int_t^\infty e^{-\theta u} dF(u)] dt}{\int_0^T (\theta t)^N [\int_t^\infty e^{-\theta u} \overline{F}(u) du] dt} \leq \frac{\int_T^\infty e^{-\theta t} dF(t)}{\int_T^\infty e^{-\theta t} \overline{F}(t) dt}.$$

Differentiating $\int_0^T (\theta t)^N [\int_t^\infty e^{-\theta u} dF(u)] dt / \int_0^T (\theta t)^N [\int_t^\infty e^{-\theta u} \overline{F}(u) du] dt$ with respect to T:

$$(\theta T)^N \int_T^\infty e^{-\theta t} dF(t) \int_0^T (\theta t)^N \left[\int_t^\infty e^{-\theta u} \overline{F}(u) du\right] dt$$

$$-(\theta T)^N \int_T^\infty e^{-\theta t} \overline{F}(t) dt \int_0^T (\theta t)^N \left[\int_t^\infty e^{-\theta u} dF(u)\right] dt$$

$$= (\theta T)^N \int_T^\infty e^{-\theta t} \overline{F}(t) dt \int_0^T (\theta t)^N \left[\int_t^\infty e^{-\theta u} \overline{F}(u) du\right] dt$$

$$\times \left\{ \frac{\int_T^\infty e^{-\theta t} dF(t)}{\int_T^\infty e^{-\theta t} \overline{F}(t) dt} - \frac{\int_0^T (\theta t)^N [\int_t^\infty e^{-\theta u} dF(u)] dt}{\int_0^T (\theta t)^N [\int_t^\infty e^{-\theta u} \overline{F}(u) du] dt} \right\} \geq 0,$$

which follows that Equation (A7.8) increases with T from $\int_0^\infty e^{-\theta t} dF(t) / \int_0^\infty e^{-\theta t} \overline{F}(t) dt$.
Forming:

$$\frac{\int_0^T (\theta t)^{N+1} [\int_t^\infty e^{-\theta u} dF(u)] dt}{\int_0^T (\theta t)^{N+1} [\int_t^\infty e^{-\theta u} \overline{F}(u) du] dt} - \frac{\int_0^T (\theta t)^N [\int_t^\infty e^{-\theta u} dF(u)] dt}{\int_0^T (\theta t)^N [\int_t^\infty e^{-\theta u} \overline{F}(u) du] dt},$$

and denoting:

$$L(T) \equiv \int_0^T (\theta t)^{N+1} \left[\int_t^\infty e^{-\theta u} dF(u)\right] dt \int_0^T (\theta t)^N \left[\int_t^\infty e^{-\theta u} \overline{F}(u) du\right] dt$$

$$- \int_0^T (\theta t)^N \left[\int_t^\infty e^{-\theta u} dF(u)\right] dt \int_0^T (\theta t)^{N+1} \left[\int_t^\infty e^{-\theta u} \overline{F}(u) du\right] dt,$$

we have $L(0) = 0$ and:

$$\frac{dL(T)}{dT} = (\theta T)^{N+1} \int_T^\infty e^{-\theta t} dF(t) \int_0^T (\theta t)^N \left[\int_t^\infty e^{-\theta u} \overline{F}(u) du \right] dt$$

$$+ (\theta T)^N \int_T^\infty e^{-\theta t} \overline{F}(t) dt \int_0^T (\theta t)^{N+1} \left[\int_t^\infty e^{-\theta u} dF(u) \right] dt$$

$$- (\theta T)^N \int_T^\infty e^{-\theta t} dF(t) \int_0^T (\theta t)^{N+1} \left[\int_t^\infty e^{-\theta u} \overline{F}(u) du \right] dt$$

$$- (\theta T)^{N+1} \int_T^\infty e^{-\theta t} \overline{F}(t) dt \int_0^T (\theta t)^N \left[\int_t^\infty e^{-\theta u} dF(u) \right] dt$$

$$= \int_T^\infty e^{-\theta t} \overline{F}(t) dt \int_0^T \left\{ (\theta T)^N (\theta t)^N (\theta T - \theta t) \left[\int_t^\infty e^{-\theta u} \overline{F}(u) du \right] \right.$$

$$\times \left. \left[\frac{\int_T^\infty e^{-\theta u} dF(u)}{\int_T^\infty e^{-\theta u} \overline{F}(u) du} - \frac{\int_t^\infty e^{-\theta u} dF(u)}{\int_t^\infty e^{-\theta u} \overline{F}(u) du} \right] \right\} dt \geq 0,$$

which follows that Equation (A7.8) increases with N to $\int_T^\infty e^{-\theta t} dF(t) / \int_T^\infty e^{-\theta t} \overline{F}(t) dt$.
Similarly, we can prove Equation (A7.9) increases with T to $\int_0^\infty e^{-\theta t} dF(t) / \int_0^\infty e^{-\theta t}$
$\overline{F}(t) dt$ and increases with N to $\int_0^\infty e^{-\theta t} dF(t) / \int_0^\infty e^{-\theta t} \overline{F}(t) dt$ and complete the proof of Equation (A7.10).

APPENDIX 7.6

Prove that:

$$\frac{\int_0^T (\theta t)^N [\int_0^t e^{-\theta u} dF(u)] dt}{\int_0^T (\theta t)^N [\int_0^t e^{-\theta u} \overline{F}(u) du] dt} \tag{A7.11}$$

increases with T from $h(0)$ and increases with N to $\int_0^T e^{-\theta t} dF(t) / \int_0^T e^{-\theta t} \overline{F}(t) dt$, and:

$$\frac{\int_T^\infty (\theta t)^N [\int_t^\infty e^{-\theta u} dF(u)] dt}{\int_T^\infty (\theta t)^N [\int_t^\infty e^{-\theta u} \overline{F}(u) du] dt} \tag{A7.12}$$

increases with T to $h(\infty)$ and increases with N to $h(\infty)$.

Proof. Note that:

$$\frac{\int_0^T (\theta t)^N \left[\int_0^t e^{-\theta u} dF(u) \right] dt}{\int_0^T (\theta t)^N \left[\int_0^t e^{-\theta u} \overline{F}(u) du \right] dt} \le \frac{\int_0^T e^{-\theta t} dF(t)}{\int_0^T e^{-\theta t} \overline{F}(t) dt} \le \frac{\int_T^\infty e^{-\theta t} dF(t)}{\int_T^\infty e^{-\theta t} \overline{F}(t) dt}$$

$$\le \frac{\int_T^\infty (\theta t)^N \left[\int_t^\infty e^{-\theta u} dF(u) \right] dt}{\int_T^\infty (\theta t)^N \left[\int_t^\infty e^{-\theta u} \overline{F}(u) du \right] dt}.$$

Using the similar discussions of Appendices 7.5 and 7.6 can be proved.

Similarly, we can prove that the failure rates in VIII and IX increase with T and N and obtain the inequalities for $0 < T < \infty$ and $N = 0,1,2,\cdots$.

APPENDIX 7.7

Prove that for $0 < T < \infty$ and $K = 0,1,2,\cdots$:

$$\frac{\int_0^T p_K(t)h(t)dt}{\int_0^T p_K(t)dt} \tag{A7.13}$$

increases with T from $h(0)$ to $1 / \int_0^\infty p_K(t)dt$ and increases with K from $F(T) / \int_0^T \overline{F}(t)dt$ to $h(T)$, and:

$$\frac{F(T)}{\int_0^T \overline{F}(t)dt} \le \frac{\int_0^T p_K(t)h(t)dt}{\int_0^T p_K(t)dt} \le h(T) \le \frac{\overline{F}(T)}{\int_T^\infty \overline{F}(t)dt}, \tag{A7.14}$$

$$\frac{\int_0^T p_K(t)h(t)dt}{\int_0^T p_K(t)dt} \le \frac{1}{\int_0^\infty p_K(t)dt}. \tag{A7.15}$$

Proof. Note that:

$$\lim_{T \to 0} \frac{\int_0^T p_K(t)h(t)dt}{\int_0^T p_K(t)dt} = h(0), \quad \lim_{T \to \infty} \frac{\int_0^T p_K(t)h(t)dt}{\int_0^T p_K(t)dt} = \frac{1}{\int_0^\infty p_K(t)dt},$$

$$\lim_{K \to 0} \frac{\int_0^T p_K(t)h(t)\mathrm{d}t}{\int_0^T p_K(t)\mathrm{d}t} = \frac{F(T)}{\int_0^T \overline{F}(t)\mathrm{d}t}, \quad \lim_{K \to \infty} \frac{\int_0^T p_K(t)h(t)\mathrm{d}t}{\int_0^T p_K(t)\mathrm{d}t} = h(T).$$

Differentiating $\int_0^T p_K(t)h(t)\mathrm{d}t / \int_0^T p_K(t)\mathrm{d}t$ with respect to T:

$$p_K(T)h(T)\int_0^T p_K(t)\mathrm{d}t - p_K(T)\int_0^T p_K(t)h(t)\mathrm{d}t$$

$$= p_K(T)\int_0^T p_K(t)[h(T) - h(t)]\mathrm{d}t \ge 0,$$

which follows that Equation (7A.13) increases with T.

Making the difference between $\int_0^T p_K(t)h(t)\mathrm{d}t / \int_0^T p_K(t)\mathrm{d}t$ for K, and letting:

$$L(T) \equiv \int_0^T p_{K+1}(t)h(t)\mathrm{d}t \int_0^T p_K(t)\mathrm{d}t - \int_0^T p_K(t)h(t)\mathrm{d}t \int_0^T p_{K+1}(t)\mathrm{d}t,$$

we have $L(0) = 0$ and:

$$L'(T) = p_{K+1}(T)\int_0^T p_K(t)[h(T) - h(t)]\mathrm{d}t$$

$$- p_K(T)\int_0^T p_{K+1}(t)[h(T) - h(t)]\mathrm{d}t$$

$$= \frac{H(T)^K}{(K+1)!}e^{-H(T)}\int_0^T p_K(t)[h(T) - h(t)][H(T) - H(t)]\mathrm{d}t \ge 0,$$

which follows that Equation (A7.13) increases with K.

From these results, we can obtain Equations (A7.14) and (A7.15).

APPENDIX 7.8

Prove that for $0 \le T < \infty$ and $K = 0,1,2,\cdots$:

$$\frac{\int_T^\infty p_K(t)h(t)\mathrm{d}t}{\int_T^\infty p_K(t)\mathrm{d}t} \tag{A7.16}$$

increases with T from $1/\int_0^\infty p_K(t)\mathrm{d}t$ to $h(\infty)$ and increases with K from $\overline{F}(T)/\int_T^\infty \overline{F}(t)\mathrm{d}t$ to $h(\infty)$, and:

$$\frac{\overline{F}(T)}{\displaystyle\int_T^\infty \overline{F}(t)\mathrm{d}t} \le \frac{\displaystyle\int_T^\infty p_K(t)h(t)\mathrm{d}t}{\displaystyle\int_T^\infty p_K(t)\mathrm{d}t}, \quad \frac{1}{\displaystyle\int_0^\infty p_K(t)\mathrm{d}t} \le \frac{\displaystyle\int_T^\infty p_K(t)h(t)\mathrm{d}t}{\displaystyle\int_T^\infty p_K(t)\mathrm{d}t}. \tag{A7.17}$$

Proof. Note that:

$$\lim_{T\to 0} \frac{\displaystyle\int_T^\infty p_K(t)h(t)\mathrm{d}t}{\displaystyle\int_T^\infty p_K(t)\mathrm{d}t} = \frac{1}{\displaystyle\int_0^\infty p_K(t)\mathrm{d}t}, \quad \lim_{T\to\infty} \frac{\displaystyle\int_T^\infty p_K(t)h(t)\mathrm{d}t}{\displaystyle\int_T^\infty p_K(t)\mathrm{d}t} = h(\infty),$$

$$\lim_{T\to 0} \frac{\displaystyle\int_T^\infty p_K(t)h(t)\mathrm{d}t}{\displaystyle\int_T^\infty p_K(t)\mathrm{d}t} = \frac{\overline{F}(T)}{\displaystyle\int_T^\infty \overline{F}(t)\mathrm{d}t}, \quad \lim_{K\to\infty} \frac{\displaystyle\int_T^\infty p_K(t)h(t)\mathrm{d}t}{\displaystyle\int_T^\infty p_K(t)\mathrm{d}t} = h(\infty).$$

By similar methods used in Appendix 7.7, we can easily prove Appendix 7.8.

APPENDIX 7.9

Prove that for $0 < T < \infty$ and $K = 0,1,2,\cdots$:

$$\frac{\displaystyle\int_0^T p_K(t)h(t)\mathrm{d}t}{\displaystyle\int_0^T p_K(t)[h(t)/Q(t)]\mathrm{d}t} \tag{A7.18}$$

increases with T from $1/\mu$ to $1/\int_0^\infty p_{K+1}(t)\mathrm{d}t$ and increases with K from: $F(T)/\int_0^T h(t)[\int_t^\infty \overline{F}(u)\mathrm{d}u]\mathrm{d}t$ to $\overline{F}(T)/\int_T^\infty \overline{F}(t)\mathrm{d}t$, and:

$$\frac{1}{\mu} \le \frac{\displaystyle\int_0^T p_K(t)h(t)\mathrm{d}t}{\displaystyle\int_0^T p_K(t)[h(t)/Q(t)]\mathrm{d}t} \le \frac{1}{\displaystyle\int_0^\infty p_{K+1}(t)\mathrm{d}t}, \tag{A7.19}$$

$$\frac{\displaystyle\int_0^T p_K(t)h(t)\mathrm{d}t}{\displaystyle\int_0^T p_K(t)[h(t)/Q(t)]\mathrm{d}t} \le \frac{\overline{F}(T)}{\displaystyle\int_T^\infty \overline{F}(t)\mathrm{d}t}, \tag{A7.20}$$

Proof. Note that:

$$\lim_{T\to 0} \frac{\displaystyle\int_0^T p_K(t)h(t)\mathrm{d}t}{\displaystyle\int_0^T p_K(t)[h(t)/Q(t)]\mathrm{d}t} = \lim_{T\to 0} \frac{\overline{F}(T)}{\displaystyle\int_T^\infty \overline{F}(t)\mathrm{d}t} = \frac{1}{\mu},$$

$$\lim_{T \to \infty} \frac{\int_0^T p_K(t)h(t)dt}{\int_0^T p_K(t)[h(t)/Q(t)]dt} = \frac{1}{\int_0^\infty p_K(t)[h(t)/Q(t)]dt} = \frac{1}{\int_0^\infty p_{K+1}(t)dt},$$

$$\lim_{K \to 0} \frac{\int_0^T p_K(t)h(t)dt}{\int_0^T p_K(t)[h(t)/Q(t)]dt} = \frac{F(T)}{\int_0^T h(t)[\int_t^\infty \overline{F}(u)du]dt},$$

$$\lim_{K \to \infty} \frac{\int_0^T p_K(t)h(t)dt}{\int_0^T p_K(t)[h(t)/Q(t)]dt} = \frac{\overline{F}(T)}{\int_T^\infty \overline{F}(t)dt}.$$

Differentiating $\int_0^T p_K(t)h(t)dt / \int_0^T p_K(t)[h(t)/Q(t)]dt$ with respect to T:

$$p_K(T)h(T)\int_0^T p_K(t)\frac{h(t)}{Q(t)}dt - p_K(T)\frac{h(T)}{Q(T)}\int_0^T p_K(t)h(t)dt$$

$$= p_K(T)\frac{h(T)}{Q(T)}\int_0^T p_K(t)\frac{h(t)}{Q(t)}[Q(T)-Q(t)]dt \geq 0,$$

which follows that Equation (A7.18) increases with T.

Making difference between $\int_0^T p_K(t)h(t)dt / \int_0^T p_K(t)[h(t)/Q(t)]dt$ for K, and letting:

$$L(T) \equiv \int_0^T p_{K+1}(t)h(t)dt\int_0^T p_K(t)\frac{h(t)}{Q(t)}dt$$

$$- \int_0^T p_K(t)h(t)dt\int_0^T p_{K+1}(t)\frac{h(t)}{Q(t)}dt,$$

we have $L(0) = 0$, and:

$$L'(T) = \frac{H(T)^K}{(K+1)!}e^{-H(T)}h(T)\int_0^T p_K(T)\left[\frac{1}{Q(t)}-\frac{1}{Q(T)}\right][H(T)-H(t)]dt \geq 0,$$

which follows that Equation (A7.18) increases with K.

From these results, we can obtain Equations (A7.19) and (A7.20).

APPENDIX 7.10

Prove that for $0 \le T < \infty$ and $K = 0,1,2,\cdots$:

$$\frac{\displaystyle\int_T^\infty p_K(t)h(t)\mathrm{d}t}{\displaystyle\int_T^\infty p_K(t)[h(t)/Q(t)]\mathrm{d}t} \tag{A7.21}$$

increases with T from $1/\int_0^\infty p_{K+1}(t)\mathrm{d}t$ to $h(\infty)$ and increases with K from: $\overline{F}(T)/\int_T^\infty [\int_t^\infty \overline{F}(u)\mathrm{d}u]\mathrm{d}t$ to $h(\infty)$, and:

$$\frac{\displaystyle\int_T^\infty p_K(t)h(t)\mathrm{d}t}{\displaystyle\int_T^\infty p_K(t)\mathrm{d}t} \le \frac{\displaystyle\int_T^\infty p_K(t)h(t)\mathrm{d}t}{\displaystyle\int_T^\infty p_K(t)h(t)[\int_t^\infty \overline{F}(u)\mathrm{d}u/\overline{F}(t)]\mathrm{d}t},$$

$$\frac{1}{\displaystyle\int_0^\infty p_{K+1}(t)\mathrm{d}t} \le \frac{\displaystyle\int_T^\infty p_K(t)h(t)\mathrm{d}t}{\displaystyle\int_T^\infty p_K(t)h(t)[\int_t^\infty \overline{F}(u)\mathrm{d}u/\overline{F}(t)]\mathrm{d}t}.$$

Proof. Note that:

$$\lim_{T\to 0}\frac{\displaystyle\int_T^\infty p_K(t)h(t)\mathrm{d}t}{\displaystyle\int_T^\infty p_K(t)[h(t)/Q(t)]\mathrm{d}t} = \frac{1}{\displaystyle\int_0^\infty p_{K+1}(t)\mathrm{d}t},$$

$$\lim_{T\to\infty}\frac{\displaystyle\int_T^\infty p_K(t)h(t)\mathrm{d}t}{\displaystyle\int_T^\infty p_K(t)[h(t)/Q(t)]\mathrm{d}t} = h(\infty),$$

$$\lim_{K\to 0}\frac{\displaystyle\int_T^\infty p_K(t)h(t)\mathrm{d}t}{\displaystyle\int_T^\infty p_K(t)[h(t)/Q(t)]\mathrm{d}t} = \frac{\overline{F}(T)}{\displaystyle\int_T^\infty h(t)[\int_t^\infty \overline{F}(u)\mathrm{d}u]\mathrm{d}t},$$

$$\lim_{K\to\infty}\frac{\displaystyle\int_T^\infty p_K(t)h(t)\mathrm{d}t}{\displaystyle\int_T^\infty p_K(t)[h(t)/Q(t)]\mathrm{d}t} = \lim_{T\to\infty}Q(T) = h(\infty).$$

Using $h(t)/Q(t) \le 1$ and similar methods in Appendix 7.9, we can easily prove Appendix 7.10.

ACKNOWLEDGMENT

This work is supported by National Natural Science Foundation of China (NO. 71801126), Natural Science Foundation of Jiangsu Province (NO. BK20180412), Aeronautical Science Foundation of China (NO. 2018ZG52080), and Fundamental Research Funds for the Central Universities (NO. NR2018003).

REFERENCES

1. Lai, C.D., Xie, M. Concepts and applications of stochastic aging in reliability. In Pham, H. (ed.), *Handbook of Reliability Engineering*. London, UK: Springer, 2003: pp. 165–180.
2. Barlow, R.E., Proschan, F. *Mathematical Theory of Reliability*. New York: John Wiley & Sons, 1965.
3. Finkelstein, M. *Failure Rate Modeling for Reliability and Risk*. London, UK: Springer, 2008.
4. Nakagawa, T. *Maintenance Theory of Reliability*. London, UK: Springer, 2005.
5. Nakagawa, T. *Random Maintenance Policies*. London, UK: Springer, 2014.
6. Nakagawa, T., Zhao, X. *Maintenance Overtime Policies in Reliability Theory*. Cham, Switzerland: Springer, 2015.
7. Zhao, X., Al-Khalifa, K.N., Hamouda, A.M.S., Nakagawa, T. What is middle maintenance policy? *Quality and Reliability Engineering International*, 2016, 32, 2403–2414.
8. Zhao, X., Al-Khalifa, K.N., Hamouda, A.M.S., Nakagawa, T. Age replacement models: A summary with new perspectives and methods. *Reliability Engineering and System Safety*, 2017, 161, 95–105.
9. Nakagawa, T. *Stochastic Processes with Applications to Reliability Theory*. London, UK: Springer, 2011.
10. Zhao, X., Qian, C., Nakagawa, T. Comparisons of maintenance policies with periodic times and repair numbers. *Reliability Engineering and System Safety*, 2017, 168, 161–170.

8 Accelerated Life Tests with Competing Failure Modes
An Overview

Kanchan Jain and Preeti Wanti Srivastava

CONTENTS

8.1 INTRODUCTION

A longer time period is necessary to test systems or components with a long expected lifetime under normal operating conditions and many units are required which is very costly and impractical. In such situations, accelerated life test (ALT) methods are used that lead to failure/degradation of systems or components in shorter time periods. Hence, failure data can be obtained during a reasonable period without changing failure mechanisms.

ALTs were introduced by Chernoff (1962) and Bessler et al. (1962). They are used during Design and Development, Design Verification, and Process Validation stages of a product life cycle. Designing of optimal test plans is a critical step for assuring that ALTs help in prediction of the product reliability accurately, quickly, and economically.

In ALT, systems or components are:

- Subjected to more severe conditions than those experienced at normal conditions (accelerated stress). Stress factors can be temperature, voltage, mechanical load, thermal cycling, humidity, and vibration.
- Put in operation more vigorously at normal operating conditions (accelerated failure). Products such as home appliances and vehicle tires are put to accelerated failure.

For accurate prediction of the reliability, the types of stresses to which systems/ components are subjected and the failure mechanisms must be understood.

Different Types of Stress are:

1. Constant
2. Step
3. Ramp-step
4. Triangular-cyclic
5. Ramp-soak-cyclic
6. Sinusoidal-cyclic

8.1.1 ACCELERATED LIFE TEST MODELS

In engineering applications, several ALT models have been proposed and used successfully. Accelerated Failure Time (AFT) models are the most widely used ALT models.

- *Partially accelerated life test model*: Degroot and Goel (1979) introduced Partially Accelerated Life Test (PALT) models wherein the items are run at normal as well as accelerated conditions.

  A PALT model consists of a life distribution and an acceleration factor for extrapolating accelerated data results to normal operating condition when the life-stress relationship cannot be specified. The acceleration factor—the

ratio of a reliability measure—for example, mean life, at use condition to that at accelerated condition provides a quantitative estimate of the relationship between the accelerated condition and the fiean condition.
* *Fully accelerated life test model*: Introduced by Bhattacharya and Soejoeti (1989), a fully ALT consists of testing the items at accelerated condition only. A fully ALT model consists of:

  1. A life distribution that represents the scatter in product life
  2. Relationship between life and stress

Some of such stress-life relationships used in the literature (Nelson 1990; Yang 2007; Elsayed 2012; Srivastava 2017) are:

* Life-Temperature models described by Arrhenius and Eyring relationships
* Life-Voltage model described by Inverse Power relationships
* Life-Usage Rate relationship
* Temperature-Humidity model
* Temperature-Nonthermal model

8.1.2 An Accelerated Life Test Procedure

An ALT is undertaken in the design and development phase as well as in the verification and validation phases of a product life cycle.

Nelson (1990) gave a comprehensive presentation of statistical models and methods for accelerated tests.

8.1.3 Competing Failure Modes

Many products have more than one cause of failure referred to as a failure mode or failure mechanism. Examples include:

* The Turn, Phase, or Ground insulation failing in motors
* A ball or the race failing in ball bearing assemblies
* A semiconductor device that fails at a junction or a lead
* A cylindrical fatigue specimen failing in the cylindrical portion, in the fillet (or radius), or in the grip
* Solar lighting device with capacitor and controller failure as two modes of failure

In these examples, the assessment of each risk factor in the presence of other risk factors is necessary and gives rise to competing risks analysis. For such an analysis, each complete observation must be composed of the failure time and the corresponding cause of failure. The causes of failure can be independent or dependent upon each other.

The procedure underlying an ALT is shown in the following flowchart (Figure 8.1).

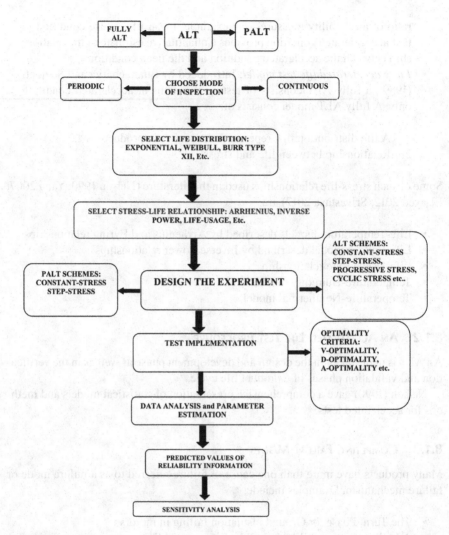

FIGURE 8.1 Flow chart.

8.2 ACCELERATED LIFE TESTS WITH INDEPENDENT CAUSES OF FAILURES

Let n identical units be put to test and suppose that a unit fails due to one of the r (>2) fatal risk factors. Let T_j be the life time of the unit due to jth risk factor with cumulative distribution function (CDF), $G_j(t)$, and probability density function (PDF), $g_j(t)$. The overall failure time of a test unit is $T = \min \{T_1, T_2, ..., T_r\}$ with CDF:

$$F(t) = 1 - \prod_{i=1}^{r} (1 - G_j(t)), \qquad (8.1)$$

and PDF:

$$f(t) = \sum_{j=1}^{r} h_j(t) \prod_{j=1}^{r} \left(1 - G_j(t)\right), \tag{8.2}$$

where $h_j(t)$, the hazard rate corresponding to the jth risk factor, is defined as:

$$h_j(t) = \frac{g_j(t)}{(1 - G_j(t))}. \tag{8.3}$$

Let C be the indicator variable for the cause of failure, then the joint distribution of (T, C) is given by:

$$f_{T,C}(t, j) = g_j(t). \tag{8.4}$$

$f_{T,C}(t, j)$ is used in the formulation of the likelihood function, which is used to estimate model parameters and obtain optimal plans using the Fisher Information Matrix. **Fisher information** measures the amount of information that an observable random variable X carries about an unknown parameter θ upon which the likelihood function depends.

8.2.1   CONSTANT-STRESS ACCELERATED LIFE TEST WITH INDEPENDENT CAUSES OF FAILURES

In a constant-stress ALT (CSALT) set-up, sub-groups of test specimens are allocated to different test chambers and, in each test chamber, the test units are subjected to different but fixed stress levels. The experiment is terminated according to a pre-specified censoring scheme. Each unit is tested under the same temperature for a fixed duration of time. For example, 10 units are tested for 100 hours at 310 K, 10 different units are tested for 100 hours at 320 K, and another 10 different units are tested for 100 hours at 330 K.

Figure 8.2 exhibits the constant-stress patterns.

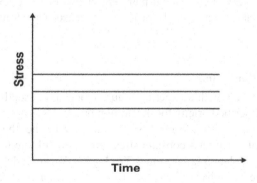

FIGURE 8.2   Constant-stress loading.

McCool (1978) presented a technique for finding interval estimates for Weibull parameters of a primary failure mode when there is a secondary failure mode with the same (but unknown) Weibull shape parameter. Moeschberger and David (1971) and David and Moeschberger (1978) gave an expression for the likelihood of competing risk data under censoring and fixed experimental conditions. Large sample properties of maximum likelihood estimators (MLEs) were discussed for Weibull and log-normal distributions. Herman and Patell (1971) discussed the MLEs under competing have causes of failure.

Klein and Basu (1981, 1982a) analyzed ALT for more than one failure mode. For independent competing failure modes for each stress level, the authors found MLEs with life times as exponential or Weibull, with common or different shape parameters under Type-I, Type-II, or progressively censored data. Using a general stress function, Klein and Basu (1982b) obtained estimates of model parameters under various censoring schemes. A dependent competing risk model was proposed by considering a bivariate Weibull distribution as the joint distribution of two competing risks.

Nelson (1990) and Craiu and Lee (2005) analyzed ALTs under competing causes of failure for semiconductor devices, ball bearing assemblies, and insulation systems. Kim and Bai (2002) analyzed ALT data with two competing risks taking a mixture of two Weibull distributions and location parameters as linear functions of stress.

Pascual (2007) considered the problem of planning ALT when the respective times to failure of competing risks are independently distributed as Weibull with a commonly known shape parameter.

Shi et al. (2013) proposed a CSALT with competing risks for failure from exponential distribution under progressive Type-II hybrid censoring. They obtained the MLE and Bayes estimators of the parameter and proved their equivalence under certain circumstances. A Monte Carlo simulation demonstrated the accuracy and effectiveness of the estimations.

Yu et al. (2014) proposed an accelerated testing plan with high and low temperatures as multiple failure modes for a complicated device. They gave the reliability function of the product and established the efficiency of the plan through a numerical example.

Wu and Huang (2017) considered planning of two or more level CSALTs with competing risk data from Type-II progressive censoring assuming exponential distribution.

8.2.1.1 Model Illustration

In this section, CSALT with competing failure modes proposed by Wu and Huang (2017) has been described briefly for illustration purpose.

Consider a CSALT with L levels of stress and let y_l be the lth stress level, $l = 1$, 2, ..., L. Each unit is run at a constant-stress and may fail due to J failure modes. Assume that at y_l, the latent failure times $X_{i1l}, X_{i2l}, ..., X_{iJl}$ are independent and exponentially distributed with hazard rate $\lambda_{jl}(> 0)$, $i = 1, 2, ..., n, l = 1, 2, ..., L$, and $j = 1$, 2, ..., J. The failure time of the ith test unit is:

$$X_{il} = \text{mim}\{X_{i1l}, X_{i2l}, \ldots, X_{iJl}\}.$$

It is assumed that at the lth stress level, the mean life time of a test unit is a log-linear function of standardized stress:

$$\log\left(\frac{1}{\lambda_{jl}}\right) = \beta_{0j} + \beta_{1j}s_l, \tag{8.5}$$

where:

$-\infty < \beta_{0j} < \infty$,
$\beta_{1j} < 0$ are unknown design parameters.

The standardized stress, s_l, is:

$$s_l = \frac{y_l - y_D}{y_L - y_D}, 0 \le s_l \le 1, l = 1, 2, \ldots, L,$$

$y_1 < y_2 < \ldots, < y_L$ are L ordered stress levels and y_D is the stress at normal operating condition. The log linear function is a common choice of life-stress relationship because it includes the power law and the Arrhenius law as special cases.

The failure density and failure distribution of the ith unit under jth risk are, respectively:

$$f(x_{il}, 1) = \lambda_{jl}\, e^{-\lambda_{+1}\, x_{il}}, x_{il} > 0 \tag{8.6}$$

$$F(x_{il}, 1) = \frac{\lambda_{jl}}{\lambda_{+1}}\left(1 - e^{-\lambda_{+1}\, x_{il}}\right), x_{il} > 0, \lambda_{+1} = \sum_{j=1}^{J} \lambda_{jl}. \tag{8.7}$$

The failure distribution at failure time x_{il} is:

$$F(x_{il}) = \left(1 - e^{-\lambda_{+1}\, x_{il}}\right), x_{il} > 0. \tag{8.8}$$

The authors have used progressive Type-II censoring scheme. Under this scheme, n_1 units are tested at stress level s_1 with $\Sigma_{l=1}^{L} n_l = n$. For each stress level l, m_1 failures are observed. The data are collected as follows:

When the first failure time, $X_{(1)l}$, and its cause of failure, δ_{1l}, are observed, r_{1l} of the surviving units are selected randomly and removed. When the second failure time, $X_{(2)l}$, and its cause of failure, δ_{2l}, are observed, r_{2l} of the surviving units are selected randomly and removed. For simplicity, X_{il} is used instead of $X_{(i)l}$. Type-II progressive censored data with competing risks at stress level s_l are:

$$\left(X_{1l}, \delta_{1l}, r_{1l}\right), \left(X_{2l}, \delta_{2l}, r_{2l}\right), \ldots, \left(X_{m_l l}, \delta_{m_l l}, r_{m_l l}\right)$$

$X_{1l} < X_{2l} < ... < X_{m_l l}$ are the m_l observed life times,
δ_{1l}, δ_{2l}, ..., $\delta_{m_l l}$ are the observable causes of failures,
r_{1l}, r_{2l},...,$r_{m_l l}$ are the number of censored units.

The likelihood function under Type-II progressive censoring scheme is:

$$L = \prod_{l=1}^{L} \prod_{i=1}^{m_l} \left\{ \prod_{j=1}^{J} \lambda_{jl}^{I_{ijl}} \right\} e^{-\lambda_{+l}\{x_{il}\,(r_{il}\,+\,1)\}}, \tag{8.9}$$

where $I_{ijl} = 1$ if $\delta_{il} = j$ and zero otherwise.

$$\log\left(\frac{1}{\lambda_{jl}}\right) = \beta_{0j} + \beta_{1j}s_l, \; \lambda_{jl} = e^{-\beta_{0j}-\beta_{1j}s_l} \text{ and } \lambda_{+l} = \sum_{j=1}^{J} e^{-\beta_{0j}-\beta_{1j}s_l}.$$

Using the likelihood function, the authors have used D-optimality, variance optimality, and A-optimality criteria to obtain the optimal stress level as well as the optimal sample allocation at each stress level. They used the real data set from Nelson (1990) on times to failure of the Class-H insulation system in motors to explain the proposed method. The design temperature is 180°C. The insulation systems are tested at high temperatures of 190°C, 220°C, 240°C, and 260°C. Turn, Phase, and Ground are three causes of failure.

8.2.2 STEP-STRESS ACCELERATED LIFE TEST WITH INDEPENDENT CAUSES OF FAILURES

Step-stress loading requires one test chamber. The stress on a specimen is increased step-by-step wherein at each step it is subject to constant stress for a specified period. The experiment is terminated according to a pre-specified censoring scheme.

Figure 8.3 exhibits the step-stress loading scheme.

To model data from step-stress test, the life distribution under step-stressing must be related to the distribution under a constant stress. Such a model, known as **Cumulative Exposure Model (CEM)**, was put forward by Nelson (1980).

The CEM assumes that the remaining life of a unit depends only on the current cumulative fraction failed and current stress irrespective of accumulation of the fraction. At the current stress, survivors fail according to the CDF for that stress but starting at the previously accumulated fraction failed. Under CEM, the step-stress life distribution is

$$G(w) = \begin{cases} G_1(w), & \tau_0 \leq w < \tau_1 \\ G_i(w - \tau_{i-1} + s_{i-1}), & \tau_{i-1} \leq w < \tau_i, \; i = 1, 2, ..., k-1 \\ G_k(w - \tau_{k-1} + s_{k-1}), & \tau_{i-1} \leq w < \tau_i, \end{cases} \tag{8.10}$$

$s_0 = \tau_0 = 0$, $s_i \, (i > 0)$ is the solution of $G_{i+1}(s_i) = G_i(\tau_i - \tau_{i-1})$, $i = 1, 2, ..., k-1$.

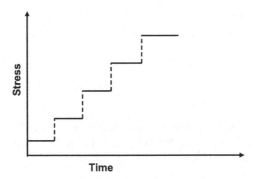

FIGURE 8.3   Step-stress loading.

Khamis and Higgens (1998) formulated the Weibull CEM, which is based on the time transformation of exponential CEM. Bai and Chun (1991) studied optimum simple step-stress accelerated life tests (SSALTs) with competing causes of failure when the distributions of each failure cause were independent and exponential.

Balakrishnan and Han (2008) and Han and Balakrishnan (2010) considered an exponential SSALT with competing risks using Type-I and Type-II censored data respectively. Donghoon and Balakrishnan (2010) studied inferential problem for exponential distribution under time constraint. Using time-censored data, Liu and Qiu (2011) devised a multiple-step SSALT with independent competing risks.

Srivastava et al. (2014) considered simple SSALT under Type-I censoring using the Khamis-Higgins model (an alternative to the Weibull CEM) with competing causes of failure. The Khamis-Higgins model is based on time transformation of the exponential model. The life distribution of each failure cause, which is independent of other, is assumed to be Weibull with the log of characteristic life as a linear function of the stress level.

Haghighi (2014) studied a step-stress test under competing risks and degradation measurements and estimated the reliability function.

8.2.2.1   Model Illustration

In this section, the design of SSALT plan is explained with competing failure modes using the methodology adopted by Srivastava et al. (2014).

Consider an SSALT with two causes of failure. There are two independent potential failure times for n test specimens corresponding to two causes of failure. The failure time of a unit is the lowest of its potential failure times. Two stress levels, x_1 and x_2 ($x_1 < x_2$), are used and x_0 is the stress level under normal operating condition. For any level of stress, i, the life time under each failure cause j, follows a Weibull distribution with shape parameter δ (known) and scale parameter θ_{ij}, i, $j = 1, 2$. Hence:

$$G_j(w) = 1 - \exp\left\{\frac{-w^\delta}{\theta_{ij}}\right\}, 0 \le w < \infty. \tag{8.11}$$

The characteristic life, which is the 63.2th percentile of the distribution, of two potential failure times are log-linear functions of stress and:

$$\log \theta_{ij}^{\frac{1}{\delta}} = \alpha_j + \beta_j x_i; \ i = 0, 1, 2; j = 1, 2. \tag{8.12}$$

$\alpha_j, \beta_j \ (<0)$ are unknown parameters depending on the nature of the product and the test method and δ is known. It can be shown as follows that $\theta_{ij}^{\frac{1}{\delta}}$ is the characteristic life of expression 8.11.

$$G_j(\xi_p) = p \Rightarrow \xi_P = (-\theta_{ij} \log (1-p))^{1/\delta}, \ \Rightarrow \theta_{ij}^{1/\delta} \approx \xi_{0.632}, \ i, j = 1, 2. \tag{8.13}$$

For each failure cause, Weibull CEM is assumed. Failure times and failure causes of test units are observed jointly and continuously.

From the CEM and Weibull distributed life assumptions, the CDF of failure cause, $j = 1, 2$, under a simple time step-stress test is the Khamis-Higgins model given by:

$$G_j(w) = G_j(w; \theta_{1j}, \theta_{2j})$$

$$= \begin{cases} 1 - \exp\left\{ \dfrac{-w^\delta}{\theta_{1j}} \right\} & \text{if } 0 < w < \tau \\[3mm] 1 - \exp\left\{ -\dfrac{\tau^\delta}{\theta_{1j}} - \dfrac{(w^\delta - \tau^\delta)}{\theta_{2j}} \right\} & \text{if } \tau \le w < \infty \end{cases} \tag{8.14}$$

Since only the smaller of W_1 and W_2 is observed, let the overall failure time of a test unit be

$$W = \min \{W_1, W_2\}.$$

The CDF and PDF of W are

$$F(w) = F(w; \theta)$$

$$= 1 - (1 - G_1(w))(1 - G_2(w))$$

$$= \begin{cases} 1 - \exp\left\{ -\left(\dfrac{1}{\theta_{11}} + \dfrac{1}{\theta_{12}} \right) w^\delta \right\} & \text{if } 0 < w < \tau, \\[3mm] 1 - \exp\left\{ -\left(\dfrac{1}{\theta_{11}} + \dfrac{1}{\theta_{12}} \right) \tau^\delta - \left(\dfrac{1}{\theta_{21}} + \dfrac{1}{\theta_{22}} \right) (w^\delta - \tau^\delta) \right\} & \text{if } \tau \le w < \infty \end{cases} \tag{8.15}$$

$$f(w) = f(w; \theta)$$

$$
= \begin{cases}
\delta w^{\delta-1} \left(\dfrac{1}{\theta_{11}} + \dfrac{1}{\theta_{12}} \right) \exp \left\{ -\left(\dfrac{1}{\theta_{11}} + \dfrac{1}{\theta_{12}} \right) w^{\delta} \right\} & \text{if } 0 < w < \tau, \\[3em]
\delta w^{\delta-1} \left(\dfrac{1}{\theta_{21}} + \dfrac{1}{\theta_{22}} \right) \exp \left\{ \begin{aligned} &-\left(\dfrac{1}{\theta_{11}} + \dfrac{1}{\theta_{12}} \right) \tau^{\delta} \\ &-\left(\dfrac{1}{\theta_{21}} + \dfrac{1}{\theta_{22}} \right) (w^{\delta} - \tau^{\delta}) \end{aligned} \right\} & \text{if } \tau \le w < \infty
\end{cases}
\tag{8.16}
$$

respectively, where $\theta = (\theta_1, \theta_2)$ with $\theta_i = (\theta_{i1}, \theta_{i2})$ for $i = 1, 2$. Furthermore, let j denote the indicator for the cause of failure. For $j, j' = 1, 2$ and $j' \neq j$, the joint PDF of (W, C) is given by:

$$f_{w,c}(w, j) = g_j(w)(1 - G_{j'}(w))$$

$$
= \begin{cases}
\dfrac{\delta w^{\delta-1}}{\theta_{1j}} \exp \left\{ -\left(\dfrac{1}{\theta_{11}} + \dfrac{1}{\theta_{12}} \right) w^{\delta} \right\} & \text{if } 0 < w < \tau, \\[3em]
\dfrac{\delta w^{\delta-1}}{\theta_{2j}} \exp \left\{ \begin{aligned} &-\left(\dfrac{1}{\theta_{11}} + \dfrac{1}{\theta_{12}} \right) \tau^{\delta} \\ &-\left(\dfrac{1}{\theta_{21}} + \dfrac{1}{\theta_{22}} \right) (w^{\delta} - \tau^{\delta}) \end{aligned} \right\} & \text{if } \tau \le w < \infty.
\end{cases}
\tag{8.17}
$$

The relative risk imposed on a test unit before τ and due to failure cause j is denoted by

$$\pi_{1j} = \Pr[C = j \mid 0 < W < \tau] = \frac{\theta_{1j}^{-1}}{\theta_{11}^{-1} + \theta_{12}^{-1}}, \ j = 1, 2. \tag{8.18}$$

Similarly, the relative risk after τ due to the cause j is denoted by

$$\pi_{2j} = \Pr[C = j \mid W \ge \tau] = \frac{\theta_{2j}^{-1}}{\theta_{21}^{-1} + \theta_{22}^{-1}}, \ j = 1, 2. \tag{8.19}$$

These equations are simply the proportion of failure rates in the given time frame. It follows from Equations 8.11 through 8.13 that W and C are independent given the time frame in which a failure has occurred. For $j = 1, 2$, let

n_{1j} = number of units failing before τ due to failure cause j,
n_{2j} = number of units failing after τ due to failure cause j.

Under the assumption of the CEM, the likelihood function of θ based on the Type-I censored sample is:

$$L(\theta) = \prod_{i=1}^{n_{11}} \left[\frac{\delta w_i^{\delta-1}}{\theta_{11}} e^{-w_i^\delta/\theta_{1\bullet}} \right] \prod_{i=1}^{n_{12}} \left[\frac{\delta w_i^{\delta-1}}{\theta_{12}} e^{-w_i^\delta/\theta_{1\bullet}} \right]$$

$$\times \prod_{i=1}^{n_{21}} \left[\frac{\delta w_i^{\delta-1}}{\theta_{21}} e^{-\frac{\tau^\delta}{\theta_{1\bullet}} - \frac{w_i^\delta - \tau^\delta}{\theta_{2\bullet}}} \right] \prod_{i=1}^{n_{22}} \left[\frac{\delta w_i^{\delta-1}}{\theta_{22}} e^{-\frac{\tau^\delta}{\theta_{1\bullet}} - \frac{w_i^\delta - \tau^\delta}{\theta_{2\bullet}}} \right] \quad (8.20)$$

$$\times \left[e^{-\left\{ \frac{n_c(T^\delta - \tau^\delta)}{\theta_{2\bullet}} + \frac{n_c \tau^\delta}{\theta_{1\bullet}} \right\}} \right],$$

where
$$\frac{1}{\theta_{1\bullet}} = \frac{1}{\theta_{11}} + \frac{1}{\theta_{12}},$$

$$\frac{1}{\theta_{2\bullet}} = \frac{1}{\theta_{21}} + \frac{1}{\theta_{22}},$$

$n_{1\bullet} = n_{11} + n_{12},$

$n_{2\bullet} = n_{21} + n_{22},$

$n = n_{1\bullet} + n_{2\bullet} + n_c,$

n is fixed and known.

The authors estimated model parameters and obtained optimum plan for the time-censored SSALTs which minimizes the sum over all causes of failure of asymptotic variances of the MLEs of the log characteristics life at design stress. The inferential procedures involving design parameters also were studied.

8.2.3 MODIFIED RAMP-STRESS ACCELERATED LIFE TEST WITH INDEPENDENT CAUSES OF FAILURES

Modified ramp-stress loading proposed by Srivastava and Gupta (2015) requires one test chamber. The stress is increased at low constant stress rate starting from the normal operating stress level, s_0 for example to the stress level, s_1, for example. Thereafter, it is increased at the higher constant stress rate until the termination of the experiment.

Modified ramp-stress ALT is designed using a generalized formulation of the CEM wherein stress $V(t) = k_1 t$, where k_1 is the rate of increase of stress.

Figure 8.4 shows this stress pattern.

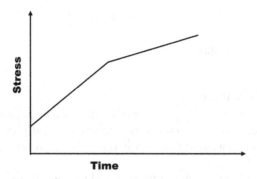

FIGURE 8.4   Modified ramp-stress loading.

Consider the life distribution $F_0(t ; \mathcal{V}, x)$ that depends on constant stress, $\mathcal{V}$ (voltage, for example) and other variables through the scale parameter $\alpha(\mathcal{V}, x)$ that is a function of $\mathcal{V}, x$, and coefficients:

$$F_0(t ; \mathcal{V}, x) = G\left(\frac{t}{\alpha(\mathcal{V}, x)}\right) \tag{8.21}$$

where the scale parameter is set equal to unity in the assumed CDF, $G(\cdot)$.

$$\Rightarrow F_0(t_I) = G(\varepsilon), \tag{8.22}$$

where

$$t_I = \Delta_1 + \Delta_2 + \dots + \Delta_i + \dots + \Delta_I \tag{8.23}$$

is the time after I steps in step-stress testing with step i at stress level $\mathcal{V}_i$ for a time:

$$\Delta_i = t_i - t_{i-1}, \tag{8.24}$$

with $t_0 = 0$, and

$$\varepsilon = \frac{\Delta_1}{\alpha(\mathcal{V}_1, x)} + \frac{\Delta_2}{\alpha(\mathcal{V}_2, x)} + \dots + \frac{\Delta_i}{\alpha(\mathcal{V}_i, x)} + \dots + \frac{\Delta_I}{\alpha(\mathcal{V}_I, x)} \tag{8.25}$$

as the cumulative exposure for the failure mode.

When $\alpha(\mathcal{V}, x) = \alpha(\mathcal{V}(t), x)$, that is, the stress, $\mathcal{V}$, is a function of time then, using Equation 8.25, the corresponding cumulative exposure function is given by

$$\varepsilon(t) = \lim_{\Delta_i \to 0}\left[\frac{\Delta_1}{\alpha(\mathcal{V}_1, x)} + \frac{\Delta_2}{\alpha(\mathcal{V}_2, x)} + \dots + \frac{\Delta_i}{\alpha(\mathcal{V}_i, x)} + \dots + \frac{\Delta_I}{\alpha(\mathcal{V}_I, x)}\right].$$

$$\tag{8.26}$$

$$= \int_0^t \frac{dt}{\alpha(\mathcal{V}(t), x)}$$

$$\Rightarrow F_0\big(t \; ; \mathcal{V}(t), x\big) = G\big(\varepsilon(t)\big) \tag{8.27}$$

is the **generalized formulation of** CEM.

8.2.3.1 Model Illustration

Srivastava and Gupta (2017) explored formulation of the optimum time-censored ALT model under modified ramp-stress loading when different failure causes have independent exponential life distributions. Their procedure is explained as follows.

Suppose each unit fails by one of the two fatal risk factors and the time to failure by each competing risk has an independent exponential life distribution obeying the linear CEM. $\varepsilon(t)$ at time t for $0 < t \le \tau_1$ and $\tau_1 < t \le \eta$ has been obtained under stress level s as:

$$\varepsilon(t) = \int_0^t \frac{1}{\theta(s(y))}\,dy = \int_0^t \frac{1}{e^{\gamma_{0j}}\left(\dfrac{s_0}{s(y)}\right)^{\gamma_{1j}}}\,dy = \int_0^t \frac{1}{e^{\gamma_{0j}}\left(\dfrac{s_0}{s_0 + y\beta_1}\right)^{\gamma_{1j}}}\,dy,$$

$$= \frac{e^{-\gamma_{0j}} s_0^{-\gamma_{1j}} \big((s_0 + \beta_1 t)^{1+\gamma_{1j}} - (s_0)^{1+\gamma_{1j}}\big)}{\beta_1(1+\gamma_{1j})} = W_{1j}(t),\ 0 < t \le \tau_1 \tag{8.28}$$

and

$$\varepsilon(t) = \int_0^t \frac{1}{\theta(s(y))}\,dy = \varepsilon(\tau_1) + \int_{\tau_1}^t \frac{1}{e^{\gamma_{0j}}\left(\dfrac{s_0}{s_1 + \beta_2(y-\tau_1)}\right)^{\gamma_{1j}}}\,dy$$

$$= \varepsilon(\tau_1) + \frac{e^{-\gamma_{0j}} s_0^{-\gamma_{1j}} \big((s_1 + \beta_2(t-\tau_1))^{1+\gamma_{1j}} - s_1^{1+\gamma_{1j}}\big)}{\beta_2(1+\gamma_{1j})} \tag{8.29}$$

$$= \varepsilon(\tau_1) + W_{2j}(t), \tau_1 < t \le \eta.$$

Then the CDF of failure cause j ($j = 1,2$) under modified ramp-stress is:

$$G_j(t) = J(\varepsilon(t)), \tag{8.30}$$

where:
$J(\cdot)$ is the exponential CDF with mean θ set equal to one and
$\varepsilon(t)$ is the cumulative exposure (damage) function.

Hence, the life distribution under modified ramp-stress loading corresponding to failure cause j ($j = 1, 2$) is:

$$G_j(t) \equiv G_j(t; \gamma_{0j}, \gamma_{1j})$$

$$= \begin{cases} 1 - \exp\{-W_{1j}(t)\}, & \text{if } 0 < t < \tau_1 \\ 1 - \exp\{-W_{1j}(\tau_1) - W_{2j}(t)\}, & \text{if } \tau_1 \le t < \infty \end{cases} \tag{8.31}$$

and the failure density is:

$$g_j(t) \equiv g_j(t; \gamma_{0j}, \gamma_{1j})$$

$$= \begin{cases} \exp\{-W_{1j}(t)\} W'_{1j}(t), & \text{if } 0 < t < \tau_1 \\ \exp\{-W_{1j}(\tau_1) - W_{2j}(t)\} W'_{2j}(t), & \text{if } \tau_1 \le t < \infty \end{cases} \tag{8.32}$$

Let $T = \min\{T_1, T_2\}$ denote the overall failure time of a test unit, then its CDF and PDF, respectively, are

$$F(t) \equiv F(t; \gamma_{0j}, \gamma_{1j})$$

$$= 1 - (1 - G_1(t))(1 - G_2(t))$$

$$= \begin{cases} 1 - \exp\{-W_{11}(t) - W_{12}(t)\}, & \text{if } 0 < t < \tau_1 \\ 1 - \exp\{-W_{11}(\tau_1) - W_{21}(t) - W_{12}(\tau_1) - W_{22}(t)\}, & \text{if } \tau_1 \le t < \infty \end{cases} \tag{8.33}$$

$$f(t) \equiv f(t; \gamma_{0j}, \gamma_{1j})$$

$$= \begin{cases} \exp\{-W_{11}(t) - W_{12}(t)\}(W'_{11}(t) + W'_{12}(t)), & \text{if } 0 < t < \tau_1 \\ \exp\{-W_{11}(\tau_1) - W_{21}(t) - W_{12}(\tau_1) - W_{22}(t)\}(W'_{21}(t) + W'_{22}(t)), & \text{if } \tau_1 \le t < \infty \end{cases} \tag{8.34}$$

Furthermore, let the indicator for the cause of failure be denoted by j. Then, under assumptions, for $j, j' = 1, 2$ and $j' \ne j$, the joint PDF of (T, C) is given by:

$$f_{T,C}(t, j) = g_j(t)(1 - G_{j'}(t))$$

$$= \begin{cases} \exp\{-W_{1j}(t)\} \exp\{-W_{1j'}(t)\}(W'_{1j}(t)), & \text{if } 0 < t < \tau_1 \\ \exp\{-W_{1j}(\tau_1) - W_{2j}(t)\} \exp\{-W_{1j'}(\tau_1) - W_{2j'}(t)\}(W'_{2j}(t)), & \text{if } \tau_1 \le t < \infty \end{cases} \tag{8.35}$$

The relative risk imposed on a test unit before τ_1 due to failure cause j for $j = 1,2$ is denoted by:

$$\pi_{1j} = \Pr[C = j \mid 0 < T < \tau_1] = \frac{\displaystyle\int_0^{\tau_1} \exp\{-W_{1j}(t)\} \exp\{-W_{1j'}(t)\} \, (W'_{1j}(t)) \, dt}{1 - \exp\{-W_{11}(\tau_1) - W_{12}(\tau_1)\}}. \quad (8.36)$$

Similarly, the relative risk after τ_1 due to the cause j for $j = 1,2$ is denoted by:

$$\pi_{2j} = \Pr[C = j \mid T \geq \tau_1]$$

$$= \frac{\displaystyle\int_{\tau_1}^{\infty} \exp\{-W_{1j}(\tau_1) - W_{2j}(t)\} \exp\{-W_{1j'}(\tau_1) - W_{2j'}(t)\} \, (W'_{2j}(t)) \, dt}{\exp\{-W_{11}(\tau_1) - W_{12}(\tau_1)\}} \quad (8.37)$$

Define for $j = 1, 2$,

n_{1j} is the number of units that fail before τ_1 due to the failure cause j,
n_{2j} is the number of units that fail after τ_1 due to the failure cause j.

Using Equation 8.36, the likelihood function is:

$$L(\gamma_{0j}, \gamma_{1j}) = \prod_{j=1}^{2} \prod_{i=1}^{n_{1j}} f(t_i, c_i) \prod_{j=1}^{2} \prod_{i=1}^{n_{2j}} f(t_i, c_i) (1 - F(\eta))^{n_c}$$

$$= \prod_{i=1}^{n_{11}} \left[\exp\{-W_{11}(t_i)\} \exp\{-W_{12}(t_i)\} \, (W'_{11}(t_i)) \right]$$

$$\prod_{i=1}^{n_{12}} \left[\exp\{-W_{12}(t_i)\} \exp\{-W_{11}(t_i)\} \, (W'_{12}(t_i)) \right] \quad (8.38)$$

$$\prod_{i=1}^{n_{21}} \left[\exp\{-W_{11}(\tau_1) - W_{21}(t_i)\} \exp\{-W_{12}(\tau_1) - W_{22}(t_i)\} \, (W'_{21}(t_i)) \right]$$

$$\prod_{i=1}^{n_{22}} \left[\exp\{-W_{12}(\tau_1) - W_{22}(t_i)\} \exp\{-W_{11}(\tau_1) - W_{21}(t_i)\} \, (W'_{22}(t_i)) \right]$$

$$e^{-n_c\{W_{11}(\tau_1) + W_{21}(\eta) + W_{12}(\tau_1) + W_{22}(\eta)\}}.$$

$n = n_{11} + n_{12} + n_{21} + n_{22} + n_c$, n is fixed and known.

The model parameters have been estimated and the optimal plan reveals relevant experimental variables, namely, stress rate and stress rate change point(s) using D-optimality criterion, which consists in finding out the optimal stress rate and the optimal stress rate change point by maximizing the logarithm of the determinant of the Fisher information matrix to the base 10. This criterion is motivated by the fact that the volume of the joint confidence region of model parameters is inversely proportional to the square root of the determinant of the Fisher information matrix. The method developed has been explained using a numerical example. The results of sensitivity analysis show that the plan is robust to small deviations from the true values of baseline parameters.

Srivastava and Gupta (2018) also formulated the **triangular cyclic-stress ALT plan** with independent competing failure modes.

8.3 ACCELERATED LIFE TESTS WITH DEPENDENT CAUSES OF FAILURES

The competing failure modes usually are dependent. The literature about dependent competing failure modes is rare in engineering but is available in biostatistics and econometrics. Models with copulas have become increasingly popular for modeling multivariate survival data. Carriere (1994) and Escarela and Carriere (2003) modeled dependence between two failure times by a two-dimensional copula. Carriere (1994) used a bivariate Gaussian copula to model the effect of complete elimination of one of two competing causes of death on human mortality. In Escarela and Carriere (2003), the bivariate Frank copula was fitted to a prostate cancer data set. Ancha and Yincai (2012) have introduced copula in reliability and analyzed ALT data with dependent multiple failure modes. Bunea and Mazucchi (2014) have applied the copula-based ALT competing risk model to Nelson's motorettes data.

8.3.1 COPULAS AND THEIR PROPERTIES

Copulas help to model dependence between two failure modes. The dependence structure relates the known marginal distributions of failure modes to their bivariate distribution (Nelsen 2006). The kind of dependence structure depends upon the choice of an appropriate copula.

A probabilistic way to define the copula is provided by Sklar (1959).

Let X, Y be random variables with continuous distributions F_1, F_2 and survival functions $S_1 = \bar{F}_1$ and $S_2 = \bar{F}_2$, respectively. The joint distribution and survival functions are $H(x, y)$ and $S(x, y)$, respectively. There exists a unique 2-dimensional copula C such that for all x in R^2:

$$H(x_1, x_2) = C(F_1(x_1), F_2(x_2))$$

and conversely, if C is a two-dimensional copula and F_1, F_2 are distribution functions, then H is a two-dimensional distribution function with marginals F_1, F_2.

Survival Copula

$$S(x, y) = P\big[X > x, Y > y\big] = 1 - F_1(x) - F_2(y) + H(x,y)$$

$$= \bar{F}_1(x) + \bar{F}_2(y) - 1 + C(F_1(x), F_2(y))$$

$$= \bar{F}_1(x) + \bar{F}_2(y) - 1 + C(1 - \bar{F}_1(x), 1 - \bar{F}_2(y)) \qquad (8.39)$$

$$= \bar{C}(\bar{F}_1(x), \bar{F}_2(y))$$

with $\bar{C}(u,v) = u + v - 1 + C(1 - u, 1 - v)$.

Sklar's Theorem leads to the following relationships:

1. $P\big[X \leq x, Y > y\big] = F_1(x) - C(F_1(x), F_2(y)),$
2. $P\big[X > y, Y \leq y\big] = F_2(x) - C(F_1(x), F_2(y)),$
3. $P\big[X \leq x | Y \leq y\big] = \dfrac{C(F_1(x), F_2(y))}{F_2(y)},$
4. $P\big[X \leq x | Y > y\big] = \dfrac{F_1(x) - C(F_1(x), F_2(y))}{1 - F_2(y)},$
5. $P\big[X \leq x | Y = y\big] = C_{1|2}(F_1(x), F_2(y)) = \dfrac{\partial\big(C(u,v)\big)}{\partial v}\Big|_{u = F_1(x), v = F_2(y)},$
6. $P\big[X > x | Y > y\big] = \dfrac{\bar{C}(\bar{F}_1(x), \bar{F}_2(y))}{\bar{F}_2(y)}.$

There are many types of copula functions, such as **Gaussian copula**, **Student's t-copula**, **Frank copula**, **Clayton copula**, and **Gumbel copula**. Different copulas produce different dependence structures and the kind of dependence structure comes from the choice of an appropriate copula.

The **Gumbel-Hougaard copula** is given by:

$$C(u,v) = exp[-((-\log_e[u])^\theta + (-\log_e[v])^\theta)^{\frac{1}{\theta}}] \qquad (8.40)$$

where $\theta \in [1, \infty]$ characterizes the association between the two variables. The Gumbel-Hougaard copula is one-parametric and symmetrical. The Gumbel-Hougaard copula belongs to the family of **Archimedean copulas**, used widely because they can be constructed easily and many families belong to this class.

8.3.2 CONSTANT-STRESS ACCELERATED LIFE TEST BASED ON COPULAS

Ancha and Yincai (2012) proposed the CSALT using the Gumbel-Hougaard copula (see Equation 8.34) with exponential marginals. Bai et al. (2018) considered a dependent-competing risk model under a constant-stress setting using the Bivariate Pareto copula function with Lomax marginals and Type-II progressive censoring.

8.3.2.1 Model Illustration

The methodology used by Ancha and Yincai (2012) is described as follows.

Consider a k CSALT with two competing failure modes. At each stress level s_i, $i = 1,2,...,k$, several n_i systems are tested until r_i of them fail. (t_{i1}, c_{i1}), $(t_{i2}, c_{i2}),...,(t_{i r_i}, c_{i r_i})$ is the failure data, where c_{il} takes any integer in the set $\{1, 2\}$. $c_{il} = 1$ and $c_{il} = 2$ indicate the failure caused by failure modes 1 and 2, respectively.

Each failure mode has an exponential life distribution with hazard rate λ_{ij}, $i = 1,2,...,k; j = 1,2$ under stress s_i. Thus, Equation (8.39) gives

$$S_i(t) = \bar{C}(e^{-\lambda_{i1}t}, e^{-\lambda_{i2}t}) = e^{-(\lambda_{i1}^{\theta} + \lambda_{i2}^{\theta})^{(1/\theta)} \cdot t}$$

Under stress level s_i, the stress-life relationship is modeled using the log-linear equation:

$$\log(\mu_{ij}) = \alpha_j + \beta_j \phi(s_i), \tag{8.41}$$

where:

$\mu_{ij} = 1/\lambda_{ij}$, α_j and β_j are unknown parameters,
$\phi(s)$ is a given function of stress s.

This is a general formulation which contains the Arrhenius and inverse power law models as special cases; defined:

$$\delta_j(c_{il}) = \begin{cases} 1 & \text{if } c_{il} = j \\ 0 & \text{if } c_{il} \neq j, j = 1, 2 \end{cases} \tag{8.42}$$

Then the likelihood function due to failure mode 1 under stress level s_i is

$$L_{i1} = \prod_{l=1}^{r_i} \left\{ \lim_{\Delta t \to 0} \frac{P\left[(T_1 < T_2) \cap (t_{il} \leq T_1 < t_{il} + \Delta t)\right]}{\Delta t} \right\}^{\delta_1(c_{il})}$$

$$\times \left\{ \lim_{\Delta t \to 0} \frac{P\left[(T_2 < T_1) \cap (t_{il} \leq T_2 < t_{il} + \Delta t)\right]}{\Delta t} \right\}^{1 - \delta_1(c_{il})}$$

$$\times \left\{ P[T_1 > t_{ir_i}, T_2 > t_{ir_i}] \right\}^{n_i - r_i}$$

$$= \left(\frac{\lambda_{i1}}{\lambda_{i2}} \right)^{\theta g_{i1}} \lambda_{i2}^{\theta - r_i} \left(\lambda_{i1}^{\theta} + \lambda_{i2}^{\theta} \right)^{r_i \left(\frac{1}{\theta} - 1 \right)} \exp\left\{ -\left(\lambda_{i1}^{\theta} + \lambda_{i2}^{\theta} \right)^{\left(\frac{1}{\theta} \right)} \left(\sum_{l=1}^{r_i} t_{il} + (n - r_i) t_{ir_i} \right) \right\}$$

$$\tag{8.43}$$

and that due to failure mode 2 under stress level s_i is

$$L_{i2} = \left(\frac{\lambda_{i2}}{\lambda_{i1}}\right)^{\theta\, g_{i2}} \lambda_{i1}^{\theta - r_i} \left(\lambda_{i1}^{\theta} + \lambda_{i2}^{\theta}\right)^{r_i\left(\frac{1}{\theta}-1\right)} \exp - \left\{\left(\lambda_{i1}^{\theta} + \lambda_{i2}^{\theta}\right)^{\left(\frac{1}{\theta}\right)} \left(\sum_{l=1}^{r_i} t_{il} + (n - r_i)t_{in}\right)\right\}$$

$$(8.44)$$

where:

$g_{ij} = \sum_{l=1}^{r_i} \delta_j(c_{il})$

$g_{i1} = r_i - g_{i2}$

Therefore, the log-likelihood function under the stress s_i is:

$\log L_i = \log L_{i1} + \log L_{i2}$

$$\Rightarrow \log L_i = 2\left[\theta\, g_{i1} \log \lambda_{i1} + \theta\left(r_i - g_i \log \lambda_{i2}\right) + \left(\frac{1}{\theta} - 1\right)\log\left(\lambda_{i1}^{\theta} + \lambda_{i2}^{\theta}\right) - \left(\lambda_{i1}^{\theta} + \lambda_{i2}^{\theta}\right)^{\frac{1}{\theta}} TTT_i\right]$$

$$(8.45)$$

where $TTT_i = \sum_{l=1}^{r_i} t_{il} + (n - r_i)t_{in}$.

Ancha and Yincai (2012) estimated the model parameters and compared via simulation the results for the dependent and the independent failure modes. They applied CSALT to the data set on insulated system of electromotors from Klein and Basu (1981) using the Gumbel copula with exponential marginals. The original data consists of three failure modes: turn failure, phase failure, and ground failure. 323 K and 423 K are the two temperatures used, and the four accelerated temperatures are 453 K, 463 K, 493 K, and 513 K. They obtained the MLEs of mean life times at normal operating conditions—323 K and 423 K.

8.3.3 Constant-Stress Partially Accelerated Life Test Based on Copulas

Srivastava and Gupta (2019) designed constant-stress PALT (CSPALT) using the Gumbel-Hougaard copula. The formulation is based on tampered failure rate model under a constant-stress set-up (Srivastava and Sharma 2014) and Type-I censoring.

The tampered failure rate (TFR) model assumes that changing the acceleration factor in different test chambers has a multiplicative effect on initial failure rate function. Thus, for $(m + 1)$ test chambers including the one in which items are tested under normal operating condition, the TFR model is:

$$h^*(y) = \begin{cases} h(y) & \text{under used condition} \\ h_j(y) = A_j h_{j-1}(y) = \prod_{i=1}^{j} A_i h(y) \text{ at } j\text{th stress level, } j = 1,2,\ldots,m \end{cases}$$

$$(8.46)$$

where the acceleration factor A_i (>1), $i = 1, 2,\ldots, m$ is assumed to be a parameter of the model. This contrasts with a fully ALT model wherein a regression model on the accelerating variable is specified (see also, Srivastava [2017]).

8.3.3.1   Model Illustration

The methodology used is described as follows.

Under the partially accelerated environmental condition using the constant-stress tampered failure rate model with $m = 1$ (Srivastava and Sharma 2014) and the fact that exponential distribution has a constant hazard rate, λ_j, the CDF of T_j, $j = 1, 2$ is:

$$G_j(t) = \begin{cases} e^{-\int_0^t h(u)du} = 1 - e^{-\lambda_j t}, \text{under normal operating condition} \\[2mm] e^{-\int_0^t Ah(u)du} = 1 - e^{-A\lambda_j t}, \text{under accelerated condition} \end{cases} \qquad (8.47)$$

with pdf $g_j(t)$.

The joint survival probability for the case of two dependent competing failure modes is

$$S(t) = P[T > t] = P[\min\{T_1, T_2\} > t] = P[T_1 > t, T_2 > t] \qquad (8.48)$$

Under the tampered failure rate model and the Gumbel-Hougaard copula with exponential marginals, $S(t)$ is given as

$$S(t) = \begin{cases} \bar{C}(e^{-\lambda_1 t}, e^{-\lambda_2 t}) = e^{-(\lambda_1^\theta + \lambda_2^\theta)^{(1/\theta)} \cdot t} \text{ , under normal operating condition} \\[2mm] \bar{C}(e^{-A\lambda_1 t}, e^{-A\lambda_2 t}) = e^{-A(\lambda_1^\theta + \lambda_2^\theta)^{(1/\theta)} \cdot t} \text{ , under accelerated condition.} \end{cases} \qquad (8.49)$$

The probabilities of failure of a unit under different failure modes over different intervals are required for the formulation of the likelihood function and:

- Probability that a product fails under failure mode j in chamber 1 (normal operating conditions) is calculated using Equation (8.49) and Section 8.3.1(5) as:

$$\lim_{\Delta t \to 0} \frac{P[(T_j < T_k) \cap (t < T_j \leq t + \Delta t)]}{\Delta t}$$

$$= g_j(t) \, dt \left. \frac{\partial \bar{C}(u,v)}{\partial u} \right|_{u=\bar{G}_1(t), v=\bar{G}_2(t)} \qquad (8.50)$$

$$= (\lambda_1^\theta + \lambda_2^\theta)^{(1/\theta)-1} \lambda_j^\theta e^{-(\lambda_1^\theta + \lambda_2^\theta)^{(1/\theta)} \cdot t} dt, \, (j,k) = 1,2, j \neq k$$

- Probability that a product fails under failure mode j in chamber 2 (accelerated condition) is:

$$\lim_{\Delta t \to 0} \frac{P[(T_1 < T_2) \cap (t < T_1 \le t + \Delta t)]}{\Delta t} = g_1(t)\, dt\, \frac{\partial \bar{C}(u,v)}{\partial u}\bigg|_{u=\bar{G}_1(t), v=u=\bar{G}_2(t)} \tag{8.51}$$

$$= A(\lambda_1^\theta + \lambda_2^\theta)^{(1/\theta)-1} \lambda_j^\theta e^{-A(\lambda_1^\theta + \lambda_2^\theta)^{(1/\theta)} \cdot t}\, dt, (j,k) = 1,2, j \ne k$$

As $n\phi_1$ and $n\phi_2$ test units are allocated to the normal operating condition and accelerated condition, respectively, the likelihood function of λ_1, λ_2 and A with censoring time η using Equations 8.50 and 8.51 is:

$$L(\lambda_1, \lambda_2, A) = L_1 L_2, \tag{8.52}$$

where:

$$L_1 = \prod_{i=1}^{n\phi_1} \begin{bmatrix} \left\{ (\lambda_1^\theta + \lambda_2^\theta)^{(1/\theta)-1} \lambda_1^\theta e^{-(\lambda_1^\theta + \lambda_2^\theta)^{(1/\theta)} t_i} \right\}^{\delta_{11}} \\ \left\{ (\lambda_1^\theta + \lambda_2^\theta)^{(1/\theta)-1} \lambda_2^\theta e^{-(\lambda_1^\theta + \lambda_2^\theta)^{(1/\theta)} t_i} \right\}^{\delta_{12}} \\ \left\{ e^{-(\lambda_1^\theta + \lambda_2^\theta)^{(1/\theta)} \cdot \eta} \right\}^{1-\delta_{11}-\delta_{12}} \end{bmatrix}$$

$$L_2 = \prod_{i=1}^{n\phi_2} \begin{bmatrix} \left\{ A(\lambda_1^\theta + \lambda_2^\theta)^{(1/\theta)-1} \lambda_1^\theta e^{-A(\lambda_1^\theta + \lambda_2^\theta)^{(1/\theta)} \cdot t_i} \right\}^{\delta_{21}} \\ \left\{ A(\lambda_1^\theta + \lambda_2^\theta)^{(1/\theta)-1} \lambda_2^\theta e^{-A(\lambda_1^\theta + \lambda_2^\theta)^{(1/\theta)} \cdot t_i} \right\}^{\delta_{22}} \\ \left\{ e^{-A(\lambda_1^\theta + \lambda_2^\theta)^{(1/\theta)} \cdot \eta} \right\}^{1-\delta_{21}-\delta_{22}} \end{bmatrix}$$

where:

$$\delta_{mj} = \begin{cases} 1, & \text{cause of failure is } j \text{ in chamber } m \text{ (normal operating condition)} \\ 0, & \text{otherwise.} \end{cases}$$

Define as
Φm as the proportion of units that are allocated in chamber m, $m = 1, 2$ and $\Phi_1 + \Phi_2 = 1$.

The authors have estimated model parameters and obtained optimal plan that consists in finding the optimal allocation, $n_1 = n\, \Phi_1$, n the first test chamber in normal conditions using D-optimality criterion. The method developed has been explained using numerical example and sensitivity analysis were carried out.

8.3.4   STEP-STRESS ACCELERATED LIFE TEST BASED ON COPULAS

Zhou et al. (2018) have addressed the statistical analysis of an SSALT in the presence of dependent competing failure modes. The dependence structure among distributions of life times is constructed by the copula function with an unknown copula parameter. Under the CEM for SSALT with two assumed copulas, namely the Gumbel and Clayton copulas, an expectation maximization (EM) algorithm is developed to obtain MLEs of model parameters and the missing information principle is used to obtain their standard errors (SEs). SSALT is applied to the Y11X-1419 type of Aerospace Electrical Connector composed of contact element, insulator, and mechanical connection. Three kinds of failure modes—contact failure, insulation failure, and mechanical connection failure have been considered. For assessing the storage reliability of electrical connectors, the data are collected in an SSALT accelerated by temperature because it is the most important environmental factor which affects the storage reliability of the electromechanical components. They used the MLE method to estimate the parameters of the candidate copula functions, Akaike's information criterion to select optimal copula functions, and verified strong dependence among failure modes. The results of the case studies show that the method proposed is valid and effective for the statistical analysis of SSALT with dependent competing failure modes.

8.4   BAYESIAN APPROACH TO ACCELERATED LIFE TEST WITH COMPETING FAILURE MODE

Zhang and Mao (1998) and Bunea and Mazzuchi (2005, 2006) considered the analysis of ALT with competing failure modes from a Bayesian viewpoint. Bunea and Mazzuchi (2005, 2006) considered two Bayesian models: Exponential Gamma and the other with prior as an ordered Dirichlet distribution. Tan et al. (2009) proposed a Bayesian method for analyzing incomplete data obtained from CSALT when there are two or more failure modes, or competing failure modes.

8.5   CONCLUSION

This chapter is a brief review on formulation of ALT models with competing failure modes—independent or dependent. The stress loading factors used in the literature are constant, step-stress, modified ramp-stress, and triangular cyclic. In case of dependent failure modes, dependence is described through copulas. In the literature, ALT models have been designed by various authors using the classical approach or the Bayesian approach. Various authors carried out data analysis using different censoring schemes such as time-censoring, failure censoring, progressive censoring, and determined optimal plans. The methods developed also were explained using numerical examples.

REFERENCES

Ancha, X. and Yincai, T. (2012). Statistical analysis of competing failure modes in accelerated life testing based on assumed copulas. *Chinese Journal of Applied Probability and Statistics*, **28**, 51–62.

Bai, D.S. and Chun, Y.R. (1991). Optimum simple step-stress accelerated life tests with competing causes of failure. *IEEE Transactions on Reliability*, **40** (5), 622–627.

Bai, X., Shi, Y., Liu, Y., and Liu, B. (2018). Statistical analysis of dependent competing risks model in constant stress accelerated life testing with progressive censoring based on copula function. *Statistical Theory and Related Fields*, **2** (1), 48–57.

Balakrishnan, N. and Han, D. (2008). Exact inference for simple step-stress model with competing risks for failure from exponential distribution under Type-II censoring. *Journal of Statistical Planning and Inference*, **138**, 4172–4186.

Bessler, S., Chernoff, H., and Marshall, A.W. (1962). An optimal sequential accelerated life test. *Technometrics*, **4** (3), 367–379.

Bhattacharya, G.K. and Soejoeti, Z. (1989). A tampered failure rate model for step-stress accelerated life test. *Communications in Statistics—Theory and Methods*, **18** (5), 1627–1643.

Bunea, C. and Mazzuchi, T.A. (2005). Bayesian accelerated life testing under competing failure modes. *Proceedings of Annual Reliability and Maintainability Symposium*, Alexandria, VA, 152–157.

Bunea, C. and Mazzuchi, T.A. (2006). Competing failure modes in accelerated life testing. *Journal of Statistical Planning and Inference*, **136**, 1608–1620.

Bunea, C. and Mazzuchi, T.A. (2014). *Accelerated Life Tests: Analysis with Competing Failure Modes*. Wiley Stats, Reference: Statistics Reference Online, pp. 1–12.

Carriere, J. (1994). Dependent decrement theory. *Transactions, Society of Actuaries*, **XLVI**, 45–65.

Chernoff, H. (1962). Optimal accelerated life designs for estimation, accelerated life test. *Technometrics*, **4** (3), 381–408.

Craiu, R.V. and Lee, T.C.M. (2005). Model selection for the competing-risks model with and without masking. *Technometrics*, **47** (4), 457–467.

David, H.A. and Moeschberger, M.L. (1978). *The Theory of Competing Risks*. Griffin, London, UK.

DeGroot, M.H. and Goel, P.K. (1979). Bayesian estimation and optimal designs in partially accelerated life testing. *Naval Research Logistic Quarterly*, **26** (20), 223–235.

Donghoon, H. and Balakrishnan, N. (2010). Inference for a simple step-stress model with competing risks for failure from the exponential distribution under time constraint. *Computational Statistics & Data Analysis*, **54** (9), 2066–2081.

Elsayed, A.E. (2012). *Reliability Engineering*. John Wiley & Sons, Hoboken, NJ.

Escarela, G. and Carriere, J. (2003). Fitting competing risks with an assumed copula. *Statistical Methods in Medical Research*, **12** (4), 333–349.

Haghighi, F. (2014). Accelerated test planning with independent competing risks and concave degradation path. *International Journal of Performability Engineering*, **10** (1), 15–22.

Han, D. and Balakrishnan, N. (2010). Inference for a simple step-stress model with competing risks for failure from the exponential distribution under time constraint. *Computational Statistics and Data Analysis*, **54**, 2066–2081.

Herman, R.J. and Patell Rusi, K.N. (1971). Maximum likelihood estimation for multi-risk model. *Technometrics*, **13** (2), 385396. doi:10.1080/00401706.1971.10488792.

Khamis, I.H. and Higgins, J.J. (1998). New model for step-stress testing. *IEEE Transactions on Reliability*, **47** (2), 131–134.

Kim, C.M. and Bai, D.S. (2002). Analysis of accelerated life test data under two failure modes. *International Journal of Reliability, Quality and Safety Engineering*, **9**, 111–125.

Klein, J.P. and Basu, A.P. (1981). Weibull accelerated life tests when there are competing causes of failure. *Communications in Statistics Theory and Methods*, **10** (20), 2073–2100.

Klein, J.P. and Basu, A.P. (1982a). Accelerated life testing under competing exponential failure distributions. *IAPQR Transactions*, **7** (1), 1–20.

Klein, J.P. and Basu, A.P. (1982b). Accelerated life tests under competing Weibull causes of failure. *Communications in Statistics—Theory and Methods*, **11** (20), 2271–2286.

Liu, X. and Qiu, W.S. (2011). Modeling and planning of step-stress accelerated life tests with independent competing risks. *IEEE Transactions on Reliability*, **60** (4), 712–720.

McCool, J. (1978). Competing risk and multiple comparison analysis for bearing fatigue tests. *Tribology Transactions*, **21**, 271–284.

Moeschberger, M.L. and David, H.A. (1971). Life tests under competing causes of failure and the theory of competing risks. *Biometrics*, **27** (4), 909–933.

Nelsen, R.B. (2006). *An Introduction to Copulas,* 2nd ed. Springer Science + Business Media, New York.

Nelson, W.B. (1990). *Accelerated Testing: Statistical Models, Test Plans, and Data Analysis.* John Wiley & Sons, Hoboken, NJ.

Pascual, F.G. (2007). Accelerated life test planning with independent Weibull competing risks with known shape parameter. *IEEE Transactions on Reliability*, **56** (1), 85–93.

Shi, Y., Jin, L., Wei, C., and Yue, H. (2013). Constant-stress accelerated life test with competing risks under progressive type-II hybrid censoring. *Advanced Materials Research*, **712–715**, 2080–2083.

Sklar, A. (1959). Fonctions de répartition à n dimensions et leurs marges. *Publications de l'Institut de statistique de l'Université de Paris*, **8**, 229–231.

Srivastava, P.W. (2017). *Optimum Accelerated Life Testing Models with Time-varying Stresses.* World Scientific Publishing Europe, London, UK.

Srivastava, P.W. and Gupta, T. (2015). Optimum time-censored modified ramp-stress ALT for the Burr Type XII distribution with warranty: A goal programming approach. *International Journal of Reliability, Quality and Safety Engineering*, **22** (3), 23.

Srivastava, P.W. and Gupta, T. (2017). Optimum modified ramp-stress ALT plan with competing causes of failure. *International Journal of Quality and Reliability Management*, **34** (5), 733–746.

Srivastava, P.W. and Gupta, T. (2018). Optimum triangular cyclic-stress ALT plan with independent competing causes of failure. *International Journal of Reliability and Applications*, **19** (1), 43–58.

Srivastava, P.W. and Gupta, T. (2019). Copula based constant-stress PALT using tampered failure rate model with dependent competing risks. *International Journal of Quality and Reliability Management*, **36** (4), 510–525.

Srivastava, P.W. and Sharma, D. (2014). Optimum time-censored constant-stress PALTSP for the Burr Type XII distribution using tampered failure rate model. *Journal of Quality and Reliability Engineering*, **2014**, 564049, 13. doi:10.1155/2014/564049.

Srivastava, P.W., Shukla, R., and Sen, K. (2014). Optimum simple step-stress test with competing risks for failure using Khamis-Higgins model under Type-I censoring. *International Journal of Operational Research/Nepal*, **3**, 75–88.

Tan, Y., Zhang, C., and Cen, X. (2009). Bayesian analysis of incomplete data from accelerated life testing with competing failure modes. *8th International Conference on Reliability, Maintainability and Safety*, pp. 1268–1272. doi:10.1109/ICRMS.2009.5270049.

Wu, S.-J. and Huang, S.-R. (2017). Planning two or more level constant-stress accelerated life tests with competing risks. *Reliability Engineering and System Safety*, **158**, 1–8.

Yang, G. (2007). *Life Cycle Reliability Engineering.* John Wiley & Sons, Hoboken, NJ.

Yu, Z., Ren, Z., Tao, J., and Chen, X. (2014). Accelerated testing with multiple failure modes under several temperature conditions. *Mathematical Problems in Engineering*, 839042, 8. doi:10.1155/2014/839042.

Zhang, Z. and Mao, S. (1998). Bayesian estimator for the exponential distribution with the competing causes of failure under accelerated life test. *Chinese Journal of Applied Probability and Statistics*, **14** (1), 91–98.

Zhou, Y., Lu, Z., Shi, Y, and Cheng, K. (2018). The copula-based method for statistical analysis of step-stress accelerated life test with dependent competing failure modes. *Proceedings of the Institution of Mechanical Engineers, Part O: Journal of Risk and Reliability*, 1–18. doi:10.1177/1748006X18793251.

9 European Reliability Standards

Miguel Angel Navas, Carlos Sancho,
and Jose Carpio

CONTENTS

9.1 INTRODUCTION

The International Electrotechnical Commission (IEC) is a standardization organization in the fields of electrical, electronic, and related technologies. It is integrated by the national standardization bodies of each member country. The IEC includes 85 countries, including those of the European Union, Japan, and the United States, among others.

The IEC has a Technical Committee, TC56, whose current name is Dependability. The purpose of TC 56 is to prepare international standards for reliability (in its broadest sense), applicable in all technological areas. Reliability can be expressed in terms of the essential attributes of support such as availability, maintainability, etc. The standards provide systematic methods and tools for evaluating the reliability and management of equipment, services, and systems throughout their life cycles. As of June 2018, TC56 has 57 published standards in this area.

The standards cover generic aspects of administration of the reliability and maintenance program, tests and analytical techniques, software and system reliability, life cycle costs, technical risk analysis, and project risk management. This list includes standards related to product problems from reliability of components to guidance for reliability of systems engineering, standards related to process issues from technological risk analysis to integrated logistics support, and standards related to management issues from program management from reliability to administration for obsolescence.

9.2 CLASSIFICATION OF THE DEPENDABILITY STANDARDS OF THE INTERNATIONAL ELECTROTECHNICAL COMMISSION

The set of standards issued by the IEC enables handling a large part of the maintenance processes, under contrasted methods and metrics, backed by the rigor and scientific level applied in its preparation and very demanding review processes, prior to its publication. That is why IEC standards must be one of the essential sources adopted by maintenance engineers in their academic, scientific, and business activities.

A classification of the 57 current standards is presented, grouped according to their main application field, noting that many of them are complementary and others are alternatives in their use. So, it is necessary to carry out a complete analysis of the process for which standards are to be applied to make an appropriate selection of them.

In Table 9.1, the 57 standards are classified into 6 clusters according to their main field of application:

- *Management procedures*: There are 19 standards that cover different processes for application in the field of maintenance (design, life cycle, maintainability, logistics, risk, etc.), which include and develop the necessary procedures for their adoption and implementation on the assets to be maintained.
- *Establishment of requirements*: The eight standards include procedures for the specification of the reliability, maintainability, and availability requirements that the systems must comply with to be established from the design phase.
- *Test methods*: These 11 standards develop the application procedures of different tests for application to the systems to obtain real operating data, and thus to evaluate practically the behavior of the systems.
- *Method selection*: There are five standards that assist in establishing measurement metrics and in selecting the most appropriate methods for evaluating the reliability of each system based on quantitative and qualitative selection criteria.

TABLE 9.1

Classification of the Dependability Standards Issued by IEC

Cluster	Number of Standards
Management procedures	19
Establishment of requirements	8
Test methods	11
Method selection	5
Reliability evaluation methods	9
Statistical methods for reliability	5

- *Reliability evaluation methods*: The nine standards present an alternative method for evaluating the reliability of a system with a different approach. It is therefore necessary to properly choose the method to be applied, taking into account the specific characteristics of each system or equipment, since each method is more appropriate for a certain type of item.
- *Statistical methods for the evaluation of reliability*: These five standards must be applied together, since the selection of the specific statistical method depends on whether the system is repairable or not. All of them are strongly linked and must be used in an integrated manner.

9.3 MANAGEMENT PROCEDURES

These 19 standards provide maintenance managers with multiple tools to perform a comprehensive management of their activities with procedures of proven academic and business validation.

Table 9.2 presents the classification of the management procedures issued by the IEC in the field of Dependability:

- *Maintenance strategies*: These eight standards pose to potential maintenance engineers basic strategies to adopt in the management of activities and operational processes.
- *Data processing*: There are two specific standards apply to the collection, analysis, and presentation of the operating data of the systems.
- *Risk*: These three standards are for the implementation of risk management procedures.
- *Logistics*: One of the processes that has the most influence on the results of maintenance management and two standards have been developed for its treatment.
- *Improvement processes*: One standard has been developed to improve the reliability of the systems in operation.
- *Life cycle*: There are three standards that address the life cycle of systems and the impact on maintenance costs.

TABLE 9.2
Classification of the Management
Procedures Standards Issued by IEC

Cluster	Number of Standards
Maintenance strategies	8
Data processing	2
Risk	3
Logistics	2
Improvement processes	1
Life cycle	3

The eight standards of maintenance strategies are:

- *IEC 60300-1:2014: Dependability management—Part 1: Guidance for management and application (Edition 3.0)* establishes a framework for dependability management. It provides guidance on dependability management of products, systems, processes, or services involving hardware, software, and human aspects or any integrated combinations of these elements. It presents guidance on planning and implementation of dependability activities and technical processes throughout the life cycle taking into account other requirements such as those relating to safety and the environment. This standard gives guidelines for management and their technical personnel to assist them to optimize dependability.
- *IEC 60300-3-10:2001: Dependability management—Part 3-10: Application guide—Maintainability (Edition 1.0)* can be used to implement a maintainability program covering the initiation, development, and in-service phases of a product, which form part of the tasks in IEC 60300-2. It provides guidance on how the maintenance aspects of the tasks should be considered to achieve optimum maintainability.
- *IEC 60300-3-11:2009: Dependability management—Part 3-11: Application guide—Reliability centered maintenance (Edition 2.0)* provides guidelines for the development of failure management policies for equipment and structures using reliability centered maintenance (RCM) analysis techniques. This part serves as an application guide and is an extension of IEC 60300-3-10, IEC 60300-3-12 and IEC 60300-3-14. Maintenance activities recommended in all three standards, which relate to preventive maintenance, may be implemented using this standard.
- *IEC 61907:2009: Communication network dependability engineering (Edition 1.0)* gives guidance on dependability engineering of communication networks. It establishes a generic framework for network dependability performance, provides a process for network dependability implementation, and presents criteria and methodology for network technology designs, performance evaluation, security consideration, and quality of service measurement to achieve network dependability performance objectives. This standard is applicable to network equipment developers and suppliers, network integrators, and providers of network service functions for planning, evaluation, and implementation of network dependability.
- *IEC 62508:2010: Guidance on human aspects of dependability (Edition 1.0)*, provides guidance on the human aspects of dependability, and the human-centered design methods and practices that can be used throughout the whole system life cycle to improve dependability performance. This standard describes qualitative approaches.
- *IEC 62628:2012: Guidance on software aspects of dependability (Edition 1.0)* addresses the issues concerning software aspects of dependability and gives guidance on achievement of dependability in software performance influenced by management disciplines, design processes, and

application environments. It establishes a generic framework on software dependability requirements, provides a software dependability process for system life cycle applications, presents assurance criteria and methodology for software dependability design and implementation, and provides practical approaches for performance evaluation and measurement of dependability characteristics in software systems.

- *IEC 62673:2013: Methodology for communication network dependability assessment and assurance (Edition 1.0)* describes a generic methodology for dependability assessment and assurance of communication networks from a network life cycle perspective. It presents the network dependability assessment strategies and methodology for analysis of network topology, evaluation of dependability of service paths, and optimization of network configurations to achieve network dependability performance and dependability of service. It also addresses the network dependability assurance strategies and methodology for application of network health check, network outage control, and test case management to enhance and sustain dependability performance in network service operation. This standard is applicable to network service providers, network designers and developers, and network maintainers and operators for assurance of network dependability performance and assessment of dependability of service.

- *IEC TS 62775:2016: Application guidelines—Technical and financial processes for implementing asset management systems (Edition 1.0)*, which is a Technical specification, shows how the IEC dependability suite of standards, systems engineering, and the IFRS and IAS standards can support the requirements of asset management, as described by the ISO 5500x suite of standards.

The most relevant aspects of the two standards of data processing are summarized as follows:

- *IEC 60300-3-2:2004: Dependability management—Part 3-2: Application guide—Collection of dependability data from the field (Edition 2.0)* provides guidelines for the collection of data relating to reliability, maintainability, availability, and maintenance support performance of items operating in the field. It deals in general terms with the practical aspects of data collection and presentation and briefly explores the related topics of data analysis and presentation of results. Emphasis is on the need to incorporate the return of experience from the field in the dependability process as a main activity. The typing of the data is done according to the attributes of Table 9.3.

- *IEC 60706-3:2006: Maintainability of equipment—Part 3: Verification and collection, analysis and presentation of data (Edition 2.0)* addresses the collection, analysis, and presentation of maintainability-related data, which may be required during, and at the completion of, design and during item production and operation.

TABLE 9.3

Attributes of the Collection of Dependability Data from the Field

Attribute	Values
Respect to time	Continuous, discontinuous, etc.
Number of data	Complete or limited
Type of population	Finite, infinite, or hypothetical
Sample size	No sampling, random sampling, or stratified sampling
Type of data	Qualitative or quantitative
Data censorship	Uncensored, lateral censorship, or censorship by interval
Data validation	In origin, by supervisor, etc.
Data screening	Without screening or with screening standards

The most important aspects of the three standards dedicated to risk management are summarized as follows:

- *IEC/ISO 31010:2009: Risk management—Risk assessment techniques (Edition 1.0)* is a dual logo IEC/ISO supporting standard for ISO 31000 and provides guidance on selection and application of systematic techniques for risk assessment. This standard is not intended for certification, regulatory, or contractual use.
- *IEC 62198:2013: Managing risk in projects—Application guidelines (Edition 2.0)* provides principles and generic guidelines on managing risk and uncertainty in projects. In particular it describes a systematic approach to managing risk in projects based on ISO 31000, Risk management—Principles and guidelines. Guidance is provided on the principles for managing risk in projects, the framework and organizational requirements for implementing risk management, and the process for conducting effective risk management. This standard is not intended for the purpose of certification.
- *IEC TR 63039:2016: Probabilistic risk analysis of technological systems—Estimation of final event rate at a given initial state (Edition 1.0)* provides guidance on probabilistic risk analysis (hereinafter referred to as risk analysis) for the systems composed of electrotechnical items and is applicable (but not limited) to all electrotechnical industries where risk analyses are performed.

The two standards for logistics management are:

- *IEC 60300-3-12:2011: Dependability management—Part 3-12: Application guide—Integrated logistic support (Edition 2.0)* is an application guide for establishing an integrated logistic support (ILS) management system. It is intended to be used by a wide range of suppliers including large and small companies wishing to offer a competitive and quality item that is optimized for the purchaser and supplier for the complete life cycle of the item. It also includes common practices and logistic data analyses that are related to ILS.

* *IEC 62550:2017: Spare parts provisioning (Edition 1.0)* describes requirements for spare parts provisioning as a part of supportability activities that affect dependability performance so that continuity of operation of products, equipment, and systems for their intended application can be sustained. This document is intended for use by a wide range of suppliers, maintenance support organizations, and users and can be applied to all items.

The existing standard for improvement processes is:

* *IEC 61160:2005: Design review (Edition 2.0)*. This International Standard makes recommendations for the implementation of design review as a means of verifying that the design input requirements have been met and stimulating the improvement of the product's design. The intention is for it to be applied during the design and development phase of a product's life cycle.

And finally, the three standards developed for the life cycle are:

* *IEC 60300-3-3:2017: Dependability management—Part 3-3: Application guide—Life cycle costing (Edition 3.0)* establishes a general introduction to the concept of life cycle costing and covers all applications. Although costs incurred over the life cycle consist of many contributing elements, this document particularly highlights the costs associated with the dependability of an item. This standard forms part of an overall dependability management program as described in IEC 60300-1. Guidance is provided on life cycle costing for use by managers, engineers, finance staff, and contractors; it is also intended to assist those who may be required to specify and commission such activities when undertaken by others.
* *IEC 60300-3-15:2009: Dependability management—Part 3-15: Application guide—Engineering of system dependability (Edition 1.0)* provides guidance for an engineering system's dependability and describes a process for realization of system dependability through the system life cycle. This standard is applicable to new system development and for enhancement of existing systems involving interactions of system functions consisting of hardware, software, and human elements.
* *IEC 62402:2007: Obsolescence management—Application guide (Edition 1.0)*. This International Standard gives guidance for establishing a framework for obsolescence management and for planning a cost-effective obsolescence management process that is applicable through all phases of the product life cycle.

9.4 ESTABLISHMENT OF REQUIREMENTS

There are eight standards that allow establishing requirements in the design phase so that the systems and equipment have established a series of indicators and values such as reliability, availability, and maintainability and checks a posteriori the degree of compliance with them:

- *IEC 60300-3-4:2007: Dependability management—Part 3-4: Application guide—Guide to the specification of dependability requirements (Edition 2.0)* gives guidance on specifying required dependability characteristics in product and equipment specifications, together with specifications of procedures and criteria for verification. The guide includes advice on specifying quantitative and qualitative reliability, maintainability, and availability requirements. The main changes from the previous edition are the concept of systems has been included and the need to specify the dependability of the system and not just the physical equipment has been stressed; the need for verification and validation of the requirement has been included; differentiation has been made between requirements that can be measured and verified and validated, and goals, which cannot; and the content on availability, maintainability, and maintenance support has been updated and expanded to similar level of detail to reliability.
- *IEC 60300-3-14:2004: Dependability management—Part 3-14: Application guide—Maintenance and maintenance support (Edition 1.0)* describes a framework for maintenance and maintenance support and the various minimal common practices that should be undertaken. The guide outlines in a generic manner the management, processes, and techniques related to maintenance and maintenance support that are necessary to achieve adequate dependability to meet the operational needs of the customer. It is applicable to items, which include all types of products, equipment, and systems (hardware and associated software). Most of these require a certain level of maintenance to ensure that their required functionality, dependability, capability, economic, safety, and regulatory requirements are achieved.
- *IEC 60300-3-16:2008: Dependability management—Part 3-16: Application guide—Guidelines for specification of maintenance support services (Edition 1.0)* describes a framework for the specification of services related to the maintenance support of products, systems, and equipment that are carried out during the operation and maintenance phase. The purpose of this standard is to outline, in a generic manner, the development of agreements for maintenance support services as well as guidelines for the management and monitoring of these agreements by the company and the service provider.
- *IEC 60706-2:2006: Maintainability of equipment—Part 2: Maintainability requirements and studies during the design and development phase (Edition 2.0).* This part of IEC 60706 examines the maintainability requirements and related design and use parameter, and discusses some activities

necessary to achieve the required maintainability characteristics and their relationship to planning of maintenance. It describes the general approach in reaching these objectives and shows how maintainability characteristics should be specified in a requirements document or contract. It is not intended to be a complete guide on how to specify or to contract for maintainability. Its purpose is to define the range of considerations when maintainability characteristics are included as requirements for the development or the acquisition of an item.

- *IEC 61014:2003: Programmes for reliability growth (Edition 2.0)* specifies requirements and gives guidelines for the exposure and removal of weaknesses in hardware and software items for the purpose of reliability growth. It applies when the product specification calls for a reliability growth program of equipment (electronic, electromechanical, and mechanical hardware as well as software) or when it is known that the design is unlikely to meet the requirements without improvement. The main changes with respect to the previous edition are: a subclause on planning reliability growth in the design phase, a subclause on management aspects covering both reliability growth in design and the test phase, and a clause on reliability growth in the field.

- *IEC 62347:2006: Guidance on system dependability specifications (Edition 1.0)*. This International Standard gives guidance on the preparation of system dependability specifications. It provides a process for system evaluation and presents a procedure for determining system dependability requirements. This International Standard is not intended for certification or to perform conformity assessment for contractual purposes. It is not intended to change any rights or obligations provided by applicable statutory or regulatory requirements.

- *IEC 62741:2015: Demonstration of dependability requirements—The dependability case (Edition 1.0)* gives guidance on the content and application of a dependability case and establishes general principles for the preparation of a dependability case. This standard is written in a basic project context where a customer orders a system that meets dependability requirements from a supplier and then manages the system until its retirement. The methods provided in this standard may be modified and adapted to other situations as needed. The dependability case is normally produced by the customer and supplier but can also be used and updated by other organizations.

- *IEC 62853:2018: Open systems dependability (Edition 1.0)* provides guidance in relation to a set of requirements placed upon system life cycles in order for an open system to achieve open systems dependability This document is applicable to life cycles of products, systems, processes, or services involving hardware, software, and human aspects or any integrated combinations of these elements. For open systems, security is especially important since the systems are particularly exposed to attack. This document can be used to improve the dependability of open systems and to provide assurance that the process views specific to open systems achieve their expected outcomes. It helps an organization define the activities and tasks that need to be undertaken to achieve dependability objectives in an open system, including dependability related communication, dependability assessment, and evaluation of dependability throughout system life cycles.

9.5 TEST METHODS

There are 11 standards designed to help methodologically in the use of testing and testing procedures to obtain field data in a controlled manner and to serve as a basis for estimating the operation indicators that systems and equipment will have:

- *IEC 60605-2:1994: Equipment reliability testing—Part 2: Design of test cycles (Edition 1.0)* applies to the design of operating and environmental test cycles referred to in 8.1 and 8.2 of IEC 605-1.
- *IEC 60706-5:2007: Maintainability of equipment—Part 5: Testability and diagnostic testing (Edition 2.0)* provides guidance for the early consideration of testability aspects in design and development, and assists in determining effective test procedures as an integral part of operation and maintenance.
- *IEC 61070:1991: Compliance test procedures for steady-state availability (Edition 1.0)* specifies techniques for availability performance testing of frequently maintained items when the availability performance measure used is either steady-state availability or steady-state unavailability.
- *IEC 61123:1991: Reliability testing—Compliance test plans for success ratio (Edition 1.0)* specifies procedures for applying and preparing compliance test plans for success ratio or failure ratio. The procedures are based on the assumption that each trial is statistically independent.
- *IEC 61124:2012: Reliability testing—Compliance tests for constant failure rate and constant failure intensity (Edition 3.0)* gives a number of optimized test plans, the corresponding operating characteristic curves, and expected test times. In addition, the algorithms for designing test plans using a spreadsheet program are given, together with guidance on how to choose test plans. This standard specifies procedures to test an observed value of failure rate, failure intensity, meantime to failure (MTTF), and mean operating time between failures (MTBF). It provides an extensive number of statistical tests.
- *IEC 61163-1:2006: Reliability stress screening—Part 1: Repairable assemblies manufactured in lots (Edition 2.0)* describes particular methods to apply and optimize reliability stress screening processes for lots of repairable hardware assemblies in cases where the assemblies have an unacceptably low reliability in the early failure period, and when other methods such as reliability growth program and quality control techniques are not applicable.
- *IEC 61163-2:1998: Reliability stress screening—Part 2: Electronic components (Edition 1.0)* provides guidance on reliability stress screening techniques

and procedures for electronic components. Is intended for use of (1) component manufacturers as a guideline, (2) component users as a guideline to negotiate with component manufacturers on stress screening requirements or plan a stress screening process in house due to reliability requirements, and (3) subcontractors who provide stress screening as a service.

- *IEC 61164:2004: Reliability growth—Statistical test and estimation methods (Edition 2.0)* gives models and numerical methods for reliability growth assessments based on failure data, which were generated in a reliability improvement program. These procedures deal with growth, estimation, confidence intervals for product reliability, and goodness-of-fit tests. In Table 9.4, the types of model developed are classified.

- *IEC 62309:2004: Dependability of products containing reused parts—Requirements for functionality and tests (Edition 1.0)* introduces the concept to check the reliability and functionality of reused parts and their usage within new products. It also provides information and criteria about the tests/analysis required for products containing such reused parts, which are declared "qualified-as-good-as-new" relative to the designed life of the product. The purpose of this standard is to ensure by tests and analysis that the reliability and functionality of a new product containing reused parts is comparable to a product with only new parts.

- *IEC 62429:2007: Reliability growth—Stress testing for early failures in unique complex systems (Edition 1.0)*. This International Standard gives guidance for reliability growth during final testing or acceptance testing of unique complex systems. It gives guidance on accelerated test conditions and criteria for stopping these tests.

- *IEC 62506:2013: Methods for product accelerated testing (Edition 1.0)* provides guidance on the application of various accelerated test techniques for measurement or improvement of product reliability. Identification of potential failure modes that could be experienced in the use of a product/item and their mitigation is instrumental to ensure dependability of an item. The object of the methods is to either identify potential design weakness or provide information on item dependability, or to achieve necessary reliability/availability improvement, all within a compressed or accelerated period of time. This standard addresses accelerated testing of non-repairable and repairable systems.

TABLE 9.4

Attributes of the Collection of Dependability Data from the Field

Type of Model	Continuous Time	Discrete Time
Classic design	Section 6.1	—
Bayesian design	Section 6.2	—
Classic tests	Section 7.1	Section 7.2
Bayesian tests	—	—

9.6 METHOD SELECTION

There are five key standards, since the selection of methods to implement in a reliability program is a very individualized process, so much so that it is not possible to make a generic suggestion for the selection of one or more of the specific methods. The choice of the appropriate method should be made with the joint effort of experts in reliability and in the field of systems engineering. The selection should be made at the beginning of the development of the program and its applicability should be reviewed. These standards help in making the selection of the most appropriate method for a system or equipment.

- *IEC 60300-3-1:2003: Dependability management—Part 3-1: Application guide—Analysis techniques for dependability—Guide on methodology (Edition 2.0)* gives a general overview of commonly used dependability analysis techniques. It describes the usual methodologies, their advantages and disadvantages, data input, and other conditions for using various techniques. This standard is an introduction to selected methodologies and is intended to provide the necessary information for choosing the most appropriate analysis methods.

The 12 methods included are briefly explained in Annex A of the standard and reference is made to the IEC standard developed by each method, if any. This standard includes a guide for the selection of the appropriate analysis method taking into account the characteristics of the system or equipment:

- Complexity of the system
- Novelty of the system
- Quantitative analysis or qualitative analysis
- Single failure or multiple failures
- Behavior dependent on time or a sequence
- Existence of dependent events
- Analysis below—up or top-down
- Suitable for reliability assignment
- Required domain
- Acceptance and common use
- Need for tool support
- Credibility checks
- Availability of tools
- Normalization, referencing the seven methods with specific IEC standards

- *IEC 60300-3-5:2001: Dependability management—Part 3-5: Application guide—Reliability test conditions and statistical test principles (Edition 1.0)* provides guidelines for the planning and performing of reliability tests and the use of statistical methods to analyze test data. It describes the tests related to repaired and non-repaired items together with tests for constant and non-constant failure intensity and constant and non-constant failure rate.

This standard establishes the methods and conditions for reliability tests and principles for the performance of statistical tests. It includes a detailed guide for the selection of the statistical methods used to analyze the data coming from reliability tests of repairable or non-repairable elements.

The requirements for a correct specification of the reliability test to be executed are established so that all the variables that may affect the test are determined and bounded prior to the application of the statistical test and contrast methods.

The following standard focuses on the analysis of trial data. For the non-repairable elements, parametric methods adjusted to the exponential distribution are proposed for failure rate $\lambda(t)$ constant and adjusted to the Weibull distribution for $\lambda(t)$ with trend.

The statistical nature of failure modes in repairable elements is described as a stochastic point process (SPP). The failure intensity $z(t)$ refers exclusively to repairable elements. This means that the failure current of a single repairable element can be estimated using the successive times between failures. It is estimated by the number of failures per unit of time or another variable.

In this case, the failures of each element happen sequentially and this is known as an SPP. It is important to maintain the traceability of the sequence of times between failures. If the times between failures are distributed exponentially, then the failure current is constant. Therefore, the time between failures can be modeled by an exponential distribution. In this case, the number of failures per unit of time can be modeled by a homogeneous Poisson process (HPP).

In many cases where there is a trend in the failure intensity, the power-law process (PLP) can be applied. This leads to a model from which the trend can be estimated. If there is a trend (intensity of increasing or decreasing failure), a non-homogeneous Poisson process (NHPP) can be applied. See classification in Table 9.5.

Attached is a list of standards for the estimation of reliability in non-repairable elements according to IEC 60300-3-5:

- Contrasts of the constant failure rate hypothesis: IEC 60605-6
- Point estimation and confidence intervals for the exponential distribution: IEC 60605-4
- Goodness of fit contrast for the Weibull distribution: IEC 61649
- Point estimation and confidence intervals for the Weibull distribution: IEC 61649
- Point estimation and confidence intervals for the binomial distribution: ISO 11453

TABLE 9.5

Appropriate Models for Data Analysis According to IEC 60300-3-5

Item	Trend	Appropriate Model
Non-repairable	Constant	Exponential distribution
Non-repairable	Non-constant	Weibull distribution
Repairable	Constant	Homogenous Poisson process (HPP)
Repairable	Non-constant	Non-homogenous Poisson process (NHPP)

Attached is a list of standards for estimating reliability in repairable items:

- Contrasts for constant failure intensity: IEC 60605-6
- Point estimation and confidence intervals for the exponential distribution: IEC 60605-4
- Estimation of the parameters and statistical contrast of the PLP: IEC 61710

- *IEC 60319:1999: Presentation and specification of reliability data for electronic components (Edition 3.0)* describes the information needed for characterizing reliability of a component and also the detailed requirements for reporting reliability data. It gives guidance to component users as to how they should specify their reliability requirements to component manufacturers. The data, derived from laboratory tests, should enable circuit and equipment designers to evaluate the reliability of circuits and systems.
- *IEC 61703:2016: Mathematical expressions for reliability, availability, maintainability and maintenance support terms (Edition 2.0)*, to account for mathematical constraints, splits the items between the individual items considered as a whole (e.g., individual components) and the systems made of several individual items. It provides general considerations for the mathematical expressions for systems as well as individual items but the individual items that are easier to model are analyzed in more detail with regard to their repair aspects. This standard is mainly applicable to hardware dependability, but many terms and their definitions may be applied to items containing software.

This standard provides the definitions related to reliability as well as the mathematical expressions that should be used in the calculations of the main variables. In this standard, the following classes of elements are considered separately:

- Non-repairable items
- Items repairable with time to zero restoration
- Repairable items with time to non-zero restoration

For non-repairable items, repairable items with time zero restoration, and repairable items with time to non-zero restoration develop and formulate the mathematical expressions:

- Reliability; $R(t)$
- Instantaneous failure rate; $\lambda(t)$ (non-repairable items)
- Instantaneous failure intensity; $z(t)$ (repairable items)
- Average failure rate; $\bar{\lambda}(t_1, t_2)$ (non-repairable items)
- Average failure intensity; $\bar{z}(t_1, t_2)$ (repairable items)
- Mean Time To Failure: MTTF (non-repairable items)
- Mean Up Time: MUT (repairable items)
- Mean Time Between Failures: MTBF (repairable items)

Likewise, and for the repairable items with time to the non-zero restoration, mathematical expressions are included for the calculation of availabilities and instantaneous, average and asymptotic unavailability, and maintainability, average repair rate, and average repair time.

- *IEC 62308:2006: Equipment reliability—Reliability assessment methods (Edition 1.0).* This International Standard describes early reliability assessment methods for items based on field data and test data for components and modules. It is applicable to mission, safety and business critical, high integrity, and complex items. It contains information on why early reliability estimates are required and how and where the assessment would be used.

9.7 RELIABILITY EVALUATION METHODS

Each of these nine standards develops a specific method for evaluating the reliability of a system or equipment. The selection of the most appropriate method and formulation must be made under the criteria specified in the selection standards of the previous section:

- *IEC 60812:2006: Analysis techniques for system reliability: Procedure for failure mode and effects analysis (FMEA) (Edition 2.0).* This International Standard describes Failure Mode and Effects Analysis (FMEA) and Failure Mode, Effects, and Criticality Analysis (FMECA) and gives guidance as to how they may be applied to achieve various objectives by providing the procedural steps necessary to perform analysis, identify appropriate terms, define basic principles, and provide examples of the necessary worksheets or other tabular forms.

The FMEA analysis is a top-down and qualitative reliability analysis method that is particularly suitable for the study of material, component, and equipment failures and their effects on the next higher functional level. The iterations of this step (identification of the single failure modes and the evaluation of their effects on the next higher level) produce the identification of all the single failure modes of the system.

The FMEA lends itself to the analysis of systems of different technologies (electrical, mechanical, hydraulic, software, etc.) with simple functional structures. The FMECA analysis extends the FMEA to include the criticality analysis, quantifying the effects of the failures in terms of probability of occurrence and their severity. The severity of the effects is assigned with respect to a specific scale.

Both FMECA and FMEA are normally carried out when a certain risk is foreseen in the program corresponding to the start of the development of a process or product. The factors that can be considered are new technology, new processes, new designs or changes in the environment, charges, or regulations. These analyses can be performed on components or systems that are part of products, processes, or manufacturing equipment. They also can be carried out on software systems.

The FMECA and FMEA methods generally follow the following steps:

- Identification of how the component of a system should work
- Identification of their potential failure modes, causes and effects

- Identification of the risk related to failure modes and their effects
- Identification of the recommended actions to eliminate or reduce the risk
- Follow up activities to close the recommended actions

- *IEC 61025:2006: Fault tree analysis (FTA) (Edition 2.0)* describes FTA and provides guidance on its application to perform an analysis, identifies appropriate assumptions, events and failure modes, and provides identification standards and symbols.

The FTA is a top-down approach to the analysis of the reliability of a product. It seeks the identification and analysis of the conditions and factors that cause, or contribute to, the occurrence of an undesired determined event and that may affect the operation, safety, economy, or other specified characteristics of the product.

The FTA also can be performed to provide a prediction model of the reliability of a system and allow cost-benefit studies in the design phase of a product. Used as a tool for the detection and quantitative evaluation of a cause of failure, the FTA represents an efficient method that identifies and evaluates the modes and causes of failure of known or suspected effects.

Taking into account the known unfavorable effects and the ability to find the respective modes and causes of failure, the FTA allows the timely mitigation of the potential failure modes, allowing the improvement of the reliability of the product in its design phase. Built to represent hardware and software architecture in addition to analyzing functionality, the FTA, developed to deal with basic events, becomes a systematic reliability modeling technique that considers the complex interactions between parts of a system through the modeling of its functional or fault dependencies, of events that trigger failures and of common cause events and allowing the representation of networks.

To estimate the reliability and availability of a system using the FTA technique, methods such as the Boolean reduction and the analysis of the cutting sets are used. The basic data that are required are the failure rates of the components, repair rates, probability of occurrence of failure modes, etc. FTA has a double application, as a means to identify a cause of a known failure and as a tool for analyzing failure modes and modeling and predicting reliability. The key elements of a fault tree are events or gates and cutting sets.

The gates represent results and the events represent entrances to the gates. The symbolic representation of some specific gates may vary from one textbook or analysis software to another; however, the representation of the basic gates is clearly universal (see Table 9.6).

- *IEC 61078:2016: Analysis techniques for dependability—Reliability block diagram and Boolean methods (Edition 3.0).* This International Standard describes the requirements to apply when RBDs are used in dependability analysis; the procedures for modeling the dependability of a system with reliability block diagrams; how to use RBDs for qualitative and quantitative analysis; the procedures for using the RBD model to calculate availability, failure frequency, and reliability measures for different types of systems with constant (or time dependent) probabilities of blocks success/failure,

TABLE 9.6

Symbols That Are Used in the Representation of the FTA Method

FTA Symbols	Event or Gate
▭ ▭	Higher or intermediate event
○	Basic event
◇	Undeveloped event
△	Transfer gate
OR gate symbol	OR gate
AND gate symbol	AND gate

and for non-repaired blocks or repaired blocks; and some theoretical aspects and limitations in performing calculations for availability, failure frequency, and reliability measures.

The RBD analysis is a method of analyzing a system by graphically representing a logical structure of a system in terms of subsystems or components. This allows the success paths of the system to be represented by the way in which the blocks (subsystems/components) connect logically (see Figure 9.1).

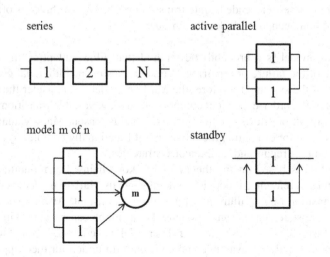

FIGURE 9.1 Elementary models for RBD analysis.

Block diagrams are among the first tasks that are completed during the definition of the product. They should be built as part of the initial conceptual development. They should be started as soon as the program definition exists, completed as part of the requirements analysis, and continuously extended to a greater level of detail, as the data becomes available to make decisions and perform cost-benefit studies.

To construct an RBD, several techniques of qualitative analysis can be used:

- Establish the definition of the success of the system
- Divide the system into appropriate functional blocks for the purpose of reliability analysis (some blocks can represent substructures of the system that, in turn, can be represented by other RBDs (system reduction))
- Carry out the qualitative analysis (there are several methods for the quantitative evaluation of an RBD. Depending on the type of structure (reducible or irreducible), simple Boolean techniques, truth tables, or analysis of cutoff or path sets can be used for the prediction of reliability and system availability values from the data of basic component)

You can evaluate more complex models in which the same block appears more than once in the diagram by using:

- The total probability theorem
- Truth Boolean tables

- *IEC 61165:2006: Application of Markov techniques (Edition 2.0).* This International Standard provides guidance on the application of Markov techniques to model and analyze a system and estimate reliability, availability, maintainability, and safety measures. This standard is applicable to all industries where systems, which exhibit state-dependent behavior, have to be analyzed. The Markov techniques covered by this standard assume constant time-independent state transition rates. Such techniques often are called homogeneous Markov techniques.

The Markov model is a probabilistic method that allows adapting the statistical dependence of the failure or repairing characteristics of the individual components to the state of the system. Therefore, the Markov model can consider the effects of the failures of the order-dependent components and the variable transition rates that change as a result of efforts or other factors. For this reason, Markov analysis is an adequate method for evaluating the reliability of functionally complex system structures and complex repair and maintenance strategies.

The method is based on the theory of Markov chains. For reliability applications, the normal reference model is the homogeneous Markov model over time that requires transition rates (failure and repair) to be constant. At the expense of the increase in the state space, non-exponential transitions can be approximated by a sequence of exponential transitions. For this model, general and efficient techniques of numerical methods are available and their only limitation for their application is the dimension of the state space.

The representation of the behavior of the system by means of a Markov model requires the determination of all possible states of the system, preferably represented graphically by means of a state transition diagram. In addition, transition rates (constants) from one state to another have to be specified (failure or repair rates of a component, event rates, etc.). The typical result of a Markov model is the probability of being in a given set of states (normally this probability is the measure of availability).

The appropriate field of application of this technique is when the transition rates (failure or repair) depend on the state of the system or vary with the load, the level of effort, the structure of the system (e.g., waiting), the policy of maintenance, or other factors. In particular, the structure of the system (cold or hot waiting, spare parts) and the maintenance policy (single or multiple repair equipment) induce dependencies that cannot be considered with other, less computationally intensive techniques. Typical applications are predictions of reliability/availability. For the application of this methodology, the following key steps must be taken into account:

- Definition of the space of the states of the system
- Assignment of transition rates between states (independent of time)
- Definition of the exit measures (group of states that lead to a system failure)
- Generation of the mathematical model (matrix of transition rates) and resolution of Markov models by using an appropriate software package
- Analysis of results

In Figure 9.2, the white circles represent operational states, while the gray circles represent non-operative states. λ_x are the transition failure rates from one state to another and μ_x are the step repair rates from one state to another.

- *IEC 61709:2017: Electric components—Reliability—Reference conditions for failure rates and stress models for conversion (Edition 3.0)* gives guidance on the use of failure rate data for reliability prediction of electric

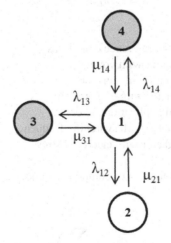

FIGURE 9.2 State transition diagram in Markov analysis.

components used in equipment. The method presented in this document uses the concept of reference conditions, which are the typical values of stresses that are observed by components in the majority of applications. Reference conditions are useful since they provide a known standard basis from which failure rates can be modified to account for differences in environment from the environments taken as reference conditions. Each user can use the reference conditions defined in this document or use their own. When failure rates stated at reference conditions are used it allows realistic reliability predictions to be made in the early design phase.

The stress models described herein are generic and can be used as a basis for conversion of failure rate data given at these reference conditions to actual operating conditions when needed and this simplifies the prediction approach. Conversion of failure rate data is only possible within the specified functional limits of the components. This document also gives guidance on how a database of component failure data can be constructed to provide failure rates that can be used with the included stress models.

Reference conditions for failure rate data are specified so that data from different sources can be compared on a uniform basis. If failure rate data are given in accordance with this document, then additional information on the specified conditions can be dispensed with. This document does not provide base failure rates for components—rather it provides models that allow failure rates obtained by other means to be converted from one operating condition to another operating condition. The prediction methodology described in this document assumes that the parts are being used within its useful life.

This international standard is intended for the prediction of reliability of components used in equipment and focuses on organizations with their own data, describing how to establish and use such data to make predictions of reliability. The failure rate of a component under operating conditions is calculated as follows:

$$\lambda = \lambda_{ref}\pi_U\pi_I\pi_T\pi_E\pi_S\pi_{ES} \tag{9.1}$$

with:

λ_b is the failure rate in the reference conditions
Π_U is the dependence factor with voltage
Π_I is the dependence factor with current
Π_T is the dependence factor with temperature
Π_E is the environmental application factor
Π_S is the dependence factor with switching frequency
Π_{ES} is the dependence factor with electrical stress

Therefore, the failure rate for sets of components under operating conditions is calculated as aggregation as follows:

$$\lambda_{Equip} = \sum_{i=1}^{n}(\lambda)_i \tag{9.2}$$

The standard develops specific stress models and values of the π factors applicable to the different types of components that must be used to convert the reference failure rates to failure rates in the operating conditions. The π factors are modifiers of the failure rate associated with a specific condition or effort. They provide a measure of the modification of the failure rate as a consequence of changes in the effort or condition.

- *IEC 61882:2016: Hazard and operability studies (HAZOP studies)—Application guide (Edition 2.0)* provides a guide for HAZOP studies of systems using guide words. It gives guidance on application of the technique and on the HAZOP study procedure, including definition, preparation, examination sessions and resulting documentation, and follow-up. Documentation examples illustrating HAZOP studies, as well as a broad set of examples encompassing various applications, are provided.

A HAZOP study is a detailed process of identification of hazards and operational problems carried out by a team. A HAZOP deals with the identification of potential deviations in the design proposal, examination of its possible causes, and evaluation of its consequences.

The basis of a HAZOP is a guide-words exam that constitutes a deliberate search for deviations in the design proposal. The proposed design contemplates the behavior of a system, its elements, and characteristics desired, or specified, by the designer. To facilitate the examination, the system is divided into parts so that the design proposal of each of the parties can be defined properly.

The design proposal of a given part of a system is expressed in terms of elements that convey the essential benefits of that part and that represent its natural divisions. The elements can be steps or steps of a procedure, individual signals and elements of equipment in a control system, equipment or components in a process or electronic system, and so on.

The identification of the deviations in the design proposal is obtained through a process of questions using predetermined guide words. The role of the word guide is to stimulate imaginative thinking, to focus the study, and to provoke ideas and discussion, thus maximizing the opportunities to achieve a more complete study.

The HAZOP is well suited in the later stages of detailed design to examine operational capabilities and when changes are made to existing facilities. The best time to conduct a HAZOP study is just before the design freezes. The HAZOP studies consist of four basic sequential steps:

- Definition of scope, objectives, responsibilities, and equipment
- Preparation of the study, registration format, and data collection
- Examination dividing the system into parts and identifying problems, causes, and consequences (identify protection mechanisms and measures)
- Documentation and follow-up with report of initial conclusions, preventive, and corrective actions taken and final results report

- *IEC 62502:2010: Analysis techniques for dependability—Event tree analysis (ETA) (Edition 1.0)* specifies the consolidated basic principles of

ETA and provides guidance on modeling the consequences of an initiating event as well as analyzing these consequences qualitatively and quantitatively in the context of dependability and risk-related measures.

The event tree considers a number of possible consequences of an initiating event or system failure. Thus, the event tree can be combined very efficiently with the fault tree. The root of an event tree can be seen as the main event of a fault tree. This combination is sometimes called analysis of causes and consequences. To evaluate the seriousness of certain consequences that derive from the initiating event, all possible consequences paths should be identified and investigated, and their probabilities determined.

The analysis with event tree is used when it is essential to investigate all the possible paths of consequent events, their sequences, and the consequences or most probable results of the initiating event. After an initiating event, there are some first events or subsequent consequences that may follow. The probability associated with the occurrence of a specific path (sequence of events) represents the product of the conditioned probabilities of all the events of that path.

The key elements in the application of the event tree are (Figure 9.3):

- The initiator (initiating event)
- Subsequent events
- Consequences of the events

- *IEC 62551:2012: Analysis techniques for dependability—Petri net techniques (Edition 1.0)* provides guidance on a Petri-net based methodology for dependability purposes. It supports modeling a system, analyzing the model, and presenting the analysis results. This methodology is oriented to dependability-related measures with all the related features, such as reliability, availability, production availability, maintainability, and safety.

Petri net is a graphical tool for the representation and analysis of complex logical interactions between the components or events of a system. The typical complex interactions that are naturally included in the language of the Petri net are concurrency, conflict, synchronization, mutual exclusion, and resource limitation.

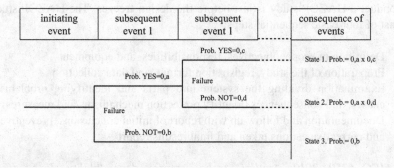

FIGURE 9.3   General outline of an event tree.

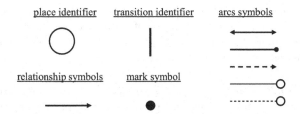

FIGURE 9.4 Petri net symbols not timed.

The static structure of the system modeled is represented by a Petri net graph, which is composed of three primary elements (see Figure 9.4):

- *Nodes or places*: usually drawn as circles, that represent the conditions in which the system can be found
- *Transitions*: usually drawn as bars, that represent events that can change one condition into another
- *Arcs*: drawn as arrows that connect nodes with transitions and transitions with nodes and that represent the admissible logical connections between conditions and event

A condition is valid in a given situation if the corresponding node is marked; that is, it contains at least one "•" mark (drawn as a black dot). The dynamics of the system is represented by the movement of the marks in the graph. A transition is allowed if its input nodes contain at least one mark.

A permitted transition can be triggered and that trigger removes a mark from each entry node and places a mark on each exit node. The distribution of the marks in the nodes is called marking.

Starting from an initial marking, the application of the activation and firing rules produces all the possible markings that constitute the attainable set of the Petri nets. This achievable set provides all the states that the system can reach from the initial state.

Standard Petri nets do not contemplate the notion of time. However, many extensions have appeared in which temporary aspects overlap the Petri network. If a trigger rate (constant) is assigned to each transition, then the dynamics of the Petri net can be analyzed by a continuous-time Markov chain whose state space is isomorphic with the attainable set of the corresponding Petri net.

The Petri net can be used as a high level language to generate Markov models and some tools used for reliability analysis are based on this methodology. Petri nets also provide a natural environment for simulation.

The use of Petri nets is recommended when complex logical interactions must be considered (concurrence, conflict, synchronization, mutual exclusion, and resource limitation). In addition, Petri net usually is an easier and more natural language to use in describing a Markov model.

The key element of a Petri net analysis is the description of the structure of the system and its dynamic behavior in terms of primary elements (nodes, transitions, arcs, and marks) typical of the Petri net language. This step requires the use of ad hoc software tools.

- *IEC 62740:2015: Root cause analysis (RCA) (Edition 1.0)* describes the basic principles of root cause analysis (RCA) and specifies the steps that a process for RCA should include. This standard identifies a number of attributes for RCA techniques that assist with the selection of an appropriate technique. It describes each RCA technique and its relative strengths and weaknesses. RCA is used to analyze the root causes of focus events with both positive and negative outcomes, but it is most commonly used for the analysis of failures and incidents.

Causes for such events can be varied in nature, including design processes and techniques, organizational characteristics, human aspects, and external events. An RCA can be used for investigating the causes of non-conformances in quality (and other) management systems as well as for failure analysis (e.g., in maintenance or equipment testing). An RCA is used to analyze focus events that have occurred; therefore, this standard only covers a posteriori analyses.

It is recognized that some of the RCA techniques with adaptation can be used proactively in the design and development of items and for causal analysis during risk assessment; however, this standard focuses on the analysis of events that have occurred. The intent of this standard is to describe a process for performing RCA and to explain the techniques for identifying root causes. These techniques are not designed to assign responsibility or liability, which is outside the scope of this standard.

9.8 STATISTICAL METHODS FOR THE EVALUATION OF RELIABILITY

These five standards should be used together, since they are all complementary. In a first phase it is necessary to classify the nature of the system or equipment subjected to the analysis in relation to its maintainability, since the methods are different if it is a repairable or non-repairable item. The IEC 60300-3-5: 2001 standard includes a complete procedure for the adequate selection of the most appropriate statistical method in each case.

- *IEC 60605-6:2007: Equipment reliability testing—Part 6: Tests for the validity and estimation of the constant failure rate and constant failure intensity (Edition 3.0)* specifies procedures to verify the assumption of a constant failure rate or constant failure intensity. These procedures are applicable whenever it is necessary to verify these assumptions. This may be a requirement or to assessing the behavior in time of the failure rate or the failure intensity. The major technical changes with respect to the previous edition concern the inclusion of corrected formulas for tests previously included in a corrigendum, and the addition of new methods for the analysis of multiple items.

The standard develops the tests to check the hypothesis of constant failure rate $\lambda(t)$ for non-repairable elements, and the tests to check the hypothesis of constant failure intensity $z(t)$ for repairable elements.

In Section 6.2 of the standard the U-test (Laplace test) is developed to analyze whether the nonrepairable equipment object of the study has a trend in its failure rate. The standard also includes three graphical methods of trend testing in Sections 6.3 through 6.5 of the standard as support to the researcher to assess whether it can be assumed that the non-repairable elements under study have a trend or not trend.

In Section 7.2 of the standard, the procedure is developed to check if a repairable element has a constant failure intensity $z(t)$, based on the calculation of the U-test (Laplace test).

For testing completed by time:

$$U = \frac{\sum_{i=1}^{r} T_i - r\dfrac{T^*}{2}}{T^* \sqrt{\dfrac{r}{12}}} \tag{9.3}$$

For testing completed by failure:

$$U = \frac{\sum_{i=1}^{r} T_i - (r-1)\dfrac{T_r}{2}}{T_r \sqrt{\dfrac{r-1}{12}}} \tag{9.4}$$

with:
   r is the total number of failures
   T^* is the total time of the test completed by time
   T_r is the total time of the test completed by failure
   T_i is the cumulative time of the test in the ith failure

With the zero growth hypothesis (i.e., the failure times follow a HPP), the U-test is roughly distributed according to a standardized exponential distribution of mean 0 and deviation 1. The U-test can be used to test whether there is evidence of reliability growth, positive or negative, independent of the reliability growth model.

A bilateral test for positive or negative growth with significance level α has critical values $u_{1-\alpha/2}$ and $-u_{1-\alpha/2}$, where $u_{1-\alpha/2}$ is the $(1-\alpha/2)100$ percent percentile of the typical normal distribution. If $-u_{1-\alpha/2} < U < u_{1-\alpha/2}$, then there is no evidence of positive or negative growth of the reliability to a significance level α. In this case, the hypothesis of an exponential distribution of times between successive failures of the HPP is accepted with significance level α:

$$-u_{1-\alpha/2} < U < u_{1-\alpha/2} \tag{9.5}$$

For the significance levels required in each test, the appropriate critical values of the percentile table of the normalized typified distribution should be chosen according to Table 9.7.

TABLE 9.7

Critical Values for a Level

of Significance α

α	$U\alpha$ Value
0.025	2.24
0.050	1.96
0.100	1.64

In Section 7.3 of the standard, the procedure is developed to check whether a set of repairable elements of the same characteristics has a constant failure intensity $z(t)$, based on the calculation of the U-test:

$$U = \frac{\sum_{i=1}^{k} \sum_{j=1}^{n} T_{ij} - 0{,}5\left(r_1 T_1^* + r_2 T_2^* + \ldots + r_k T_k^*\right)}{\sqrt{\frac{1}{12}\left(r_1 T_2^{*2} + r_2 T_2^{*2} + \ldots + r_k T_k^{*2}\right)}} \tag{9.6}$$

with:

r_i is the total number of failures to consider from the ith item

T_i^* is the total time of the test for the ith item

T_{ij} is the time accumulated at the jth failure of the ith item

k is the total number of items

As in the case of Section 7.2 of the standard, a bilateral test for positive or negative growth with significance level α has critical values $u_{1-\alpha/2}$ and $-u_{1-\alpha/2}$, where $u_{1-\alpha/2}$ is the $(1-\alpha/2)100$ percent percentile of the typical normal distribution

In Section 7.4 of the standard, the graphical procedure $M(t)$ plot is developed to check whether one or a set of repairable elements of the same characteristics has constant failure intensity. It is a more qualitative than quantitative test.

- *IEC 60605-4:2001: Equipment reliability testing—Part 4: Statistical procedures for exponential distribution—Point estimates, confidence intervals, prediction intervals and tolerance intervals (Edition 2.0)* provides statistical methods for evaluating point estimates, confidence intervals, prediction intervals, and tolerance intervals for the failure rate of items whose time to failure follows an exponential distribution.

This standard develops the statistical procedure for the exponential distribution and allows estimating the value of constant failure rate $\lambda(t)$ for non-repairable elements and the constant failure intensity $z(t)$ value for non-repairable elements. It also includes the formulation for the calculation of confidence intervals, tolerances, and so on.

This norm must apply complementary to IEC 60605-6 in such a way that if the result of the application of U-test accepts the hypothesis of exponential distribution of the times between successive failures (or a HPP), it is possible to calculate directly the value of constant failure rate $\lambda(t)$ or constant failure intensity $z(t)$.

For testing completed by time and non-repairable items, the point estimate of the failure rate:

$$\hat{\lambda} = \frac{r}{T^*} \tag{9.7}$$

For test terminated by failure:

$$\hat{\lambda} = \frac{r}{T^*} \tag{9.8}$$

with:
  r is the total number of failures in test
  T^* is the total time of the test completed by time or by failure

For testing completed by time and repairable elements, the point estimate of the failure intensity:

$$\hat{Z} = \frac{r}{T^*} \tag{9.9}$$

For test terminated by failure:

$$\hat{Z} = \frac{r}{T^*} \tag{9.10}$$

with:
  r is the total number of failures in test
  T^* is the total time of the test completed by time or by failure

The standard includes the calculation of bilateral confidence intervals, for example, for tests completed by time for repairable items:

$$Z_{L2} = \lambda_{L2} = \frac{X_{\alpha/2}^2 2r}{2T^*} \tag{9.11}$$

$$Z_{U2} = \lambda_{U2} = \frac{X_{1-\frac{\alpha}{2}}^2 (2r+2)}{2T^*} \tag{9.12}$$

with:
  X^2 is the fractile table value of the X^2 distribution for the 90 percent confidence interval.

In addition, the standard allows for prediction intervals for failures for a future period in Section 9.6 and a procedure for assigning tolerance intervals in Section 9.7.

- *IEC 61649:2008: Weibull analysis (Edition 2.0)* provides methods for ana-lyzing data from a Weibull distribution using continuous parameters such as time to failure, cycles to failure, mechanical stress, and so on. This standard is applicable whenever data on strength parameters such as times to fail-ure, cycles, and stress are available for a random sample of items operating under test conditions or in-service to estimate measures of reliability per-formance of the population from which these items were drawn. The main changes with respect to the previous edition are as follows: the title has been shortened and simplified to read "Weibull analysis" and provision of meth-ods for both analytical and graphical solutions has been added.

In non-repairable items, when the failure rate $\lambda(t)$ does not have a constant behavior over time, usually the Weibull distribution is tried:

$$f(t) = \beta\alpha(\alpha t)^{\beta-1} e^{-(\alpha t)^{\beta}} \tag{9.13}$$

$$R(t) = e^{-(\alpha t)^{\beta}} \tag{9.14}$$

$$\lambda(t) = \beta\alpha(\alpha t)^{\beta-1} \tag{9.15}$$

where:
  α is the scale parameter
  β is the shape parameter
  $f(t)$ is the probability density function of the failure
  $R(t)$ is the reliability function

The Weibull distribution is used to model data without considering whether the fail-ure rate is increasing, decreasing, or constant. The Weibull distribution is flexible and can be adapted to a wide variety of data.

The standard contemplates the Weibull distribution with two and three param-eters, graphical methods, and goodness of fit. It also includes a section for the inter-pretation of the resulting probability graph.

It develops computational methods for the point estimation of parameters by means of maximum likelihood estimation (MLE), confidence intervals, as well as the Weibayes approach, and the "sudden death" method.

- *IEC 61710:2013: Power law model—Goodness-of-fit tests and estimation meth-ods (Edition 2.0)* specifies procedures to estimate the parameters of the power law model, to provide confidence intervals for the failure intensity, to provide prediction intervals for the times to future failures, and to test the goodness-of-fit of the power law model to data from repairable items. It is assumed that the time to failure data have been collected from an item or some identical items operating under the same conditions (e.g., environment and load).

This standard develops the statistical procedure for an NHPP by means of PLP and allows estimating the value of the failure intensity $z(t)$ for tests of one or more repairable items in tests terminated by time or by failure. It also allows the estimation of the $z(t)$ in tests for groups of failures in time intervals.

This standard must be applied in a complementary way to IEC 60605-6 so that if the result of the application of the U-test is rejected, there is a trend (intensity of increasing or decreasing failure) and may be applicable PLP:

$$E\left[N\left(t\right)\right] = \lambda t^{\beta} \tag{9.16}$$

Then $z(t)$ can be calculated as follows:

$$Z\left(t\right) = \frac{d}{dT} E\left[N\left(t\right)\right] = \lambda \beta t^{\beta-1} \tag{9.17}$$

with:
$E[N(t)]$ is the expected accumulated number of failures up to time t
λ is the scale parameter
β is the shape parameter

The methods of estimating $z(t)$ differ according to the type of test carried out:

- One or more repairable items observed in the same space time (the statistics of Section 7.2.1 of the standard are applied)
- Multiple repairable items observed in different time intervals (the statistics of Section 7.2.2 of the standard are applied)
- Groups of failures in time intervals (the statistics of Section 7.2.3 of the standard are applied)

For one or multiple repairable items observed in the same period of time, Section 7.2.1 the summation is calculated:

$$S_1 = \sum_{j=1}^{N} \ln\left(\frac{T^*}{t_j}\right); \text{ for tests completed on time} \tag{9.18}$$

$$S_2 = \sum_{j=1}^{N} \ln\left(\frac{t_N}{t_j}\right); \text{ for tests completed to failure} \tag{9.19}$$

with:
T^* is the total time of the test completed by time
t_N is the total time of the test completed by failure
t_j is the cumulative time of test in jth failure

The unbiased estimation of the parameter of form β is calculated:

$$\hat{\beta} = \frac{N-1}{S_1}; \text{ for tests completed on time} \tag{9.20}$$

$$\hat{\beta} = \frac{N-2}{S_2}; \text{ for tests completed to failure} \tag{9.21}$$

The unbiased estimation of the scale parameter λ is calculated:

$$\hat{\lambda} = \frac{N}{k(T^*)^{\beta}}; \text{ for tests completed on time} \tag{9.22}$$

$$\hat{\lambda} = \frac{N}{k(t_N)^{\beta}}; \text{ for tests completed to failure} \tag{9.23}$$

with:
   N is the total number of failures accumulated in test
   k is the total number of test items

$z(t)$, therefore, according to the PLP:

$$\hat{Z}(t) = \hat{\lambda}\hat{\beta}t^{\hat{\beta}-1} \tag{9.24}$$

For multiple repairable items observed in different time intervals (Section 7.2.2), the parameter of form β is calculated iteratively:

$$\frac{N}{\hat{\beta}} + \sum_{i=1}^{N} \ln t_i - \frac{N \sum_{j=1}^{k} T_j^{\beta} \ln T_j}{\sum_{j=1}^{k} T_j^{\beta}} = 0 \tag{9.25}$$

And the estimation of the scale parameter λ is calculated:

$$\hat{\lambda} = \frac{N}{\sum_{J=1}^{k} T_j^{\hat{\beta}}} \tag{9.26}$$

with:
   N is the total number of failures accumulated in the test
   k is the total number of items
   t_i is time to the ith failure $(i = 1, 2, ..., N)$
   T_j is the total time of observation for item $j = 1, 2, ..., k$

The goodness-of-fit test given in IEC 61710 (2013) is the Cramér–von Mises statistic C^2, with $M = N$ and $T = T^*$ for testing completed based on time, and $M = N - 1$ and $T = T_N$ for tests completed to failure:

$$C^2 = \frac{1}{12M} + \sum_{j=1}^{M}\left[\left(\frac{t_j}{T}\right)^{\beta} - \left(\frac{2j-1}{2M}\right)\right]^2 \tag{9.27}$$

A critical value of $C^2_{0.90}(M)$ is selected, with a level of significance of 10 percent of the tabulated value. If C^2 exceeds the critical value $C^2_{0.90}(M)$, $C^2 > C^2_{0.90}(M)$, then the hypothesis that the PLP model fits the test data must be rejected.

In Annex C of IEC 61710 of 2013 a Bayesian estimate for PLP is included. The methods reflected in the main body of this standard are based on the classic approach to make statistical estimates. This means that the parameters of PLP, λ, and β are assumed to be fixed, but unknown and a classical method such as maximum likelihood is used to estimate the values of both parameters, using the observed data of the accumulated times until the failure of a repairable item or items.

An alternative approach is that of the Bayesian estimate. This approach deals with the parameters of PLP, λ, and β as random variables not observed. This affects the stages of the estimation process. A Bayesian approach to estimating the PLP can be summarized in the following steps:

1. Choose a probability distribution that reflects the degree of knowledge of each of the parameters, λ and β, before collecting any data. This distribution is called the a priori distribution
2. Collect the observed data of the accumulated failure times for the repairable items in question
3. Estimate the parameters of PLP with a posteriori distribution calculated using the Bayes theorem and reflects what is known about the parameters after observing the data

- *IEC 61650:1997: Reliability data analysis techniques—Procedures for comparison of two constant failure rates and two constant failure (event) intensities (Edition 1.0)* specifies procedures to compare two observed failure rates, failure intensities, rates/intensities of relevant events. The procedures are used to determine whether an apparent difference between the two sets of observations can be considered statistically significant. Numerical methods and a graphical procedure are prescribed. Simple practical examples are provided to illustrate how the procedures can be applied.

9.9 CONCLUSIONS

The IEC standards published in the field of reliability provide maintenance engineers with tools, procedures, and methods to deal with a large part of the management and control activities that they have to develop, in a standardized and auditable manner and that have the support from official, business, and scientific community organizations.

However, these standards are not being used systematically or generalized by maintenance organizations, nor are they cross-referenced in indexed scientific publications.

It is estimated that this lack of knowledge and use may have part of its origin in the form of organization and structuring of the contents published in the different standards, which hinder their understanding and practical application.

In this chapter an attempt has been made to classify and present the 57 published standards grouped according to their main field of application to improve their dissemination, understanding, and orientation of use:

- *Management procedures*: They are standards for the management of different maintenance activities: design, life cycle, maintainability, logistics, risk, etc.
- *Establishment of requirements*: These standards develop procedures for the specification of design requirements for systems and equipment such as reliability, maintainability, availability, etc.
- *Test methods*: They are norms that present the procedures for the design and application of different tests in order to evaluate the behavior of the systems and equipment in operation.
- *Method selection*: These standards establish metrics for the selection of the most appropriate method for evaluating the reliability of a system or equipment.
- *Reliability evaluation methods*: They are the standards that develop each one of the methods for the evaluation of the reliability of a system or equipment, with a different estimation approach.
- *Statistical methods for the evaluation of reliability*: They are the standards developed by the statistical method for evaluating the reliability of repairable and non-repairable systems or equipment (these standards must be applied jointly and in an integrated manner).

Researchers, organizations, and maintenance professionals are encouraged to use these standards as procedures or as a reference guide, particularly in the field of reliability assessment because they provide mathematical methods and metrics that have the consensus and support of international standardization bodies.

REFERENCES

IEC/ISO 31010:2009 Edition 1.0, *Risk Management: Risk Assessment Techniques*, International Electrotechnical Commission (IEC), Geneva, Switzerland.

IEC 60300-1:2014 Edition 3.0, *Dependability Management: Part 1: Guidance for Management and Application*, International Electrotechnical Commission (IEC), Geneva, Switzerland.

IEC 60300-3-1:2003 Edition 2.0, *Dependability Management: Part 3-1: Application Guide— Analysis Techniques for Dependability—Guide on Methodology*, International Electrotechnical Commission (IEC), Geneva, Switzerland.

IEC 60300-3-2:2004 Edition 2.0, *Dependability Management: Part 3-2: Application Guide— Collection of Dependability Data from the Field*, International Electrotechnical Commission (IEC), Geneva, Switzerland.

IEC 60300-3-3:2017 Edition 3.0, *Dependability Management: Part 3-3: Application Guide—Life Cycle Costing*, International Electrotechnical Commission (IEC), Geneva, Switzerland.

IEC 60300-3-4:2007 Edition 2.0, *Dependability Management: Part 3-4: Application Guide—Guide to the Specification of Dependability Requirements*, International Electrotechnical Commission (IEC), Geneva, Switzerland.

IEC 60300-3-5:2001 Edition 1.0, *Dependability Management: Part 3-5: Application Guide—Reliability Test Conditions and Statistical Test Principles*, International Electrotechnical Commission (IEC), Geneva, Switzerland.

IEC 60300-3-10:2001 Edition 1.0, *Dependability Management: Part 3-10: Application Guide—Maintainability*, International Electrotechnical Commission (IEC), Geneva, Switzerland.

IEC 60300-3-11:2009 Edition 2.0, *Dependability Management: Part 3-11: Application Guide—Reliability Centred Maintenance*, International Electrotechnical Commission (IEC), Geneva, Switzerland.

IEC 60300-3-12:2011 Edition 2.0, *Dependability Management: Part 3-12: Application Guide—Integrated Logistic Support*, International Electrotechnical Commission (IEC), Geneva, Switzerland.

IEC 60300-3-14:2004 Edition 1.0, *Dependability Management: Part 3-14: Application Guide—Maintenance and Maintenance Support*, International Electrotechnical Commission (IEC), Geneva, Switzerland.

IEC 60300-3-15:2009 Edition 1.0, *Dependability Management: Part 3-15: Application Guide—Engineering of System Dependability*, International Electrotechnical Commission (IEC), Geneva, Switzerland.

IEC 60300-3-16:2008 Edition 1.0, *Dependability Management: Part 3-16: Application Guide—Guidelines for Specification of Maintenance Support Services*, International Electrotechnical Commission (IEC), Geneva, Switzerland.

IEC 60319:1999 Edition 3.0, *Presentation and Specification of Reliability Data for Electronic Components*, International Electrotechnical Commission (IEC), Geneva, Switzerland.

IEC 60605-2:1994 Edition 1.0, *Equipment Reliability Testing: Part 2: Design of Test Cycles*, International Electrotechnical Commission (IEC), Geneva, Switzerland.

IEC 60605-4:2001 Edition 2.0, *Equipment Reliability Testing: Part 4: Statistical Procedures for Exponential Distribution—Point Estimates, Confidence Intervals, Prediction Intervals and Tolerance Intervals*, International Electrotechnical Commission (IEC), Geneva, Switzerland.

IEC 60605-6:2007 Edition 3.0, *Equipment Reliability Testing: Part 6: Tests for the Validity and Estimation of the Constant Failure Rate and Constant Failure Intensity*, International Electrotechnical Commission (IEC), Geneva, Switzerland.

IEC 60706-2:2006 Edition 2.0, *Maintainability of Equipment: Part 2: Maintainability Requirements and Studies During the Design and Development Phase*, International Electrotechnical Commission (IEC), Geneva, Switzerland.

IEC 60706-3:2006 Edition 2.0, *Maintainability of Equipment: Part 3: Verification and Collection, Analysis and Presentation of Data*, International Electrotechnical Commission (IEC), Geneva, Switzerland.

IEC 60706-5:2007 Edition 2.0, *Maintainability of Equipment: Part 5: Testability and Diagnostic Testing*, International Electrotechnical Commission (IEC), Geneva, Switzerland.

IEC 60812:2006 Edition 2.0, *Analysis Techniques for System Reliability—Procedure for Failure Mode and Effects Analysis (FMEA)*, International Electrotechnical Commission (IEC), Geneva, Switzerland.

IEC 61014:2003 Edition 2.0, *Programmes for Reliability Growth*, International Electrotechnical Commission (IEC), Geneva, Switzerland.

IEC 61025:2006 Edition 2.0, *Fault Tree Analysis (FTA)*, International Electrotechnical Commission (IEC), Geneva, Switzerland.

IEC 61070:1991 Edition 1.0, *Compliance Test Procedures for Steady-State Availability*, International Electrotechnical Commission (IEC), Geneva, Switzerland.

IEC 61078:2016 Edition 3.0, *Reliability Block Diagrams*, International Electrotechnical Commission (IEC), Geneva, Switzerland.

IEC 61123:1991 Edition 1.0, *Reliability Testing: Compliance Test Plans for Success Eatio*, International Electrotechnical Commission (IEC), Geneva, Switzerland.

IEC 61124:2012 Edition 3.0, *Reliability Testing: Compliance Tests for Constant Failure Rate and Constant Failure Intensity*, International Electrotechnical Commission (IEC), Geneva, Switzerland.

IEC 61160:2005 Edition 2.0, *Design Review*, International Electrotechnical Commission (IEC), Geneva, Switzerland.

IEC 61163-1:2006 Edition 2.0, *Reliability Stress Screening: Part 1: Repairable Assemblies Manufactured in Lots*, International Electrotechnical Commission (IEC), Geneva, Switzerland.

IEC 61163-2:1998 Edition 1.0, *Reliability Stress Screening: Part 2: Electronic Components*, International Electrotechnical Commission (IEC), Geneva, Switzerland.

IEC 61164:2004 Edition 2.0, *Reliability Growth: Statistical Test and Estimation Methods*, International Electrotechnical Commission (IEC), Geneva, Switzerland.

IEC 61165:2006 Edition 2.0, *Application of Markov Techniques*, International Electrotechnical Commission (IEC), Geneva, Switzerland.

IEC 61649:2008 Edition 2.0, *Weibull Analysis*, International Electrotechnical Commission (IEC), Geneva, Switzerland.

IEC 61650:1997 Edition 1.0, *Reliability Data Analysis Techniques: Procedures for Comparison of Two Constant Failure Rates and Two Constant Failure (event) Intensities*, International Electrotechnical Commission (IEC), Geneva, Switzerland.

IEC 61703:2016 Edition 2.0, *Mathematical Expressions for Reliability, Availability, Maintainability and Maintenance Support Terms*, International Electrotechnical Commission (IEC), Geneva, Switzerland.

IEC 61709:2017 Edition 3.0, *Electric Components: Reliability: Reference Conditions for Failure Rates and Stress Models for Conversion*, International Electrotechnical Commission (IEC), Geneva, Switzerland.

IEC 61710:2013 Edition 2.0, *Power Law Model: Goodness-of-fit Tests and Estimation Methods*, International Electrotechnical Commission (IEC), Geneva, Switzerland.

IEC 61882:2016 Edition 2.0, *Hazard and Operability Studies (HAZOP studies): Application Guide*, International Electrotechnical Commission (IEC), Geneva, Switzerland.

IEC 61907:2009 Edition 1.0, *Communication Network Dependability Engineering*, International Electrotechnical Commission (IEC), Geneva, Switzerland.

IEC 62198:2013 Edition 2.0, *Managing Risk in Projects: Application Guidelines*, International Electrotechnical Commission (IEC), Geneva, Switzerland.

IEC 62308:2006 Edition 1.0, *Equipment Reliability: Reliability Assessment Methods*, International Electrotechnical Commission (IEC), Geneva, Switzerland.

IEC 62309:2004 Edition 1.0, *Dependability of Products Containing Reused Parts: Requirements for Functionality and Tests*, International Electrotechnical Commission (IEC), Geneva, Switzerland.

IEC 62347:2006 Edition 1.0, *Guidance on System Dependability Specifications*, International Electrotechnical Commission (IEC), Geneva, Switzerland.

IEC 62402:2007 Edition 1.0, *Obsolescence Management: Application Guide*, International Electrotechnical Commission (IEC), Geneva, Switzerland.

IEC 62429:2007 Edition 1.0, *Reliability Growth: Stress Testing for Early Failures in Unique Complex Systems*, International Electrotechnical Commission (IEC), Geneva, Switzerland.

IEC 62502:2010 Edition 1.0, *Analysis Techniques for Dependability: Event Tree Analysis (ETA)*, International Electrotechnical Commission (IEC), Geneva, Switzerland.

IEC 62506:2013 Edition 1.0, *Methods for Product Accelerated Testing*, International Electrotechnical Commission (IEC), Geneva, Switzerland.

IEC 62508:2010 Edition 1.0, *Guidance on Human Aspects of Dependability*, International Electrotechnical Commission (IEC), Geneva, Switzerland.

IEC 62550:2017 Edition 1.0, *Spare Parts Provisioning*, International Electrotechnical Commission (IEC), Geneva, Switzerland.

IEC 62551:2012 Edition 1.0, *Analysis Techniques for Dependability: Petri Net Techniques*, International Electrotechnical Commission (IEC), Geneva, Switzerland.

IEC 62628:2012 Edition 1.0, *Guidance on Software Aspects of Dependability*, International Electrotechnical Commission (IEC), Geneva, Switzerland.

IEC 62673:2013 Edition 1.0, *Methodology for Communication Network Dependability Assessment and Assurance*, International Electrotechnical Commission (IEC), Geneva, Switzerland.

IEC 62740:2015 Edition 1.0, *Root Cause Analysis (RCA)*, International Electrotechnical Commission (IEC), Geneva, Switzerland.

IEC 62741:2015 Edition 1.0, *Demonstration of Dependability Requirements: The Dependability Case*, International Electrotechnical Commission (IEC), Geneva, Switzerland.

IEC TS 62775:2016 Edition 1.0, *Application Guidelines: Technical and Financial Processes for Implementing Asset Management Systems*, International Electrotechnical Commission (IEC), Geneva, Switzerland.

IEC 62853:2018 Edition 1.0, *Open Systems Dependability*, International Electrotechnical Commission (IEC), Geneva, Switzerland.

IEC TR 63039:2016 Edition 1.0, *Probabilistic Risk Analysis of Technological Systems: Estimation of Final Event Rate at a Given Initial State*, International Electrotechnical Commission (IEC), Geneva, Switzerland.

10 Time-Variant Reliability Analysis Methods for Dynamic Structures

Zhonglai Wang and Shui Yu

CONTENTS

10.1   INTRODUCTION

Reliability analysis aims to estimate the probability that products perform their intended performance under the specified working conditions during their lifecycle. For highly reliable products, it is difficult to collect enough data to conduct reliability analysis using the statistics-based method. From the aspect of failure mechanism of products, the physics-based method will be a proper choice for reliability analysis with insufficient data. Traditional physics-based static (time-invariant) reliability analysis methods have been developed extensively such as the First Order Reliability Method (FORM) [1], the Second Order Reliability Method (SORM) [2], the moment-based method [3], and surrogate models [4], which only consider the static performance or simplify the dynamic performance to be the static performance. For most products, the performance is usually dynamic because of various time-varying loadings, working conditions, and inherent motion. Time-invariant reliability analysis methods have shown poor capability in satisfying the reliability accuracy requirements for time-varying and high nonlinear performance functions of products [5]. Therefore, such engineering requirements have fostered the development of time-variant reliability methods and several time-variant reliability analysis methods have been developed.

Time-variant reliability analysis aims to estimate the probability that products successfully complete the intended performance during a given time interval. There are typically two categories of time-variant reliability analysis methods: simulation and analytical. Typical analytical time-variant reliability analysis methods include Gamma process method [6], extreme value method [7], composite limit state method [8], compound random processes method [9], and crossing-rate based methods [10,11]. The high model error would be produced due to the model approximation since system parameters or performance functions are usually assumed to follow a certain distribution in the Gamma process, extreme value, and compound random processes methods. When handling high nonlinear limit state functions, the computational accuracy of the composite limit state method may be unsatisfactory. After the crossing-rate method was first proposed [10,11], many crossing-based methods were developed further: e.g., differential Gaussian process method [12], the rectangular wave renewal process method [13], Laplace integration method [14], PHI2 method [15], and PHI2+ method [16]. The differential Gaussian process method, rectangular wave renewal process method, and Laplace integration method are suitable mainly for the crossing-rate calculation for some specific random processes. The developed PHI2 and PHI2+ methods based on the crossing-rate method use the parallel reliability framework to improve the computational accuracy and further broaden the application range of the crossing rate methods. However, the PHI2 and PHI2+ methods show lower computational accuracy when dealing with the time-variant reliability analysis of non-monotonic systems [16].

The other branch of the time-variant reliability analysis is the simulation methods. The typical simulation methods are MCS, importance sampling (IS), and subset simulation (SS) methods. MCS is a direct and easy-to-use method, regardless of the dimensions and nonlinearity of limit state functions, but the computational efficiency is usually forbidden for high reliability estimation for complicated engineering systems with implicit expression of the limit state function. The combination of MCS and analytical methods could improve the computational efficiency [17]. IS technique can improve the computational efficiency by introducing the importance density function, but difficulties exist in acquiring the prior failure domain information and determining the proper importance sampling density [18]. SS method is another branch of the simulation method that transforms a small failure probability into the product of some bigger conditional failure probabilities. Subset simulation with Markov Chain Monte Carlo (SS/MCMC) and subset simulation with splitting (SS/S) are the two typical sampling branches [19]. However, the nonlinearity of the limit state function affects the computational accuracy [20].

In this chapter, the expression of the time-variant reliability is described in Section 10.2. Developed three time-variant reliability analysis methods are elaborated in Section 10.3. In Section 10.4, two examples are used to illustrate and compare the proposed methods. Conclusions are provided in Section 10.5.

10.2   TIME-VARIANT RELIABILITY

Time-variant reliability is defined as the probability that products complete their intended performance under the practical working conditions during the given time interval. The typical expression of the time-variant reliability is provided as:

$$R_t\left(t_{lb},t_{ub}\right)=\Pr\left\{g\left(\mathbf{d},\mathbf{X},\mathbf{Y}\left(t\right),t\right)>0,\forall t\in\left[t_{lb},t_{ub}\right]\right\} \qquad (10.1)$$

where:
   $g(\bullet)$ is the time-variant limit state function for a certain structure
   $\mathbf{d}$ denotes the vector of deterministic design variables
   $\mathbf{X}$ defines the vector of random design variables and parameters
   $\mathbf{Y}(t)$ expresses the vector of time-variant random design variables and parameters,
        actually stochastic process
   t_{lb} and t_{ub} are lower and upper boundaries of the time interval

When $\mathbf{Y}(t)$ is a stochastic process with the autocorrelation function, the stochastic process can be decomposed into the general stochastic processes $\mathbf{Y}(\mathbf{N},t)$ with the stochastic process discretization method [21], where $\mathbf{N}=\left[N_1,...,N_r\right]$ is a vector of independent standard normal random variables. The decomposed process of a scalar Gaussian process with the mean value $m(t)$, standard deviation $\sigma(t)$, and exponential autocorrelation coefficient function $\rho(t_1,t_2)$ is provided in the appendix to this chapter. Therefore, the time-variant reliability can be rewritten as:

$$R_t\left(t_{lb},t_{ub}\right)=\Pr\left\{g\left(\mathbf{d},\mathbf{Z},t\right)>0,\forall t\in\left[t_{lb},t_{ub}\right]\right\} \qquad (10.2)$$

where $\mathbf{Z}=\left[\mathbf{X},\mathbf{N}\right]$.

To simplify the computational process, normalization is conducted for the time interval $[t_{lb}, t_{ub}]$.

$$T = \frac{t - t_{lb}}{t_{ub} - t_{lb}} \qquad (10.3)$$

Since $t \in [t_{lb}, t_{ub}]$, T belongs to $[0,1]$ in Equation 10.3. With the normalization in Equation 10.3, the expression of the time-variant reliability in Equation (10.2) can be rewritten as:

$$R_T(0,1) = \Pr\left\{ g^T(\mathbf{d}, Z, T) > 0, \forall T \in [0,1] \right\} \qquad (10.4)$$

10.3 THE PROPOSED THREE TIME-VARIANT RELIABILITY ANALYSIS METHODS

In this section, three developed time-variant reliability analysis methods will be elaborated. The first method is called the failure processes decomposition (FPD) method, which possesses the advantage for handling the time-variant reliability problems with the high order of time parameters [21]. The second method is based on the combination of the extreme value moment and improved maximum entropy (EVM-IME), which can effectively deal with the time-variant reliability problems with multiple failure modes and temporal parameters [7]. The third method is proposed based on the probability density function (PDF) estimation of the first-passage time point (P-FTP).

10.3.1 FAILURE PROCESSES DECOMPOSITION METHOD

In this subsection, the procedure of the time-variant reliability analysis based on the FPD method will be illustrated. For more details, please refer to [21]. First, the time point where the mean value of the limit state function possesses the minimal value (FMTP) is searched. With the acquired FMTP, the time-variant limit state function with high order temporal parameters is then transformed to a quadratic function of time, which is called by the first-stage failure processes decomposition. Based on the property of the quadratic function and reliability criterion, the time-variant reliability is transformed to the time-invariant system reliability, which is called by the second-stage FPD. Finally, the kernel density estimation (KDE) method is implemented to calculate the time-invariant system reliability. For a clear illustration, the flowchart of the FPD method is provided in Figure 10.1.

10.3.1.1 The FMTP Search for the Time-Variant Limit State Function

The expression of the time-variant reliability based on the extreme value theory can be given by:

$$R_T(0,1) = \Pr\left\{ g_{\min}^T(\mathbf{d}, Z, T) > 0, T \in [0,1] \right\} \qquad (10.5)$$

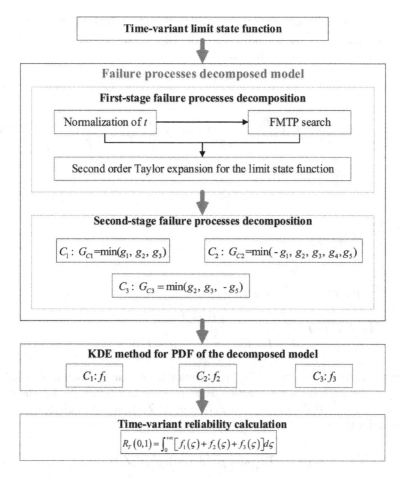

FIGURE 10.1 Flowchart of the FPD method.

For a trajectory of the stochastic process representing the time-variant uncertain limit state function, $g^T(d,Z,T)$ is actually a deterministic function of T. The minimal value $g^T_{min}(d,Z,T)$ can be achieved when $T=T^*$. The time point $\overline{T}^*$ where the mean value of the limit state function $g^T(\mathbf{d},\mathbf{Z},T)$ is minimal can be searched with the optimization model and $\overline{T}^*$ is the so-called FMTP.

$$\begin{cases} \text{find:} & T \\ \text{minimize:} & \overline{g}^T(\mathbf{d},\mathbf{Z},T) \\ \text{subject to:} & T \in [0,1] \end{cases} \quad (10.6)$$

10.3.1.2   Failure Processes Decomposition Based on Taylor Expansion

In the first-stage FPD, the second-order Taylor expansion is performed for the time-variant limit state function at the FMTP:

$$g^T\left(\mathbf{d},\mathbf{Z},T\right)$$

$$\approx g^T\left(\mathbf{d},\mathbf{Z},\bar{T}^*\right)+\frac{\partial g^T}{\partial T}\bigg|_{T=\bar{T}^*}\left(T-\bar{T}^*\right)+\frac{1}{2}\frac{\partial^2 g^T}{\partial T^2}\bigg|_{T=\bar{T}^*}\left(T-\bar{T}^*\right)^2 \quad (10.7)$$

$$= aT^2 + bT + c$$

where

$$a = \frac{1}{2}\frac{\partial^2 g^T}{\partial T^2}\bigg|_{T=\bar{T}^*}$$

$$b = \frac{\partial g^T}{\partial T}\bigg|_{T=\bar{T}^*} - \bar{T}^*\frac{\partial^2 g^T}{\partial T^2}\bigg|_{T=\bar{T}^*}$$

$$c = g^T\left(\mathbf{d},\mathbf{Z},\bar{T}^*\right) - \bar{T}^*\frac{\partial g^T}{\partial T}\bigg|_{T=\bar{T}^*} + \frac{\bar{T}^{*2}}{2}\frac{\partial^2 g^T}{\partial T^2}\bigg|_{T=\bar{T}^*}$$

When the second derivative of the approximate limit state function to T equals 0, the approximate limit state function is a monotonic function of T. Therefore, $g^T_{\min}\left(\mathbf{d},\mathbf{Z},T\right) = g^T\left(\mathbf{d},\mathbf{Z},0\right)$ or $g^T_{\min}\left(\mathbf{d},\mathbf{Z},T\right) = g^T\left(\mathbf{d},\mathbf{Z},1\right)$, and the reliability can be obtained for this case:

$$R_T\left(0,1\right) = \Pr\left\{\min\left[g^T\left(\mathbf{d},\mathbf{Z},0\right), g^T\left(\mathbf{d},\mathbf{Z},1\right)\right] > 0\right\} \quad (10.8)$$

10.3.1.3   Failure Processes Decomposition Based on Case Classification

Since the trajectory $g^T\left(d,Z,T\right)$ of the stochastic process $g^T\left(\mathbf{d},\mathbf{Z},T\right)$ is a quadratic function of T, the stochastic process $g^T\left(\mathbf{d},\mathbf{Z},T\right)$ could be considered as a collection of quadratic functions. According to the property of a quadratic function and safety criterion, three cases are classified to represent the safety of the structure in the second-stage FPD, shown in Figure 10.2a–c, respectively.

In Figure 10.2, the green curve represents the shape of the quadratic function $g^T\left(\mathbf{d},\mathbf{Z},T\right)$, the dashed red line is the symmetric axis $T_s = -\frac{b}{2a}$, and the green dots, respectively, denote $T=0$ and $T=1$. The corresponding properties for the three cases are provided in Table 10.1.

For **Case 1**, **Case 2**, and **Case 3**, three events C_1, C_2, and C_3 are provided:

$$C_1 = \left\{T_s < 0, g^T\left(\mathbf{d},\mathbf{Z},0\right) > 0, g^T\left(\mathbf{d},\mathbf{Z},1\right) > 0\right\}$$

$$= \left\{-\frac{b}{2a} < 0, c > 0, a+b+c > 0\right\} \quad (10.9)$$

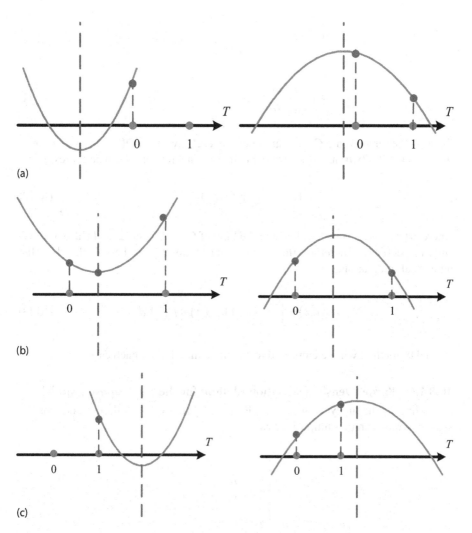

FIGURE 10.2 The geometrical relationship between $g^T(\mathbf{d},\mathbf{Z},T)$ and T. (a) case 1 of the safety situation, (b) case 2 of the safety situation, (c) case 3 of the safety situation.

TABLE 10.1

Properties for the Three Cases

Cases	Position of T_s	Location of Minimum Point
Case 1	$T_s < 0$	$T = 0$ or 1
Case 2	$0 \leq T_s < 1$	$T = 0$ or T_s or 1
Case 3	$T_s \geq 1$	$T = 0$ or 1

$$C_2 = \left\{ 0 < -\frac{b}{2a} < 1, c - \frac{b^2}{4a} > 0, c > 0, a + b + c > 0 \right\}$$ (10.10)

$$C_3 = \left\{ -\frac{b}{2a} > 1, c > 0, a + b + c > 0 \right\}$$ (10.11)

Because the three events $C_1 \sim C_3$ are mutually exclusive, the PDF of the system time-invariant reliability transformed from the time-variant reliability can be expressed by:

$$f(\varsigma) = f_1(\varsigma) + f_2(\varsigma) + f_3(\varsigma)$$ (10.12)

where $f_1(\varsigma)$, $f_2(\varsigma)$, and $f_3(\varsigma)$ denote the PDF of **Case 1**, **Case 2**, and **Case 3** occurring, respectively. Therefore, the time-variant reliability can be calculated by the numerical integration:

$$R_T(0,1) = \int_0^{+\infty} \left[f_1(\varsigma) + f_2(\varsigma) + f_3(\varsigma) \right] d\varsigma$$ (10.13)

The KDE method will be employed to calculate the PDF for each case.

10.3.1.4 Kernel Density Estimation Method for the Decomposed Model

According to the analysis in Section 10.3.1.3, there are five failure modes for the system, which can be summarized as:

$$\begin{cases} g_1 = \dfrac{b}{2a} \\[2mm] g_2 = a + b + c \\[2mm] g_3 = c \\[2mm] g_4 = c - \dfrac{b^2}{4a} \\[2mm] g_5 = 1 + \dfrac{b}{2a} \end{cases}$$ (10.14)

Because of the similar procedure for calculating the system reliability for each case, **Case 1** is taken for an example. The limit state function for the event C_1 is:

$$G_{C_1}(\mathbf{Z}) = \min \left[g_1, g_2, g_3 \right] = \min \left[\frac{b}{2a}, a + b + c, c \right]$$ (10.15)

M samples are directly drawn from the limit state function $G_{C_1}(\mathbf{Z})$, and the vector of samples can be obtained as $\mathbf{G}_{C_1} = \{\varsigma_1^{C_1}, \varsigma_2^{C_1}, \cdots, \varsigma_M^{C_1}\}$. Then the PDF $f_{G_{C_1}}(\varsigma)$ for the event C_1 is:

$$f_{C_1}(\varsigma) = \frac{1}{Mh_{C_1}} \sum_{i=1}^{M} K\left(\frac{\varsigma - \varsigma_i^{C_1}}{h_{C_1}}\right) \tag{10.16}$$

where:

$K(\bullet)$ is the kernel function and the Gaussian kernel function in this model is considered: i.e., $K(u) = \frac{1}{\sqrt{2\pi}}\exp\left(-\frac{u^2}{2}\right)$

h is the bandwidth of the kernel function

The bandwidth of the kernel function is important for the prediction accuracy and the optimal value of h is:

$$h_{opt} = \left(\frac{4}{3M}\right)^{0.2} \sigma(\mathbf{G}) \tag{10.17}$$

PDF $f_1(\varsigma)$ for **Case 1** can be estimated by:

$$\begin{aligned} f_1(\varsigma) &= f_{C_1}(\varsigma) \\ &= \frac{1}{Mh_{C_1}} \sum_{i=1}^{M} K\left(\frac{\varsigma - \varsigma_i^{C_1}}{h_{C_1}}\right) \end{aligned} \tag{10.18}$$

With the same procedure, PDFs are estimated for events C_2 and C_3 based on the KDE method. Using the estimated PDFs, the time-invariant system reliability is obtained:

$$\begin{aligned} R_T(0,1) &= \int_0^{+\infty} \left[f_1(\varsigma) + f_2(\varsigma) + f_3(\varsigma)\right]d\varsigma \\ &= \frac{1}{M} \sum_{k=1}^{3} \sum_{i=1}^{M} \Phi\left(\frac{\varsigma_i^{C_k}}{h_{C_k}}\right) \end{aligned} \tag{10.19}$$

10.3.2 THE COMBINATION OF THE EXTREME VALUE MOMENT AND IMPROVED MAXIMUM ENTROPY (EVM-IME) METHOD

In this subsection, the second method called EVM-IME will be elaborated. For more details, please refer to [7]. The accuracy of the maximum entropy first is improved by introducing the scaling function. The extreme value moments of the limit state functions are then estimated using the sparse grid stochastic collocation method. The PDF and corresponding time-variant reliability are finally estimated. The flow-chart of the EVM-IME method is provided in Figure 10.3.

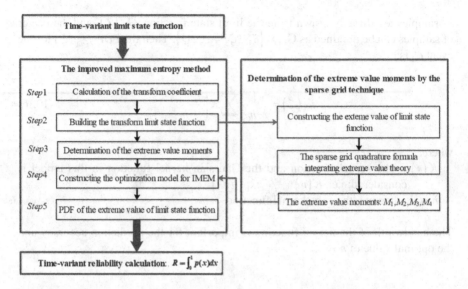

FIGURE 10.3   Flowchart of the EVM-IME method.

10.3.2.1   Determination of the Extreme Value Moments by the Sparse Grid Technique

Several numerical approximation methods could be employed for the multi-dimensional integration, such as full factorial numerical integration, univariate dimension reduction method, and sparse grid based stochastic collocation method. Full factorial numerical integration method can obtain accurate results, but suffers from the curse of the dimensionality, and is suitable mainly for solving the low dimensional problems. In the univariate dimension reduction method, a larger error may be produced due to the neglect of the interaction terms in the additive decomposition process. The sparse grid-based stochastic collocation method can avoid the curse of the dimensionality of the full factorial numerical integration method and have higher accuracy than the univariate dimension reduction method. Therefore, the sparse grid-based stochastic collocation method is used in this section. To simplify the computation process, the limit state function is transformed to be the only one including mutually independent standard normal random variables based on the Rosenblatt transformation.

Then the expression of the time-variant reliability is:

$$R_T(0,1) = \Pr\left\{ g^N(\mathbf{d}, \mathbf{N}, T) > 0, \forall T \in [0,1] \right\} \tag{10.20}$$

where $\mathbf{N} = [N_1, N_2, ..., N_k]$ is a vector of mutually independent standard normal random variables. The lth raw moments of extreme values in the time-variant reliability model can be expressed as:

$$M_l = \int_{-\infty}^{+\infty} \cdots \int_{-\infty}^{+\infty} \left\{ g^N \left(\mathbf{d}, \mathbf{N}, T \right) \right\}^l \varphi(x_1) \cdots \varphi(x_k) dx_1 \cdots dx_k \qquad (10.21)$$

where $\varphi(\cdot)$ is the PDF of a standard normal random variable.

Based on the Smolyak algorithm, the sparse grid quadrature method is employed to compute the multivariate integration and the lth raw moments of extreme values can be expressed by:

$$M_l = \sum_{i \in H(q,k)} (-1)^{q+k-|i|} \binom{k-1}{q+k-|i|} \times \sum_{j_1=1}^{m_{i_1}} \cdots \sum_{j_1=1}^{m_{i_k}} \left[g^N \left(\mathbf{d}, x_{j_1}^{i_1}, \cdots, x_{j_k}^{i_k}, T_{opt} \right) \right]^l p_{j_1}^{i_1} \cdots p_{j_k}^{i_k}$$

$$(10.22)$$

where the abscissas and weights are $x_{j_i}^{i_i} = \sqrt{2}\xi_{j_i}^{i_i}$ and $p_{j_i}^{i_i} = \frac{1}{\sqrt{\pi}}\zeta_{j_i}^{i_i}$, and $\xi_{j_i}^{i_i}$ and $\zeta_{j_i}^{i_i}$ are the abscissas and weights in the Gauss-Hermite quadrature formula; $j_i = 1, \ldots, m_{i_i}$; the multi-index $\mathbf{i} = (i_1, \ldots, i_k) \in N_+^k$; and the set $H(q,k)$ is defined by:

$$H(q,k) = \left\{ \mathbf{i} = (i_1, \ldots, i_k) \in N_+^k, \mathbf{i} \geq 1: q+1 \leq \sum_{r=1}^{k} i_r \leq q+k \right\} \qquad (10.23)$$

q will affect the computational accuracy of $H(q,k)$ and the selection of q is based on the nonlinearity of the limit state function. To balance the computational accuracy and efficiency, $2 \leq q \leq 4$ and $m_1 = 1$ for $i = 1$ and $m_i = 2^{i-1} + 1$ for $i > 1$ are provided in engineering applications.

In Equation 10.22, the input variable nodes $x_{j_1}^{i_1}, \ldots, x_{j_k}^{i_k}$ can be generated by the sparse grid technique. $g^N \left(\mathbf{d}, x_{j_1}^{i_1}, \ldots, x_{j_k}^{i_k}, T_{opt} \right) = \min \left[g^N \left(\mathbf{d}, x_{j_1}^{i_1}, \cdots, x_{j_k}^{i_k}, T \right) \right]$ is regarded as an extreme value, and the corresponding optimization model is provided by:

$$\begin{cases} \text{find:} & T \\ \text{minimize:} & g^N \left(\mathbf{d}, x_{j_1}^{i_1}, \cdots, x_{j_k}^{i_k}, T \right) \\ \text{subject to:} & T \in [0,1] \end{cases} \qquad (10.24)$$

10.3.2.2 The Improved Maximum Entropy Method Based on the Raw Moments

The maximum entropy method can obtain a relatively accurate result of the PDF based on the known moments. The typical formulation for the PDF of the time-invariant limit state function is defined as:

$$\begin{cases} \text{find:} & p(x) \\ \text{maximize:} & H = -\int p(x) \ln p(x) dx \\ \text{subject to:} & \int x^i p(x) dx = M_i, \quad i = 0,1,\cdots,l \end{cases} \qquad (10.25)$$

where:

$p(x)$ is the PDF of the time-invariant limit state function $g(\mathbf{X})$

H is the entropy of the PDF $p(x)$

M_i is the ith raw moment and $M_0 = 1$

l is the number of the given moment constraints, which is defined to be 4 here

For the optimization problem, a Lagrangian multiplier $\lambda_i, i = 0,1,\ldots,4$ is introduced into the structure Lagrangian function L and:

$$L = -\int p(x)\ln p(x)dx - \sum_{i=0}^{4} \lambda_i \left[\int x^i p(x)dx - M_i\right] \qquad (10.26)$$

$\frac{\delta L}{\delta p(x)} = 0$ is satisfied for calculating the optimal solution, and therefore the analytical expression of $p(x)$ can be easily obtained by:

$$p(x) = \exp\left[-\sum_{i=0}^{4} \lambda_i x^i\right]. \qquad (10.27)$$

The objective function to be minimized based on the Kullback–Leibler (K–L) divergence between the true PDF and estimated PDF can be provided by:

$$I(\lambda) = \lambda_0 + \sum_{i=1}^{4} \lambda_i M_i \qquad (10.28)$$

where $\lambda_0 = \ln\left[\int \exp\left(-\sum_{i=1}^{4} \lambda_i x^i\right)dx\right]$. The optimization with equality constraints in Equation 10.25 can be converted into an unconstrained optimization:

$$\begin{cases} \text{find:} & \lambda_1, \ \lambda_2, \ \lambda_3, \ \lambda_4 \\[2mm] \text{minimize:} & I = \ln\left[\int \exp\left(-\sum_{i=1}^{4} \lambda_i x^i\right)dx\right] + \sum_{i=1}^{4} \lambda_i M_i \end{cases} \qquad (10.29)$$

With the obtained raw moments, the PDF $p(x)$ of the limit state function $g(\mathbf{X})$ can be acquired from the optimization model in Equation 10.29:

$$p(x) = \exp\left[-\ln\left[\int \exp\left(-\sum_{i=1}^{4} \lambda_i x^i\right)dx\right] - \sum_{i=1}^{4} \lambda_i x^i\right] \qquad (10.30)$$

Reliability is then calculated based on the PDF $p(x)$ from the maximum entropy method:

$$R = \int_{0}^{+\infty} p(x)dx \qquad (10.31)$$

The reliability results from this method may be not accurate due to the trunca-tion error of the integral. Addressing this issue, the monotonic scaling function is introduced to improve the computational accuracy of the maximum entropy method. The truncation error of the numerical integration can be greatly reduced by changing the definition domain of the PDF from an infinite interval to a lim-ited interval.

The scaling function is expressed as:

$$g^s(\mathbf{X}) = -\frac{1 - \exp\left[\dfrac{g(\mathbf{X})}{c}\right]}{1 + \exp\left[\dfrac{g(\mathbf{X})}{c}\right]} \tag{10.32}$$

where c is a conversion coefficient and $c > 0$.

The scaling function $g^s(\mathbf{X})$ is a monotonic increasing function for the limit state function $g(\mathbf{X})$, and $\lim_{g(\mathbf{X}) \to \infty} g^s(\mathbf{X}) = 1$, $\lim_{g(\mathbf{X}) \to -\infty} g^s(\mathbf{X}) = -1$. Therefore, the defini-tion domain is changed from the infinite interval [−inf, +inf] to the limited inter-val [−1, 1]. c is an important coefficient which affects the relationship between $g(\mathbf{X})$ and $g^s(\mathbf{X})$, shown in Figure 10.4. From Figure 10.4, it is possible to see that the greater coefficient c will lead to the gentle curve. In this subsection, $c = \left|\mu_{g(\mathbf{x})}\right|$ is chosen.

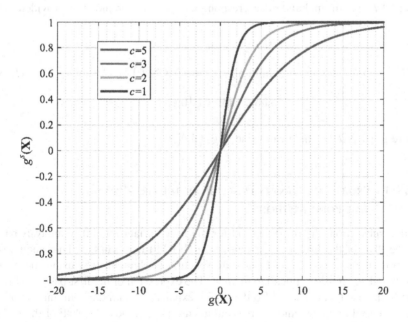

FIGURE 10.4   Conversion relationship between $g(\mathbf{x})$ and $g^s(\mathbf{x})$ for different c.

With the scaling function $g^s(\mathbf{X})$, the unconstrained optimization could be rewritten as:

$$\begin{cases} \text{find:} \quad \lambda_1, \ \lambda_2, \ \lambda_3, \ \lambda_4 \\[2mm] \text{minimize:} \quad I = \ln\left[\int_{-1}^{1}\exp\left(-\sum_{i=1}^{4}\lambda_i x^i\right)dx\right] + \sum_{i=1}^{4}\lambda_i M_i^s \end{cases} \tag{10.33}$$

and the corresponding reliability is:

$$R = \int_0^1 p(x)\,dx \tag{10.34}$$

where M_i^s is the ith raw moment of $g^s(\mathbf{X})$, $p(x)$ is the PDF obtained from M_i^s.

For the time-variant reliability analysis, the scaling function can be expressed by:

$$g^s(\mathbf{d},\mathbf{N},T) = -\frac{1-\exp\left[\dfrac{g^T(\mathbf{d},\mathbf{N},T)}{c}\right]}{1+\exp\left[\dfrac{g^T(\mathbf{d},\mathbf{N},T)}{c}\right]} \tag{10.35}$$

where c is the conversion coefficient.

To obtain the conversion coefficient c, the mean value of the limit state function $g^T(\mathbf{d},\mathbf{N},T)$ is minimal and the corresponding optimization model is provided by:

$$\begin{cases} \text{find:} \quad c \\[2mm] \text{minimize:} \quad c = \mu_g(\mathbf{d},\mathbf{N},T) \\[2mm] \text{subject to:} \quad T \in [0,1] \end{cases} \tag{10.36}$$

where $\mu_g(\mathbf{d},\mathbf{N},T)$ is the mean value of $g^T(\mathbf{d},\mathbf{N},T)$.

10.3.3 Probability Density Function of the First-Passage Time Point Method

In this subsection, the third method for the time-variant reliability analysis based on the PDF of the first-passage time point (F-PTP) is discussed. The mean value function of the time-variant limit state function is obtained first using the sparse grid based stochastic collocation method. The expression of the first-passage time is then built based on the second-order Taylor expansion. With the combination of the fourth central moments and the maximum entropy methods, the PDF of the F-PTP is obtained and the time-variant reliability can be calculated with the integration.

10.3.3.1   Time-Variant Reliability Model Based on PDF of F-PTP

According to the reliability criterion, the reliability situation and failure situation can be described by the relationship between the realization $g^T(\mathbf{d},Z,T)$ of the stochastic process representing the time-variant limit state function and the time interval [0,1], shown in Figure 10.5. For example, there are three intersections t_1, t_2, and t_3, between $g^T(\mathbf{d},Z,T)$ and horizontal coordinate axes, where t_1 is the F-PTP. When $t_1 \in [0,1]$, the failure occurs, shown in Figure 10.5a. When $t_1 > 1$, the system operates successfully, shown in Figure 10.5b.

Actually, the F-PTP, t_1, is a function of random input vector $\mathbf{Z}$, denoted as $\mathbf{t_1} = t(\mathbf{Z})$. Therefore, the failure probability during the time interval [0,1] can be expressed as:

$$P(0,1) = \Pr\{\mathbf{t_1} \in [0,1]\} \tag{10.37}$$

If the PDF $f(\tau)$ of the F-PTP function $\mathbf{t_1} = t(\mathbf{Z})$ is available, the failure probability can be calculated by:

$$P(0,1) = \Pr\{0 < \mathbf{t_1} < 1\}$$
$$= \int_0^1 f(\tau)\,d\tau \tag{10.38}$$

10.3.3.2   Establish $f(\tau)$ by Using the Maximum Entropy Method Combined with the Moment Method

The mean value function of the time-variant random limit state function $g^T(\mathbf{d},Z,T)$ can be expressed by:

$$\mu(T) = \int g^T(\mathbf{d},Z,T)p(\mathbf{Z})\,d\mathbf{Z} \tag{10.39}$$

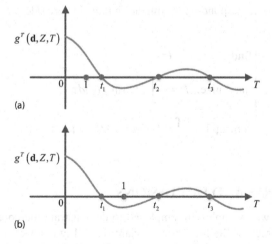

(a)

(b)

FIGURE 10.5   The geometrical relationship between $g^T(\mathbf{d},Z,T)$ and T, (a) the safe situation on the life cycle [0,T], (b) the failure situation on the life cycle [0,T].

The mean value function $\mu(T)$ can be calculated by the sparse grid-based stochastic collocation method, which is elaborated in Section 10.3.2. The F-PTP, T^*, of the mean value function can be acquired via the optimization model. The second-order Taylor expansion is used to approximate the time-variant limit state function at T^*:

$$g^T(\mathbf{d},\mathbf{Z},T) \approx g^T(\mathbf{d},\mathbf{Z},T^*) + \frac{\partial g^T}{\partial T}\bigg|_{T=T^*}(T-T^*)$$

$$+\frac{1}{2}\frac{\partial^2 g^T}{\partial T^2}\bigg|_{T=T^*}(T-T^*)^2 \tag{10.40}$$

$$= AT^2 + BT + C$$

where:

$$A = \frac{1}{2}\frac{\partial^2 g^T}{\partial T^2}\bigg|_{T=T^*}\frac{1}{2},$$

$$B = \frac{\partial g^T}{\partial T}\bigg|_{T=T^*} - T^*\frac{\partial^2 g^T}{\partial T^2}\bigg|_{T=T^*},$$

$$C = g^T(\mathbf{d},\mathbf{Z},T^*) - T^*\frac{\partial g^T}{\partial T}\bigg|_{T=T^*} + \frac{(T^*)^2}{2}\frac{\partial^2 g^T}{\partial T^2}\bigg|_{T=T^*}.$$

The approximate limit state function is a quadratic function of T. Therefore, the F-PTP function of the limit state function is $\mathbf{t_1} = t(\mathbf{Z}) = \frac{-B-\sqrt{B^2-4AC}}{2A}$. The fourth raw moments of $\mathbf{t_1}$ can be also computed by the sparse grid-based stochastic collocation method, denote as M_i^F, $i = 1,2,3,4$. The maximum entropy method is used to establish $f(\tau)$, and the corresponding optimization model is provided:

$$\begin{cases} \text{find:} & f(\tau) \\[2mm] \text{maximize:} & H = -\displaystyle\int f(\tau)\ln f(\tau)d\tau \\[2mm] \text{subject to:} & \displaystyle\int \tau^i f(\tau)d\tau = M_i^F, \quad i = 0,1,\cdots,l \end{cases} \tag{10.41}$$

10.4   EXAMPLES AND DISCUSSIONS

In this section, two examples are employed to demonstrate the computational efficiency and accuracy of the three time-variant reliability analysis methods. For the accuracy comparison, the results from the MCS method are provided as the benchmark. The error on the failure probability is:

$$\text{Err} = \frac{|P - P_{MCS}|}{P_{MCS}} \times 100\% \tag{10.42}$$

where P is the failure probability from the three proposed methods, P_{MCS} is failure probability from the MCS method.

10.4.1 NUMERICAL EXAMPLE

The time-variant limit state function is:

$$g(\mathbf{d}, \mathbf{X}, t) = x_1^2 x_2 - 5x_3 t + (x_4 + 1)\exp(x_5 t^2) - \Delta \tag{10.43}$$

where Δ is the threshold, X_i $(i = 1, 2, 3, 4, 5)$ are input random variables to be independent normally distributed, and $X_i \sim N(5, 0.5^2)$ $(i = 1, 2, 3, 4, 5)$; the time interval is defined as $t \in [0, 1]$.

Table 10.2 provides the failure probability results from the MCS method, the FPD method, the EVM-IME method, and the F-PTP method for different threshold. Furthermore, the relative errors of the FPD method (Err1), the EVM-IME method (Err2), and the F-PTP method (Err3) are also given in Table 10.2.

From Table 10.2, the error of the proposed methods is very small for different threshold. Compared with the three proposed method, the order of the computational accuracy is approximately the EVM-IME method > the FPD method > the F-PTP method.

For this example, 10^5 trajectories are generated in the MCS method. The optimization algorithm is used to estimate the minimum value of a trajectory of a limit state function. For a given trajectory, nearly 20 functions are called for the global minimum value. Therefore, the total function calls 2×10^6 are used in the MCS method. Similarly, nearly 20 function calls are used to obtain FMTP, and 10,000 samples are used to build the kernel density function. Therefore, 10,020 function calls are used in the FPD method. In the EVM-IME method, $q = 2$ is set and 61 samples are drawn for each limit state function using the sparse grid stochastic technique. For each limit state function, there are nearly 20 function calls via the optimization. Therefore, there are $20 + 20 \times 61 + 20 = 1,260$ function calls in the

TABLE 10.2
Failure Probability Results for Example 1

Δ	MCS	FPD	EVM-IME	F-PTP	Err1 (%)	Err2 (%)	Err3 (%)
70	0.00823	0.00897	0.00854	0.00893	8.99	3.77	8.51
72	0.01149	0.01099	0.01126	0.01117	4.35	2.00	2.79
74	0.01449	0.01506	0.01458	0.01626	3.93	0.6211	12.22
76	0.01798	0.01900	0.01880	0.01936	5.67	4.56	7.68
78	0.02312	0.02453	0.02386	0.02103	6.88	3.20	9.04

EVM-IME method. In the F-PTP method, 101 function calls are used. Therefore, the order of the computational efficiency is the F-PTP method > the EVM-IME method > the FPD method.

10.4.2 A Corroded Simple Supported Beam Under Random Loadings

A corroded simple supported beam subjected to random loadings is used as an engineering case to illustrate the three proposed methods. As shown in Figure 10.6, the parameters for the beam are $L = 5$ m, $b_0 = 0.2$ m, and $h_0 = 0.04$ m. The uniform load p and the time-varying random loading $F(t)$ are simultaneously applied to the beam. The uniform load p is:

$$p = \sigma b_0 h_0 \, (\text{N/m}) \tag{10.44}$$

where:

$$\sigma = 78,500 \, (\text{N/m}).$$

The time-varying random loading $F(t)$ is a Gaussian process with the mean value of 3,500 N, standard deviation 700 N, and autocorrelation function $\rho(t_1, t_2) = \exp\left(-\left(\frac{t_2 - t_1}{0.0833}\right)^2\right)$. The time-varying random loading $F(t)$ can be expressed by the random process discretization method:

$$F(t) = 3,500 + 700 \sum_{i=1}^{r} \frac{N_i}{\sqrt{\tau_i}} \phi_i^T \rho(t) \tag{10.45}$$

where:
  τ_i, ϕ_i^T, and $\rho(t)$ can be obtained within different time interval according to the appendix in this chapter
  N_i are independently standard normal random variables

With the effect of $F(t)$ and p, the bending moment at the mid-span section is:

$$M(t) = \frac{F(t)L}{4} + \frac{pL^2}{8}. \tag{10.46}$$

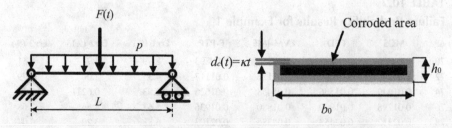

FIGURE 10.6 A corroded simple supported beam under random loadings.

Corrosion is assumed to happen around the cross-section of the beam isotropically and the growth is linear with time progression. Then the surplus area of the cross-section is provided as:

$$A(t) = b(t) \times h(t) \qquad (10.47)$$

where $b(t) = b_0 - 2\kappa t$, $h(t) = h_0 - 2\kappa t$, and $\kappa = 0.00003\,(\text{m/year})$. κ is a parameter to control the corrosion velocity. The ultimate bending moment is:

$$M_u(t) = \frac{A(t)h(t)}{4} f_y. \qquad (10.48)$$

where f_y is the steel yield stress.

The time-variant limit state function could be provided as:

$$g(\mathbf{X}, \mathbf{Y}(t), t) = M_u(t) - M(t) \qquad (10.49)$$

In this case, the time intervals [0,15] and [0,20] years are considered. The related information of random parameters is given in Table 10.3.

The reliability results for the four methods are provided in Table 10.4. From Table 10.4, it is seen that the order of the accuracy remains the same as that in example 1. Since the expression of the limit state function has little impact on the computational efficiency, the computational efficiency keeps the same order.

TABLE 10.3

Information of Random Parameters in Example 2

Parameter	Distribution Type	Mean	Standard Deviation
f	Lognormal	240	24
b_0	Lognormal	0.2	0.01
h_0	Lognormal	0.04	0.004
$F(t)$	Gaussian process	3,500	700

TABLE 10.4

Time-Variant Reliability Results in Example 2

Time Interval	MCS	FPD	EVM-IME	F-PTP	Err 1	Err 2	Err 3
[0,20]	0.00178	0.00182	0.00175	0.00174	2.28	1.69	2.25
[0,15]	0.00121	0.00119	0.00122	0.00116	1.65	0.826	4.13

10.5 CONCLUSIONS

In this chapter, three time-variant reliability analysis methods including the FPD method, the EVM-IME method, and the F-PTP method are discussed. From the procedure and examples, the following conclusions can be reached: (1) the three time-variant reliability analysis methods have the high computational accuracy, which satisfy the engineering requirements; (2) the three time-variant reliability analysis methods have the high computational efficiency, which provide the feasibility for solving complex engineering problems; (3) the order of the computational accuracy is approximately the EVM-IME method > the FPD method > the F-PTP method; and (4) the order of the computational efficiency is the F-PTP method > the EVM-IME method > the FPD method.

In the further research, the intelligent technique will be used for time-variant reliability analysis to further improve the computational efficiency under the satisfaction of high computational accuracy.

APPENDIX: DISCRETIZATION OF RANDOM PROCESSES

Consider a Gaussian process $Y(t)$ with mean value $m(t)$, standard deviation $\sigma(t)$, and autocorrelation coefficient function $\rho(t_1, t_2)$. In the time interval $[0, T]$, r time points $t_i, i = 1, ..., r$ are selected to decompose the process and $t_1 = 0$, $t_r = T$. The Gaussian process $Y(t)$ is decomposed into:

$$\hat{Y}(t) = m(t) + \sigma(t) \sum_{i=1}^{r} \frac{\xi_i}{\sqrt{\tau_i}} \phi_i^T \rho(t) \tag{A10.1}$$

where $\xi_i \sim N(0, 1)$, $i = 1, ..., r$ are independent standard normal random variables, and (τ_i, ϕ_i) are the eigenvalues and eigenvectors of the correlation matrix $\mathbf{C}$, and $C_{ij} = \rho(t_i, t_j)$, $i, j = 1, ..., r$, $\rho(t) = [\rho(t, t_1), ..., \rho(t, t_r)]^T$ is a time-variant vector. The corresponding decomposition error is given by:

$$e(t) = 1 - \sum_{i=1}^{r} \frac{\left(\phi_i^T \rho(t)\right)^2}{\tau_i} \tag{A10.2}$$

REFERENCES

1. Du X. Unified uncertainty analysis by the first order reliability method. *Journal of Mechanical Design*, 2008, 130(9): 091401.
2. Wang Z, Huang HZ, Liu Y. A unified framework for integrated optimization under uncertainty. *Journal of Mechanical Design*, 2010, 132(5): 051008.
3. Zhao YG, Ono T. Moment methods for structural reliability. *Structural Safety*, 2001, 23(1): 47–75.
4. Xiao NC, Zuo MJ, Zhou C. A new adaptive sequential sampling method to construct surrogate models for efficient reliability analysis. *Reliability Engineering & System Safety*, 2018, 169: 330–338.

5. Yu S, Wang Z, Zhang K. Sequential time-dependent reliability analysis for the lower extremity exoskeleton under uncertainty. *Reliability Engineering & System Safety*, 2018, 170: 45–52.

6. Van Noortwijk JM, van Der Weide JAM, Kallen MJ, Pandey MD. Gamma processes and peaks-over-threshold distributions for time-dependent. *Reliability Engineering & System Safety*, 2007, 92(12): 1651–1658.

7. Yu S, Wang Z, Meng D. Time-variant reliability assessment for multiple failure modes and temporal parameters. *Structural and Multidisciplinary Optimization*, 2018, 58(4): 1705–1717.

8. Majcher M, Mourelatos Z, Tsianika V. Time-dependent reliability analysis using a modified composite limit state approach. *SAE International Journal of Commercial Vehicles*, 2017, 10(2017-01-0206): 66–72.

9. Gnedenko BV, Belyayev YK, Solovyev AD. *Mathematical Methods of Reliability Theory*. Burlington, NY: Academic Press, 2014.

10. Jiang C, Wei X, Huang Z, Liu J. An outcrossing rate model and its efficient calculation for time-dependent system reliability analysis. *Journal of Mechanical Design*, 2017, 139(4): 041402.

11. Yan M, Sun B, Liao B et al. FORM and out-crossing combined time-variant reliability analysis method for ship structures. *IEEE Access*, 2018, 6: 9723–9732.

12. Sundar VS, Manohar CS. Time variant reliability model updating in instrumented dynamical systems based on Girsanov's transformation. *International Journal of Nonlinear Mechanics*, 2013, 52: 32–40.

13. Breitung K, Rackwitz R. Nonlinear combination of load processes. *Journal of Structural Mechanics*, 1982, 10(2): 145–166.

14. Zayed A, Garbatov Y, Soares CG. Time variant reliability assessment of ship structures with fast integration techniques. *Probabilistic Engineering Mechanics*, 2013, 32: 93–102.

15. Andrieu-Renaud C, Sudret B, Lemaire M. The PHI2 method: A way to compute time-variant reliability. *Reliability Engineering & System Safety*, 2004, 84(1): 75–86.

16. Singh A, Mourelatos Z P. On the time-dependent reliability of non-monotonic, non-repairable systems. *SAE International Journal of Materials and Manufacturing*, 2010, 3(1): 425–444.

17. Wang Z, Zhang X, Huang HZ et al. A simulation method to estimate two types of time-varying failure rate of dynamic systems. *Journal of Mechanical Design*, 2016, 138(12): 121404.

18. Singh A, Mourelatos Z, Nikolaidis E. Time-dependent reliability of random dynamic systems using time-series modeling and importance sampling. *SAE International Journal of Materials and Manufacturing*, 2011, 4(1): 929–946.

19. Wang Z, Mourelatos ZP, Li J, Baseski I, Singh A. Time-dependent reliability of dynamic systems using subset simulation with splitting over a series of correlated time intervals. *Journal of Mechanical Design*, 2014, 136(6): 061008.

20. Ching J, Au SK, Beck JL. Reliability estimation for dynamical systems subject to stochastic excitation using subset simulation with splitting. *Computer Methods in Applied Mechanics and Engineering*, 2005, 194(12–16): 1557–1579.

21. Yu S, Wang Z. A novel time-variant reliability analysis method based on failure processes decomposition for dynamic uncertain structures. *Journal of Mechanical Design*, 2018, 140(5): 051401.

11 Latent Variable Models in Reliability

Laurent Bordes

CONTENTS

11.1 INTRODUCTION

A latent variable is a variable that is not directly observable and is assumed to affect the response variables. There are many statistical models that involve latent variables. Such models are called latent variable models. Surprisingly, there are few monographs specifically dedicated to latent variable models (see, e.g., [1–4]). Latent variables typically are encountered in econometric, reliability, and survival statistical model with different aims. A latent variable may represent the effect of unobservable covariates or factors and then it allows accounting for the unobserved heterogeneity between subjects, it may also account for measurement errors assuming that the latent variables represent the "true" outcomes and the manifest variables represent their "disturbed" versions, it may also summarize different measurements of the same (directly) unobservable characteristics (e.g., quality of life), so that sample units may be easily ordered or classified based on these traits (represented by the latent variables). Hence, latent variable models now have a wide range of applications, especially in the presence of repeated observations, longitudinal/panel data, and multilevel data.

In this chapter, we propose to select a few latent variable models that have proved to be useful in the domain of reliability. We do not pretend to have an exhaustive view of such models but we try to show that these models lead to various estimation methodologies that require various mathematical tools if we want to derive large sample properties. Basic mathematical tools are based on empirical processes theory (see [5–7]), or martingale methods for counting processes theory (see [8]), or again Expectation-Maximization (EM) algorithms for parametric models (see [9]).

This chapter is organized into three parts. The first part is Section 11.2, which is devoted to incomplete data including right censored data, partly right and left censored data, and competing risk data. Then in the second part, Section 11.3, we consider models that allow consideration of heterogeneity in data, including frailty models, finite mixture models, cure models as well as excess risk models. The last part in Section 11.4 deals with models for time-dependent phenomena. Indeed, we consider degradation processes for which the latent variable is either a random duration; this is the case for the Gamma degradation processes with random initial time or a frailty scale parameter. We also consider bivariate degradation processes obtained from trivariate construction that requires a third latent Gamma process.

11.2 LATENT VARIABLE MODEL FOR HANDLING INCOMPLETE DATA

11.2.1 RIGHT CENSORING

Let T be a duration of interest with probability density function (PDF) f_T, survival function S_T, hazard rate function $\lambda_T = f_T/S_T$, and cumulative hazard rate function Λ_T. Let C be a censoring time with PDF f_C, survival function S_C, hazard rate function $\lambda_C = f_C/S_C$, and cumulative hazard rate function Λ_C. One of the basic latent variable model in reliability (or survival analysis) assumes that the duration of interest T is right censored if instead of observing T we observe the couple (X, Δ) where $X = T \wedge C$ and $\Delta = 1(T \leq C)$. Here we use the notations $T \wedge C = \min(T, C)$ and $1(A)$ denotes the set indicator function equal to 1 if A is true and 0 otherwise. Assuming that the random variables T and C are independent and defining for $x \geq 0$ and $\delta \in \{0,1\}$:

$$H_\delta(x) = \Pr(X \leq x; \Delta = \delta),$$

it is straightforward to check that:

$$dH_1(x) = f_T(x)S_C(x)dx$$

and:

$$H(x) \equiv H_0(x) + H_1(x) = \Pr(X \leq x) = 1 - S_T(x)S_C(x).$$

Thus, we observe that:

$$\Lambda_T(x) = \int_0^x \frac{dH_1(y)}{\bar{H}(y)}. \tag{11.1}$$

where $\bar{H}(x) = 1 - H(x)$. Then, given n independent identically distributed (i.i.d.) copies $\{(X_i, \Delta_i)\}_{1 \le i \le n}$ of (X, Δ), we naturally derive the well-known Nelson–Aalen [10,11] estimator based on the following empirical processes:

$$H_{1n}(x) \equiv \frac{1}{n} \sum_{i=1}^n 1(X_i \le x; \Delta_i = 1)$$

and:

$$\bar{H}_n(x) \equiv \frac{1}{n} \sum_{i=1}^n 1(X_i \ge x).$$

Indeed, replacing H_1 and $\bar{H}$ by their empirical counterpart just defined we obtain:

$$\hat{\Lambda}_T(x) = \int_0^x \frac{dH_{1n}(y)}{\bar{H}_n(y)} = \sum_{i=1}^n \Delta_i \frac{1(X_i \le x)}{\bar{H}_n(X_i)}.$$

The most important thing allowed by this latent variable representation is that we have the possibility to represent $\hat{\Lambda}_T$ as a functional of the two–dimensional empirical process $x \mapsto (H_{1n}(x), \bar{H}_n(x))$. Then the powerful empirical processes tools (continuous mapping theorem, functional delta–method, etc.) allow the transfer to asymptotic properties of $(H_{1n}, \bar{H}_n)$ to $\hat{\Lambda}_T$. This example is handled in several textbooks such as van der Vaart and Wellner, van der Vaart, and Korosok (see [5–7]). Since the Kaplan–Meier [12] estimator is linked to the Nelson–Aalen estimator by the product-limit operator, it is also a functional of the empirical process $x \mapsto (H_{1n}(x), \bar{H}_n(x))$ and again its asymptotic properties can be derived by the tools mentioned previously. Another point is that the latent variable representation of (X, Δ) is also useful for simulating data and trying alternative estimation methods.

11.2.2 PARTLY OBSERVED CURRENT STATUS DATA

Let T denote the random lifetime of interest. We consider the case where instead of T we observe independent copies of a finite nonnegative duration X and of a discrete variable $A \in \{0, 1, 2\}$, such that:

$$\begin{cases} X = T & \text{if} \quad A = 0 \\ X < T & \text{if} \quad A = 1 \\ X > T & \text{if} \quad A = 2 \end{cases}$$

The aim is to estimate the distribution of T based on n i.i.d. copies $\{(X_i, A_i)\}_{1 \le i \le n}$ of (X, A). Let us point out that the limit case where the event $A = 2$ (resp. $A = 1$) has zero probability corresponds to the usual random right-censoring (resp. left-censoring) setup we discussed in the previous section. If $A = 0$ has zero probability, we obtain current status data which means that each observation is either a right censoring time or a left censoring time. This observation is a special case of Turnbull [13] grouped, censored, and truncated data. Note that the Non-Parametric Maximum Likelihood Estimator (NPMLE) for the distribution of duration data partly interval censored has been studied by Huang [14]. There are many papers focusing on the derivation of the NPMLE in situations where the observation scheme does not allow us to obtain an explicit version of this quantity at the contrary to the right censoring case of the previous section.

To derive an explicit version of the distribution of interest, the authors in [15] propose the following latent variable model for (X, A). Let us introduce a non-negative variable C and a Bernoulli variable Δ such that:

$$
\begin{cases}
X = T \text{ and } A = 0 & \text{if} \quad T \le C \text{ and } \Delta = 1 \\
X = C \text{ and } A = 1 & \text{if} \quad C < T \\
X = C \text{ and } A = 2 & \text{if} \quad T \le C \text{ and } \Delta = 0
\end{cases}
$$

and for purposes of identification it is assumed that the random variables T, C, and Δ are independent. As in the previous section, distributions functionals of T and C are indexed by T and C, respectively, while $\Pr(\Delta = 1) = p \in [0,1]$. Note that $p = 1$ corresponds to the right censoring case and that $p = 0$ corresponds to current status data. However, for identification it is assumed that $p \in (0,1]$, which guaranties that a proportion of durations of interest will be observed. For the sake of simplicity we assume that T and C admit PDF functions. Thus, defining:

$$
H_a(x) = \Pr(X \le x; A = a),
$$

for $a = 0, 1, 2$ it is easy to check that:

$$
\begin{cases}
dH_0(x) & = & pS_C(x)f_T(x)dx \\
dH_1(x) & = & f_C(x)S_T(x)dx \\
dH_2(x) & = & (1-p)f_C(x)F_T(x)dx
\end{cases}
$$

and using the notation $\bar{H}_a(x) = \int_{[x,+\infty)} dH_a(y)$ simple calculations lead to:

$$
\bar{H}_0(x) + p\bar{H}_1(x) = pS_T(x)S_C(x)
$$

from which we obtain the following representation for the hazard rate function λ_T:

$$\lambda_T(x)dx = \frac{dH_0(x)}{\overline{H}_0(x) + p\overline{H}_1(x)}.$$

In addition we have:

$$p = \frac{\overline{H}_0(0)}{\overline{H}_0(0) + \overline{H}_2(0)} = \Pr(\Delta = 1 \mid T \leq C) \equiv \frac{\Pr(A = 0)}{\Pr(A = 0) + \Pr(A = 2)}.$$

Then, given n i.i.d. copies $\{(X_i, A_i)\}_{1 \leq i \leq n}$ of (X, A), [15] derives a Nelson–Aalen type estimator of Λ_T defined by:

$$\hat{\Lambda}_T(x) = \int_{[0,x]} \frac{dH_{0n}(x)}{\overline{H}_{0n}(x) + \hat{p}\overline{H}_{1n}(x)}$$

where:

$$H_{an}(x) = \frac{1}{n}\sum_{i=1}^{n} 1(X_i \leq x; A_i = a) \text{ and } \overline{H}_{an}(x) = \frac{1}{n}\sum_{i=1}^{n} 1(X_i \geq x; A_i = a)$$

and:

$$\hat{p} = \frac{\sum_{i=1}^{n} 1(A_i = 0)}{\sum_{i=1}^{n} 1(A_i \neq 1)}.$$

Alternatively the estimator $\hat{\Lambda}_T(x)$ can be written:

$$\hat{\Lambda}_T(x) = \sum_{i=1}^{n} \frac{1(A_i = 0)1(X_i \leq x)}{\sum_{j=1}^{n} \{1(X_j \geq X_i; A_j = 0)\hat{p} + 1(X_j \geq X_i; A_j = 1)\}}.$$

In addition to the fact that this estimator is explicit, it is easily seen that it can be written as functional of the three-dimensional empirical process $x \mapsto (H_{0n}(x), H_{1n}(x), H_{2n}(x))$, which allows us to derive its asymptotic behavior by standard empirical processes tools.

11.2.3 COMPETING RISKS MODELS

Latent variable models are useful for modeling missing data phenomena. Let us consider a very simple example of latent variable model useful for handling the fact that sometimes the cause of failure is unknown. In reliability, competing risk models correspond to series component systems that fail whenever one of the components is down. Thus, if the system is made of p components and if $\tilde{X}_i$ denotes the lifetime of the ith component, the lifetime of the whole system is nothing but $X = \min_{1 \leq j \leq p} \tilde{X}_j = \tilde{X}_1 \wedge \cdots \wedge \tilde{X}_p$ and we note S_X its reliability function. Let us consider several model assumptions (A1, A2, and so on):

A1. $\tilde{X}_1 \ldots \tilde{X}_p$ are i.i.d. and write S the common reliability function of these random variables. Because the reliability function S_X of X verifies:

$$S_X(x) = \Pr(X \geq x) = \left[S(x) \right]^p,$$

we have $S(x) = \left[S_X(x) \right]^{1/p}$. Thus, based on n i.i.d. copies $(X_i)_{1 \leq i \leq n}$ and defining the observable empirical process:

$$\bar{H}_n(x) = \frac{1}{n} \sum_{j=1}^{n} 1(X_j \geq x)$$

it is straightforward to estimate S by:

$$\hat{S}(x) = \left[\bar{H}_n(x) \right]^{1/p},$$

and the asymptotic properties of $\hat{S}$ are inherited from those of $\bar{H}_n$.

A2. $\tilde{X}_1, \ldots, \tilde{X}_p$ are i.n.i.d. (independent but non identically distributed) and S_i denotes the reliability function of $\tilde{X}_i$. Because $S_X(x) = \prod_{j=1}^{p} S_j(x)$, the reliability functions $S_1, \ldots, S_p$ cannot be recovered from S_X. This result is an identifiability issue that disappears if in addition to X we observe the cause of failure; that is, $\Delta = \sum_{k=1}^{p} k 1(X = X_k)$, which is well defined if $\Pr(\exists k \neq k' : \tilde{X}_k = \tilde{X}_{k'}) = 0$, which holds when the distribution of $\tilde{X}_1, \ldots, \tilde{X}_p$ are absolutely continuous with respect to the Lebesgue measure. Based on n i.i.d. copies $\{(X_i, \Delta_i)\}_{1 \leq i \leq n}$ of (X, Δ) and defining for $\delta \in \{1, \ldots, p\}$ the observable empirical processes:

$$H_{\delta,n} = \frac{1}{n} \sum_{i=1}^{n} 1(X_i \leq x; \Delta_i = \delta) \text{ and } \bar{H}_{\delta,n} = \frac{1}{n} \sum_{i=1}^{n} 1(X_i \geq x; \Delta_i = \delta)$$

the j-th cumulative hazard rate function can be consistently estimated by:

$$\hat{\Lambda}_j(x) = \int_0^x \frac{dH_{j,n}(y)}{\bar{H}_n(y)},$$

where $\bar{H}_n = \sum_{k=1}^{p} \bar{H}_{k,n}$ and the asymptotic properties of $(\hat{\Lambda}_1, \ldots, \hat{\Lambda}_p)$ are inherited from those of $(H_{1,n}, \ldots, H_{p,n})$.

A3. We consider now the case where the cause of failure Δ may be missing completely at random in the previous i.n.i.d. setup. Considering that $\tilde{X}_1, \ldots, \tilde{X}_p$

are absolutely continuous with respect to the Lebesgue measure, we note λ_j the failure rate of $\tilde{X}_j$ for $1 \leq j \leq p$. We introduce a binary random variable A, independent of (X, Δ) and we observe $(X, D) \equiv (X, A\Delta)$ instead of (X, Δ) such that the cause of failure is known only when $A = 1$, otherwise it is unknown. We write $\alpha \in (0, 1]$ the probability of the event $\{A = 1\}$. Note that if $\alpha = 0$, the failure causes are never observed and the model parameters are no longer identifiable. Let us define the sub-reliability functions $\bar{H}_d(x) = \Pr(X \geq x; D = d)$ for $0 \leq j \leq p$. It is straightforward to obtain:

$$-d\bar{H}_j(x) = \alpha \lambda_j(x) S_X(x) dx \text{ for } 1 \leq j \leq p, \qquad (11.2)$$

$$-d\bar{H}_0(x) = -(1 - \alpha) \sum_{j=1}^{p} \lambda_j(x) S_X(x) dx. \qquad (11.3)$$

Again, based on n i.i.d. copies $\{(X_i, D_i)\}_{1 \leq i \leq n}$ of (X, D), we can define for $d = 0, \ldots, p$ the observable empirical processes:

$$H_{d,n}(x) = \frac{1}{n} \sum_{i=1}^{n} 1(X_i \leq x; D_i = d),$$

$$\bar{H}_n(x) = \frac{1}{n} \sum_{i=1}^{n} 1(X_i \geq x).$$

and a natural estimator for α is $\hat{\alpha} = n^{-1} \sum_{i=1}^{n} 1(D_i > 0)$.

Using Equation 11.2, we see that for $1 \leq j \leq p$ the cumulative hazard rate function λ_j are consistently estimated by:

$$\hat{\Lambda}_j(x) = \frac{1}{\hat{\alpha}} \int_{[0,x]} \frac{dH_{j,n}(x)}{\bar{H}_n(x)} = \frac{1}{\hat{\alpha}} \sum_{i=1}^{n} \frac{1(X_i \leq x; D_i = j)}{\sum_{k=1}^{n} 1(X_k \geq X_i)}.$$

Neglecting the information coming from Equation 11.3 may be harmful. In [16], the authors propose the following strategy. Let us define $\Lambda_0 = \sum_{j=1}^{p} \Lambda_j$. Using Equation 11.3, Λ_0 can be estimated by:

$$\hat{\Lambda}_0(x) = \frac{1}{1 - \hat{\alpha}} \int_{[0,x]} \frac{dH_{0,n}(x)}{\bar{H}_n(x)} = \frac{1}{1 - \hat{\alpha}} \sum_{i=1}^{n} \frac{1(X_i \leq x; D_i = 0)}{\sum_{k=1}^{n} 1(X_k \geq X_i)}.$$

Let us define $\Lambda^* = (\Lambda_0, \ldots, \Lambda_p)$ and $\hat{\Lambda}^* = (\hat{\Lambda}_0, \ldots, \hat{\Lambda}_p)$. First they show that $\sqrt{n}(\hat{\Lambda}^* - \Lambda^*)$ converges weakly to a centered Gaussian process $\mathbf{G}$ in $(\ell[0, \tau])^{p+1}$ where $[0, \tau]$ is the study interval and $\mathbf{G}$ has a covariance function that satisfies $\mathbb{E}[\mathbf{G}(x)\mathbf{G}(y)] = \Sigma(x, y)$ for $(x, y) \in [0, \tau]^2$. Then the authors look for a linear transformation of $\hat{\Lambda}^*$, which will give an optimal estimator of $\Lambda = (\Lambda_1, \ldots, \Lambda_p)$.

To this end, they define H as the set of $p \times (p+1)$ real valued matrices such that $Ha = a^*$ for all $a^* = (a_1, \ldots, a_p)^T \in \mathbb{R}^p$ and $a = (a^{*T}, \sum_{j=1}^{p} a_j)^T \in \mathbb{R}^{p+1}$.

Then, for a consistent estimator $\hat{\Sigma}(x)$ of $\Sigma(x) \equiv \Sigma(x, x)$, the authors define:

$$\hat{H}(x) = \arg\min_{H \in \mathcal{H}} \operatorname{trace}\left(H \hat{\Sigma}(x) H^T \right)$$

where a close form expression is available for $\hat{H}(x)$ and where $\hat{H}(x)$ has to be calculated at points $X_i \in [0, \tau]$ such that $D_i > 0$. Then $\hat{\Lambda}(x) = \hat{H}(x)\hat{\Lambda}^*(x)$ is a new estimator of $\Lambda(x)$ asymptotically T-optimal in the sense that among all the estimators obtained by linear transformation of $\hat{\Lambda}$, this one has the smallest asymptotic variance trace.

11.3 LATENT VARIABLE MODEL FOR HANDLING HETEROGENEITY

11.3.1 Frailty Models

The frailty models have been introduced in the biostatistical literature in [17] to account for missing covariates or heterogeneity and have been extensively discussed in [18–20]. These models can be viewed as an extension of the Cox proportional hazard model in the sense that it is generally assumed that the hazard rate function of the duration of interest depends upon an unobservable random quantity that acts multiplicatively on it. These unobservable random quantities, varying from an individual to another, are called frailties.

In [21] frailty models are described as random effects models for time variables, where the random effect (the frailty) has a multiplicative effect on the hazard rate function.

In the univariate case we consider now, if T is the duration of interest with hazard rate function λ_T (cumulative hazard function Λ_T and survival function S_T), then we assume that we observe X having a random hazard rate function $\lambda_X(x) \equiv \lambda_{X|Z}(x) = Z\lambda_T(x)$ where Z is an unobserved positive random variable. Z is considered as a random mixture variable, varying across the population. It also means that two individuals have the same hazard rate function up to an unknown factor as in the famous proportional hazard model. Nearly all arguments in favor of assuming a Gamma distribution for Z are based on mathematical and computational aspects. However, for identification reasons the condition $\mathbb{E}(Z) = 1$ is required, then $Z \sim \Gamma(\alpha, \alpha)$ for $\alpha > 0$ where the PDF f of the Gamma distribution $\Gamma(\alpha, \beta)$ is defined by:

$$f(z) \equiv f_{\Gamma(\alpha,\beta)}(z) = \frac{\beta^\alpha z^{\alpha-1}}{\Gamma(\alpha)} e^{-\beta z} \mathbf{1}(z > 0).$$

Using the Bayes inversion formula it is easy to show that conditionally on $X = x$, the frailty Z is distributed according $\Gamma(\alpha, \alpha + \Lambda_T(x))$. We also derive the unconditional PDF f_X of X since:

$$f_X(x) = \int_0^{+\infty} \lambda_{X|Z}(x \mid z) S_{X|Z}(x \mid z) f_Z(z) dz$$

$$= \frac{\alpha^{\alpha+1} \lambda_T(x)}{(\alpha + \Lambda_T(x))^{\alpha+1}}.$$

Thus, if the model is fully parametric (i.e, if λ_T belongs to a parametric family of hazard functions $\mathbf{H} = \{\lambda(\cdot \mid \theta); \theta \in \Theta \in \mathbb{R}^d\}$), then (α, θ) are estimated by:

$$(\hat{\alpha}, \hat{\theta}) = \arg\max_{(\alpha, \theta) \in (0, +\infty) \times \Theta} \Pi_{i=1}^n \frac{\alpha^{\alpha+1} \lambda(X_i \mid \theta)}{(\alpha + \Lambda(X_i \mid \theta))^{\alpha+1}}.$$

The asymptotic properties of the estimators of α and θ are studied in [8] using the theory of martingales for counting processes in the right censoring setup. The semi-parametric joint estimation of (α, Λ_T) has been studied in [22,23]. Frailty models are interesting ways to consider population heterogeneity. By introducing a known correlation structure between the frailty random variable, it is possible to construct some homogeneity test based on an approximation of the score function (see, e.g., [24]).

In the case where Z is a positive discrete random variable belonging to $\{z_1, \ldots, z_d\} \in (0, +\infty)^d$ for some $2 \le d \in \mathbb{N}$ and $\Pr(Z = z_i) = p_i \in (0,1)$, then the reliability function S_X of X is defined by:

$$S_X(x) = \sum_{i=1}^d p_i \left[S_T(x) \right]^{z_i}.$$

It means that the PDF f_X is a convex linear combination of d PDF that are nothing but the conditional PDF of X given $Z = z_i$. This model is a special case of finite mixture models that we discuss in the next section.

11.3.2 Finite Mixture Models

Finite mixture models have been discussed widely in the literature and for an overview on theory and applications of modeling via finite mixture distributions, we refer to [9]. Basically a duration of interest T has a finite mixture distribution if its PDF can be written:

$$f_T(x) = \sum_{j=1}^d p_j f_j(x)$$

where the p_is are non-negative and sum to one and the f_js are PDF. A latent variable representation of T is possible in the sense that if $\tilde{T}_1, \ldots, \tilde{T}_d$ and Z are $p+1$ random variables such that $\tilde{T}_j$ has PDF f_j for $1 \le j \le d$, $Z \in \{1, \ldots, d\}$ with $\Pr(Z = z_j) = p_j$ for $1 \le j \le d$, then if Z and $(\tilde{T}_1, \ldots, \tilde{T}_d)$ are independent T and $\tilde{T}_Z$ have the same PDF and thus the same distribution. T can be seen as the lifetime of an individual chosen at random within d populations where the proportion of individuals coming from the ith population is p_i and the lifetimes coming from the ith population are homogeneous with PDF f_i. Sometimes we are interested in estimating the distributions of the d sub-populations, that is, the distributions of the $\tilde{T}_i$s.

Of course, if the latent variable Z is observed and if S_j denotes the reliability function of $\tilde{T}_j$ for some $1 \le j \le d$ then based on n i.i.d. copies $\{(T_i, Z_i)\}_{1 \le i \le n}$ of (T, Z):

$$\hat{S}_j(x) = \frac{\sum_{i=1}^{n} 1(T_i \ge x; Z_i = j)}{\sum_{i=1}^{n} 1(Z_i = j)} \quad \text{and} \quad \hat{p}_j = \frac{\sum_{i=1}^{n} 1(Z_i = j)}{n},$$

are obviously consistent nonparametric estimators of both S_j and p_j for $1 \le j \le d$. The problem becomes more intricate when the component information Z is no longer available. The first problem we have to face is an identifiability issue: can we recover $\boldsymbol{p} = (p_1, \ldots, p_d)$ and $\boldsymbol{f} = (f_1, \ldots, f_d)$ from the knowledge of f_T? Without additional constraints on the f_i values the answer is generally negative; indeed, for $d = 2$ suppose that there exist two PDF f_1 and f_2 such that:

$$f_T = pf_1 + (1-p)f_2$$

with $p \in (0,1)$, $f_1 = \alpha g_1 + (1-\alpha)g_2$ and $f_2 = \beta g_1 + (1-\beta)g_2$ where $g_i \ne f_j$ for $1 \le i, j \le 2$. Then the PDF admits another representation:

$$f_T = (p\alpha + (1-p)\beta)g_1 + (p(1-\alpha) + (1-p)(1-\beta))g_2$$

which shows that the semi-parametric identifiability fails. It is not possible to obtain identifiability in the semi-parametric setup without additional constraints on the sub-distribution functions f_i. See [25] for the discussion of this problem in the setup of right-censored data. In the case of mixture of parametric lifetime distributions, that is when:

$$\forall j \in \{1, \ldots, d\}, \quad f_j \in \mathbf{F} = \{f(\cdot \mid \theta); \theta \in \Theta \subset \mathbb{R}^k\},$$

we have $f_T(x) \equiv f_T(x; \boldsymbol{p}, \boldsymbol{\theta})$ with $\boldsymbol{\theta} = (\theta_1, \ldots, \theta_d)$ and for all $x \in \mathbb{R}$

$$f_T(x; \boldsymbol{p}, \boldsymbol{\theta}) = \sum_{j=1}^{d} p_j f(x \mid \theta_j)$$

Hence, the identifiability condition becomes: $f_T(x; \boldsymbol{p}, \boldsymbol{\theta}) = f_T(x; \boldsymbol{p}', \boldsymbol{\theta}')$ for all $x \in \mathbb{R}$ implies $(\boldsymbol{p}, \boldsymbol{\theta}) = (\boldsymbol{p}', \boldsymbol{\theta}')$. Classical identifiability conditions may be found in [26,27]. Additional identifiability conditions related to mixture of classical parametric lifetime distributions are given in [28].

In the case of right censoring and left truncation, that is when instead of observing T we observe (L, X, Δ) where $X = \tilde{T}_Z \wedge C = T \wedge C \geq L$ and $\Delta = 1(T \leq C)$ with C a right censoring time and L a left truncation time both independent of the label random variable Z and the lifetime T. The authors in [29] have shown that it is possible to use the EM–algorithm to estimate the unknown model parameters based on n i.i.d. realizations of (L, X, Δ). However, in the discussion of the this paper, [30] mentioned that the EM–algorithm may be trapped by a local maximum and as proposed in [31], as an alternative estimation method, to use the stochastic EM–algorithm. Here we recall the stochastic EM–algorithm principle and we show that it can be easily extended to the case of parametric mixtures when data are right censored and left truncated. Let us write $l = (l_1, \ldots, l_n)$, $x = (x_1, \ldots, x_n)$ and $\delta = (\delta_1, \ldots, \delta_n)$ where $(l, x, \delta) = ((l_1, x_1, \delta_1), \ldots, (l_n, x_n, \delta_n))$ are n i.i.d. realizations of (L, X, Δ) and for $1 \leq i \leq n$ we have $x_i = t_i \wedge c_i$. Let us write $t = (t_1, \ldots, t_n)$. For the sake of simplicity we note for $1 \leq k \leq d$, $S(\cdot \mid \theta_k)$ the reliability function of $\tilde{T}_k$ and $\lambda(\cdot \mid \theta_k)$ its hazard rate function, then it is not difficult to check that for $1 \leq k \leq d$ we have:

$$h(k, l, x, \delta; p, \theta) = \Pr(Z = k \mid (L, X, \Delta) = (l, x, \delta))$$

$$= \frac{p_k \left(\lambda(x \mid \theta_k) \right)^\delta S(x \mid \theta_k) / S(l \mid \theta_k)}{\sum_{j=1}^{p} p_j \left(\lambda(x \mid \theta_j) \right)^\delta S(x \mid \theta_j) / S(l \mid \theta_j)}.$$

It is important to note that the above probability does not depend on the distribution of L and C, thus it is possible to estimate both p and θ following the method of [25].

As the EM–algorithm, the stochastic–EM algorithm is an iterative algorithm which requires an initial value for the unknown parameter θ, for example, θ^0, and which allows us to obtain iterates $(p^s, \theta^s)_{s \geq 1}$. Indeed let (p^s, θ^s) be the current value of the unknown parameters, the next value (p^{s+1}, θ^{s+1}) is derived in the following way:

1. *Expectation step:* For $i = 1, \ldots, n$ and $j = 1, \ldots, d$ calculate:

$$p_{ij}^s = h(j, l_i, x_i, \delta_i; p^s, \theta^s).$$

2. *Stochastic step:* For $i = 1, \ldots, n$ simulate a realization z_i^s of a random variable taking the value $j \in \{1, \ldots, d\}$ with probability p_{ij}^s, and define for $j = 1, \ldots, d$:

$$X_j^s = \{i \in \{1, \ldots, n\}; z_i^s = j\}.$$

3. *Maximization step:* For $j = 1, \ldots, d$ we set:

$$p_j^{s+1} = \frac{\text{Card}(X_j^s)}{n},$$

and for $j = 1, \ldots, d$ we have:

$$\theta_j^{s+1} = \arg\max_{\theta \in \Theta} \ell_j(\theta \mid (l, x, \delta)),$$

where:

$$\ell_j(\theta \mid (l, x, \delta)) = \sum_{i \in X_j^s} \left\{ \delta_i \log \lambda(x_i \mid \theta) - \int_{l_i}^{x_i} \lambda(x \mid \theta) dx \right\}.$$

In the special case of mixture of exponential distributions, that is when $F = \{x \mapsto f(x \mid \theta) = \theta \exp(-\theta x) 1(x > 0); \theta \in (0, +\infty)\}$, it is straightforward to see that for $j = 1, \ldots, d$ we have:

$$\theta_j^{s+1} = \frac{\sum_{i \in X_j^s} \delta_i}{\sum_{i \in X_j^s} (x_i - l_i)}.$$

Obtaining an initial guess θ^0 may be a tricky problem, see [25] for discussion and comments about initialization of the stochastic EM–algorithm. There are several ways to construct the final estimate based on K iterations of the algorithm. The most classical one, because the sequence $(p^s, \theta^s)_{s \geq 1}$ is a Markov chain, consists in taking the ergodic mean of iterates, that is:

$$\hat{p} = \frac{1}{K} \sum_{s=1}^{K} p^s \quad \text{and} \quad \hat{\theta} = \frac{1}{K} \sum_{s=1}^{K} \theta^s.$$

The asymptotic properties of these estimators have been studied in [32]. It may be more stable to replace the current value of the parameters obtained at step s of the preceding stochastic EM–algorithm by the average of the estimates obtained along the $s-1$ first iterations. However, this method is at the cost of losing the Markov's property of $(p^s, \theta^s)_{s \geq 1}$.

11.3.3 CURE MODELS

Cure models are special cases of duration models; Boag, [33] was among the first to consider a population of patients containing a cured fraction. He used a mixture model to fit a data set of follow-up study of breast cancer patients and estimated the cured fraction by maximum likelihood method. As previously stated, the specificity of cure models comes from the fact that a fraction of subjects in the population will never experience the event of interest. This outcome is the reason why most of cure models are special cases of mixture models where the time of interest T has the following distribution:

$$T \sim (1-p)P_0 + p\delta_\infty,$$

where $p \in [0,1]$, P_0 is the probability measure of a non-negative random variable and δ_∞ is the Dirac measure at $\{+\infty\}$. It means that if Y is a Bernoulli latent random variable with probability of success p, T_0 a non-negative random variable, independent of Y, with cumulative distribution function $F_0(t) = P_0([0,t])$, then T has the same distribution as $(1-Y) \times T_0 + Y \times \{+\infty\}$. Hence, the distribution of T is degenerated in the sense that its cumulative distribution function F_T verifies: $F_T(t) = (1-p)P_0([0,t]) + p\delta_{+\infty}([0,t]) = (1-p)F_0(t) \to 1-p$ as t tends to $+\infty$. The parameter p corresponds to the fraction of cured patients.

Now considering right censoring means that we observe $(X,\Delta) = (T \wedge C, 1(T \le C))$ instead of T where C is a random or deterministic right censoring time. In addition to (X,Δ) an $\mathbb{R}^p$-valued covariate vector $\mathbf{Z} = (Z_1, \ldots, Z_p)$ maybe observed, and if we consider that T_0 and C are independent conditionally on $\mathbf{Z}$ but that conditionally on $\mathbf{Z} = z$ we have $Y \sim B(p(z))$ then the conditional cumulative distribution function of T given $\mathbf{Z}$ is defined by:

$$F_{T|\mathbf{Z}}(t \mid z) = (1-p(z))P_0([0,t] \mid z) + p(z)\delta_{+\infty}([0,t]) = (1-p(z))F_0(t \mid z).$$

Because $\Pr(C < +\infty) = 1$, the event $\{T = +\infty\}$ will never be observed since $X \le C$ with probability one. Concerning the probability of being cured a logistic regression model is generally assumed (see [34]):

$$p(z \mid \gamma_0, \gamma) = \frac{\exp(\gamma_0 + \gamma^T z)}{1 + \exp(\gamma_0 + \gamma^T z)}.$$

Concerning the distribution of T_0 conditionally on $\mathbf{Z} = z$ parametric and semi-parametric approaches are available. A review of most standard models and softwares is available in [35]. Let us look at the general principle of implementation of a stochastic EM–algorithm for a cure model. First let us write θ the model parameter that may include functional parameters and we note $(x_i, \delta_i, z_i)_{1 \le i \le n}$ the observed data:

Step 1: Find an initial guess $\theta^{(0)}$ for θ
Step 2: Update the current value $\theta^{(k)}$ to $\theta^{(k+1)}$
    a. For $i \in \{1, \ldots, n\}$ simulate a realization $y_i^{(k)}$ using the law of Y conditionally on $(X,\Delta,\mathbf{Z}) = (x_i, \delta_i, z_i)$ and $\theta = \theta^{(k)}$;
    b. Based on the augmented data $(x_i, \delta_i, y_i^{(k)}, z_i)_{1 \le i \le n}$ calculate $\theta^{(k+1)}$ by an appropriate method;
Step 3: Based on iterates $\theta^{(1)}, \theta^{(2)}, \ldots, \theta^{(K)}$ obtained by using repeatedly the above step 2 derive a estimate $\hat{\theta}$ of θ.

    Writing $S_0(x \mid z)$ the survival function of T_0 conditionally on $\mathbf{Z} = z$ it is easy to check that:

$$q_\theta(x,\delta,z) \equiv P_\theta\big(Y=1\,|\,(X,\Delta,\mathbf{Z})=(x,\delta,z)\big)$$

$$= \frac{p(z\,|\,\gamma_0,\gamma)(1-\delta)}{p(z\,|\,\gamma_0,\gamma)+\big(1-p(z\,|\,\gamma_0,\gamma)\big)S_0(x\,|\,z)}.$$

It is important to note that this conditional probability does not depend on the distribution of the censoring variable. This fact is essential because it allows considering the distribution of C as a nuisance parameter in the model.

Example 11.1

Let us assume that the covariate Z is real-valued and that conditionally on $Z=z$ the random time T_0 follows an exponential proportional hazards model with conditional hazard rate functions defined by $\lambda_0(z|\,\beta) = \exp(\beta_0 + \beta_1 z)$. Then setting $\gamma = (\gamma_0, \gamma_1)$, $\beta = (\beta_0, \beta_1)$, and $\theta = (\gamma_0, \gamma_1, \beta_0, \beta_1) \in \mathbb{R}^4$, and

$$q_\theta(x,\delta,z) = \frac{(1-\delta)\exp(\gamma_0 + \gamma_1 z)}{\exp(\gamma_0 + \gamma_1 z) + \exp\big(-xe^{\beta_0 + \beta_1 z}\big)}.$$

Thus, given $\theta^{(k)} = (\gamma_0^{(k)}, \gamma_1^{(k)}, \beta_0^{(k)}, \beta_1^{(k)})$, the kth iterate of θ, for the simulation Step 2a we have for $1 \le i \le n$:

$$y_i^{(k)} \sim B\big(q_{\theta^{(k)}}(x_i, \delta_i, z_i)\big),$$

while the updating Step 2b is as follows:

$$\gamma^{(k+1)} = \underset{\gamma \in \mathbb{R}^2}{\operatorname{argmax}} \sum_{i=1}^{n} \big(y_i^{(k)}(1-\delta_i)\big)\log\big(p(z_i\,|\,\gamma)\big) + \big(1-y_i^{(k)}\big)\log\big(1-p(z_i\,|\,\gamma)\big),$$

And:

$$\beta^{(k+1)} = \underset{\beta \in \mathbb{R}^2}{\operatorname{argmax}} \sum_{i=1}^{n} \big(\big(1-y_i^{(k)}\big)\delta_i\big)\big(\beta_0 + \beta_1 z_i\big) - \big(1-y_i^{(k)}\big)x_i e^{\beta_0 + \beta_1 z_i}.$$

Assuming that K iterates have been obtained, final estimate of Step 3 may be obtained by averaging the iterates, that is $\hat{\theta} = K^{-1}\sum_{k=1}^{K}\theta^{(k)}$.

11.3.4 Excess Hazard Rate Models

Excess hazard rate models are used in cancer epidemiology studies to evaluate the excess of risk due to the disease. Generally, considering that an individual is diagnosed at age $a > 0$ its risk or hazard function $\lambda_{obs}(t)$ at time $t \ge 0$ is $\lambda_{pop}(a+t) + \lambda_{exc}(t)$ where λ_{pop} is the known population risk given by life tables while λ_{exc} is an unknown additional risk due to the disease. In addition, the probability p that the individual does not die from the disease is generally not null resulting in an improper excess risk function λ_{exc} connected to p through the relationship:

$$p = \exp\left(-\int_0^{+\infty} \lambda_{exc}(s)ds\right).$$

Of course, in such a model the population risk and the excess risk may depend on covariates and data are generally incomplete including, for instance, right censoring. For example, a proportional hazards model on the excess risk function allows us to include covariates effects (see [36] for an efficient semi-parametric estimator).

Let us see that it is possible to obtain a latent variable representation for a time to event T the hazard rate function of which is λ_{obs}. Indeed, let us introduce the random variable A corresponding to the age at which the individual is diagnosed. Then let Z be a Bernoulli random variable with probability of success $p \in [0,1]$, $T_\infty = +\infty$, T_p a positive random variable with hazard rate function λ_{pop}, and T_0 a positive random variable with hazard rate function λ_0. Assume, in addition, that conditionally on A the random variables Z, T_p, and T are independent, then conditionally on $A = a$, the hazard rate of the random variable:

$$\{Z \times T_\infty + (1-Z) \times T_0\} \wedge \{T_p - A\}$$

is λ_{obs} whenever we have for all $t \geq 0$:

$$\Lambda_0(t) = \int_0^t \lambda_0(s)ds = -\log\left(\frac{e^{-\Lambda_{exc}(t)} - p}{1-p}\right),$$

where $\Lambda_{exc}(t) = \int_0^t \lambda_{exc}(s)ds$. It is interesting to note that the excess hazard rate model is close to the competing risk model. Indeed, if $T_1 = Z \times T_\infty + (1-Z) \times T_0$ and $T_2 = T_p - A$, we observe the smallest lifetime $T = T_1 \wedge T_2$ and the lack of information about the component failure (here $1(T_1 \leq T_2)$ is not observed) is compensated by the assumption that conditionally on A, the distribution of T_2 is known.

There is a large amount of literature about parametric, semi-parametric, and non-parametric estimation of these models. In addition, a major difficulty comes from the heterogeneity of the observed T_p which generally depend on covariates that include the age at diagnostic. See, for example [37] for recent discussion about this issue.

Here, for simplicity, we consider that λ_{pop} is homogeneous, more precisely it means that it does not depend on the age at diagnosis. Let S_{obs} (resp. S_{exc} and S_{pop}) be the survival function associated to the hazard rate function λ_{obs} (resp. λ_{exc} and λ_{pop}). It is straightforward to check that if $A = a$:

$$S_{obs}(t) = S_{exc}(t) \times S_{pop}(t+a) \quad \text{for all } t \geq 0.$$

Hence, based on n i.i.d. copies $(T^{(i)})_{i=1,...,n}$ of T and assuming that all the individuals are diagnosed at the same age a, the empirical estimator of S_{obs} is defined by:

$$\hat{S}_{obs}(t) = \frac{1}{n}\sum_{i=1}^{n} Y_i(t),$$

where $Y_i(t) = 1(T^{(i)} \geq t)$ and thus S_{exc} is naturally estimated by:

$$\hat{S}_{exc}(t) = \frac{\hat{S}_{obs}(t)}{S_{pop}(a+t)}.$$

In this very simple case the asymptotic properties of $\hat{S}_{exc}$ are easy to obtain. Suppose now that the age at diagnosis varies from one individual to another, and let us write a_i the age at diagnosis of the ith individual. It is well known (see [8]) that the intensity process of the counting process $N(t) = \sum_{i=1}^{n} 1(T^{(i)} \leq t)$ is:

$$\sum_{i=1}^{n} Y_i(t)\left(\lambda_{exc}(t) + \lambda_{pop}(a_i + t)\right).$$

By a method-of-moment approach we derive an estimator Λ_{exc} defined by:

$$\hat{\Lambda}_{exc}(t) = \sum_{i=1}^{n}\int_0^t \frac{dN_i(s) - Y_i(s)\lambda_{pop}(a_i+s)ds}{\sum_{i=1}^{n} Y_i(s)},$$

This estimator is known as the Ederrer II estimator of the cumulative excess risk function (see [38] for more general non-parametric estimators of the cumulative excess risk function). Note that in the case of right censoring, that is, if instead of observing $(T^{(i)}, a_i)_{i=1,...,n}$, we observe $(X^{(i)}, \Delta_i, a_i)_{i=1,...,n}$ where $\Delta_i = 1$ if $X^{(i)} = T^{(i)}$, and $\Delta_i = 0$ if $X^{(i)} < T^{(i)}$, then this estimator is still valid simply replacing $N_i(t)$ by $\tilde{N}_i(t) = 1(T^{(i)} \leq t; \Delta_i = 1)$ for $1 \leq i \leq n$.

11.4   LATENT VARIABLE OR PROCESS MODELS FOR HANDLING SPECIFIC PHENOMENA

In this section, we give a few examples of time-dependent models that describe the degradation of a system and where the latent variable may depend on the time. There are a large number of stochastic degradation models, here we focus on Gamma processes. To be more specific $X = (X_t)_{t \geq 0}$ is a Gamma process with scale parameter $b > 0$ and continuous and non-decreasing shape function $a : \mathbb{R}^+ \to \mathbb{R}^+$ with $a(0) = 0$ if X is a random process with independent Gamma distributed increments with common scale parameter $b > 0$ such that $X_0 = 0$ almost surely and $X_t - X_s \sim \Gamma(a(t) - a(s), b)$ for every $0 \leq s < t$ where for $\alpha > 0$ and $\beta > 0$ we note $\Gamma(\alpha, \beta)$ the Gamma distribution with PDF:

$$f_{\Gamma(\alpha,\beta)}(x) = \frac{\beta^\alpha x^{\alpha-1} \exp(-\beta x)}{\Gamma(\alpha)} 1(x \geq 0).$$

Note that if the shape function satisfies $a(t) = at$, then the Gamma process X is homogeneous since for $s \geq 0$ and $t \geq 0$, the distribution of $X_{t+s} - X_t$ is nothing but the $\Gamma(as,b)$ distribution which hence does not depend on t.

11.4.1 GAMMA DEGRADATION MODEL WITH RANDOM INITIAL TIME

In [39] a stochastic model is introduced for a component that deteriorates over time. The deterioration is due to defects which appear one by one and next independently propagate over time for passive components within electric power plants, where (measurable) flaw indications first initiate (one at a time) and next grow over time. The available data come from inspections at discrete times, where only the largest flaw indication is measured together with the total number of indications on each component. As a consequence the model of [39] can be seen as a competing degradation model. Let us describe here a simpler model with a single degradation trajectory that initiates at the random time T, which is a latent variable. For example, we may consider that Y_t is the length of a crack at time t that appears at time $T \geq 0$, and we consider that at time $t + T$ the length of the crack is X_t where $X = (X_t)_{t \geq 0}$ is a Gamma process with scale parameter $b > 0$ and continuous and non-decreasing shape function $a(\cdot; \theta_1): \mathbb{R}^+ \to \mathbb{R}^+$ where $a(0; \theta_1) = 0$ with $\theta_1 \in \Theta_1 \subset \mathbb{R}^p$. Then $Y_t = X_{(t-T)^+}$ where $x^+ = \max(0, x)$. Assuming that T has a PDF $f_T(\cdot; \theta_2)$ with $\theta_2 \in \Theta_2 \subset \mathbb{R}^q$ the random variable Y_t has the PDF:

$$f_{Y_t}(y; \theta) = \left(1 - F_T(t; \theta_2)\right)\delta_0(y) + \int_0^t \frac{b^{a(t-s;\theta_1)} y^{a(t-s;\theta_1)-1} \exp(-by)}{\Gamma(a(t-s;\theta_1))} f_T(s; \theta_2) ds dy$$

with respect to the sum of the Dirac measure δ_0 at 0 and the Lebesgue measure dy on $\mathbb{R}$ where $\theta = (\theta_1, \theta_2, b)$. When N i.i.d. copies $(Y^{(k)})_{k=1,\ldots,N}$ of the delayed degradation process $Y = (Y_t)_{t \geq 0}$ are observed at times $0 = t_{00} < t_{k1} < \cdots < t_{kn_k}$ for $k = 1, \ldots, N$, it is possible to derive the joint distribution of $\left(Y_{t_{k1}}^{(k)}, \ldots, Y_{t_{kn_k}}^{(k)}\right)$ to apply a maximum likelihood principle. However, due to numerical instabilities the maximization of the associated log-likelihood function is a tricky problem. An alternative estimation method based on the pseudo-likelihood (or composite likelihood) can be (see, e.g., [40]) an alternative method. It simply consists in maximizing:

$$(\theta) \mapsto \sum_{k=1}^N \sum_{i=1}^{n_k} \log\left(f_{Y_{t_{ki}}}(y_{ki}; \theta)\right),$$

where for $1 \leq i \leq n$ and $1 \leq j \leq N$, y_{ki} is the observation of $Y_{t_{ki}}^{(k)}$. In other words, the pseudo-likelihood method consists in doing as if the random variables $Y_{t_{ki}}^{(k)}$ were independent, this simplifies the calculation of the log-likelihood at the price of a loss of efficiency. See [39] for an application to competing degradation processes.

11.4.2 GAMMA DEGRADATION MODEL WITH FRAILTY SCALE PARAMETER

In [41] a fatigue crack propagation is considered where the crack growth is described by a non-homogeneous Gamma process where the scale parameter is a frailty variable. Let us consider a degradation process $X = (X_t)_{t \geq 0}$ observed at times $0 = t_0 < t_1 < \cdots < t_n$. Let $a : \mathbb{R}^+ \to \mathbb{R}^+$ be a continuous and non-decreasing shape function with $a(0) = 0$ and B a non-negative random variable with f_B as PDF. Here we assume that conditionally on $B = b > 0$ the random process X is a Gamma process with shape function a and scale parameter b. It means that conditionally on $B = b > 0$ the increments $\Delta X_i = X_{t_i} - X_{t_{i-1}}$ for $i = 1, \ldots, n$ are independent with PDF:

$$f_{\Delta X_1, \ldots, \Delta X_n | B}(\delta x_1, \ldots, \delta x_n \mid b) = \prod_{i=1}^{n} \frac{(\delta x_i)^{\Delta a_i - 1} b^{\Delta a_i} \exp(-b \delta x_i)}{\Gamma(\Delta a_i)},$$

where $\Delta a_i = a(t_i) - a(t_{i-1})$. As a consequence the unconditional distribution of $\Delta X_i = X_{t_i} - X_{t_{i-1}}$ for $i = 1, \ldots, n$ when B has a PDF f_B is given by:

$$f_{\Delta X_1, \ldots, \Delta X_n}(\delta x_1, \ldots, \delta x_n) = \prod_{i=1}^{n} \int_0^{+\infty} \frac{(\delta x_i)^{\Delta a_i - 1} b^{\Delta a_i} \exp(-b \delta x_i)}{\Gamma(\Delta a_i)} f_B(b) db.$$

For the special case of Gamma frailties, that is when $B \sim \Gamma(\alpha, \beta)$ we obtain:

$$f_{\Delta X_1, \ldots, \Delta X_n}(\delta x_1, \ldots, \delta x_n) = \prod_{i=1}^{n} \left(\frac{\delta x_i}{\delta x_i + \beta} \right)^{\Delta a_i - 1} \left(\frac{\beta}{\delta x_i + \beta} \right)^{\alpha} \frac{\Gamma(\Delta a_i + \alpha)}{\Gamma(\Delta a_i)\Gamma(\alpha)}.$$

Now suppose that the shape function a depends on an Euclidean parameter $\theta \in \Theta \subset \mathbb{R}^p$, then based on N i.i.d. copies $\left(X^{(k)} \right)_{k=1,\ldots,N}$ of X and setting δx_{ij} the observation of $\Delta X_i^{(j)} = X_{t_i}^{(j)} - X_{t_{i-1}}^{(j)}$ for $1 \leq j \leq N$ and $1 \leq i \leq n$, the likelihood function is therefore defined by:

$$\mathcal{L}(\theta, \alpha, \beta) = \prod_{j=1}^{N} \prod_{i=1}^{n} \left(\frac{\delta x_{ij}}{\delta x_{ij} + \beta} \right)^{\Delta a_i(\theta) - 1} \left(\frac{\beta}{\delta x_{ij} + \beta} \right)^{\alpha} \frac{\Gamma(\Delta a_i(\theta) + \alpha)}{\Gamma(\Delta a_i(\theta))\Gamma(\alpha)},$$

where $\Delta a_i(\theta) = a(t_i; \theta) - a(t_{i-1}; \theta)$.

11.4.3 BIVARIATE GAMMA DEGRADATION MODELS

In [42] the intervention scheduling of a railway track is discussed based on the observation of two dependent randomly increasing deterioration indicators modeled through a bivariate Gamma process $Y = \left(Y_t^{(1)}, Y_t^{(2)} \right)_{t \geq 0}$ constructed by trivariate reduction (see [43]). As we will see next this construction is based on the properties that the sum of two independent Gamma processes with common scale is still a Gamma process from one hand, and, on the other hand, that the components of the bivariate process share a common Gamma latent process which allows obtaining correlation between the two marginal processes.

Now let us consider three independent Gamma processes $X^{(i)}$ for $0 \leq i \leq 2$ with scale parameter one and shape functions $\alpha_i : \mathbb{R}^+ \to \mathbb{R}^+$. The bivariate Gamma process Y is defined by:

$$\begin{cases} Y_t^{(1)} & = & \left(X_t^{(0)} + X_t^{(1)} \right) / b_1 \\ Y_t^{(2)} & = & \left(X_t^{(0)} + X_t^{(2)} \right) / b_2 \end{cases}$$

where b_1 and b_2 are two positive scale parameters. As a consequence Y has independent increments and for $i = 1, 2$ the marginal process $\left(Y_t^{(i)} \right)_{t \geq 0}$ is a Gamma process with scale parameter b_i and shape function $\alpha_0 + \alpha_i$. In addition it is straightforward to check that we have for $i = 1, 2$:

$$\mathbb{E}\left(Y_t^{(i)} \right) = \frac{\alpha_0(t) + \alpha_i(t)}{b_i} \quad \text{and} \quad \text{var}\left(Y_t^{(i)} \right) = \frac{\alpha_0(t) + \alpha_i(t)}{b_i^2}.$$

and:

$$\text{cov}\left(Y_t^{(1)}, Y_t^{(2)} \right) = \frac{\alpha_0(t)}{b_1 b_2}.$$

Now let us consider that the process Y is homogeneous (i.e., $\alpha_i(t) = \alpha_i t$ for $0 \leq i \leq 2$) and observed at times $0 = t_0 < t_1 < t_2 < \cdots < t_n$ and let us define the increments $\Delta t_j = t_j - t_{j-1}$ and $\Delta Y_j = \left(\Delta Y_j^{(1)}, \Delta Y_j^{(2)} \right) = \left(Y_{t_j}^{(1)} - Y_{t_{j-1}}^{(1)}, Y_{t_j}^{(2)} - Y_{t_{j-1}}^{(2)} \right)$ for $1 \leq j \leq n$. It is easy to check that the bivariate Gamma process Y can be parametrized equivalently by $(\alpha_0, \alpha_2, \alpha_2, b_1, b_2)$ or $(\mu_1, \mu_2, \sigma_1^2, \sigma_2^2, \rho)$ where for $1 \leq j \leq n$:

$$\mu_i = \mathbb{E}\left(\frac{\Delta Y_j^{(i)}}{\Delta t_j} \right) = \frac{\alpha_0 + \alpha_i}{b_i} \quad \text{for } i = 1, 2,$$

$$\sigma_i^2 = \text{var}\left(\frac{\Delta Y_j^{(i)}}{\sqrt{\Delta t_j}} \right) = \frac{\alpha_0 + \alpha_i}{b_i^2} \quad \text{for } i = 1, 2,$$

$$\rho = \text{cov}\left(\frac{Y_{t_j}^{(1)} - Y_{t_{j-1}}^{(1)}}{\sqrt{\Delta t_j}}, \frac{Y_{t_j}^{(2)} - Y_{t_{j-1}}^{(2)}}{\sqrt{\Delta t_j}} \right) = \frac{\alpha_0}{b_1 b_2}.$$

Then by the moment method $(\mu_1, \mu_2, \sigma_1^2, \sigma_2^2, \rho)$ is estimated by $(\hat{\mu}_1, \hat{\mu}_2, \hat{\sigma}_1^2, \hat{\sigma}_2^2, \hat{\rho})$ where:

$$\hat{\mu}_i = \frac{\sum_{j=1}^{n} \left(Y_{t_j}^{(i)} - Y_{t_{j-1}}^{(i)} \right)}{\sum_{j=1}^{n} \Delta t_j} \quad \text{for } i = 1, 2,$$

$$\hat{\sigma}_i^2 = \frac{\sum_{j=1}^n \left(Y_{t_j}^{(i)} - Y_{t_{j-1}}^{(i)} - \hat{\mu}_i \Delta t_j \right)^2}{\sum_{j=1}^n \Delta t_j - \frac{\sum_{j=1}^n (\Delta t_j)^2}{\sum_{j=1}^n \Delta t_j}} \quad \text{for } i = 1,2,$$

$$\hat{\rho} = \frac{\sum_{j=1}^n \left(\Delta Y_j^{(1)} - \hat{\mu}_1 \Delta t_j \right)\left(\Delta Y_j^{(2)} - \hat{\mu}_2 \Delta t_j \right)}{\sum_{j=1}^n \Delta t_j - \frac{\sum_{j=1}^n (\Delta t_j)^2}{\sum_{j=1}^n \Delta t_j}},$$

are unbiased estimators of $(\mu_1, \mu_2, \sigma_1^2, \sigma_2^2, \rho)$. Now, since we have:

$$\begin{cases} \alpha_0 & = & \dfrac{\sigma_1^2 \sigma_2^2 \rho}{\mu_1 \mu_2} \\[2ex] b_i & = & \dfrac{\mu_i}{\sigma_i^2} \text{ for } i = 1,2 \\[2ex] \alpha_i & = & \dfrac{\mu_i^2}{\sigma_i^2} - \dfrac{\sigma_1^2 \sigma_2^2 \rho}{\mu_1 \mu_2} \text{ for } i = 1,2 \end{cases}$$

we obtain:

$$\begin{cases} \hat{\alpha}_0 & = & \dfrac{\hat{\sigma}_1^2 \hat{\sigma}_2^2 \hat{\rho}}{\hat{\mu}_1 \hat{\mu}_2} \\[2ex] \hat{b}_i & = & \dfrac{\hat{\mu}_i}{\hat{\sigma}_i^2} \text{ for } i = 1,2 \\[2ex] \hat{\alpha}_i & = & \dfrac{\hat{\mu}_i^2}{\hat{\sigma}_i^2} - \dfrac{\hat{\sigma}_1^2 \hat{\sigma}_2^2 \hat{\rho}}{\hat{\mu}_1 \hat{\mu}_2} \text{ for } i = 1,2 \end{cases}$$

In [42] an alternative estimation method is proposed based on the maximum likelihood principle. Indeed, based on marginal observations $\left(\Delta Y_j^{(i)} \right)_{1 \le j \le n}$ parameters b_i are estimated using the maximum likelihood principle. Then, considering increments $\Delta X_j^{(0)} = X_{t_j}^{(0)} - X_{t_{j-1}}^{(0)}$ for $1 \le j \le n$ as hidden data, the authors develop an EM algorithm to estimate $(\alpha_0, \alpha_1, \alpha_2)$.

11.5   CONCLUDING REMARKS

In this chapter we presented a panel of latent variable models that are useful in reliability and survival analysis studies. We showed that a large variety of parametric, semi-parametric, or nonparametric estimation methods can be used to estimate the models parameters. In addition, when direct calculation of the likelihood function is mathematically too complicated, or numerically unachievable, estimation methods based on EM or stochastic EM algorithms, or estimation methods based on the pseudo-likelihood principle, may be interesting alternatives to classical estimation methods.

REFERENCES

1. B. Everett. *An Introduction to Latent Variable Models.* Springer Monographs on Statistics and Applied Probability, Chapman & Hall, London, UK, 2011.
2. D. Bartholomew, M. Knott, and I. Moustaki. *Latent Variable Models and Factor Analysis: A Unified Approach.* Wiley Series in Probability and Statistics, 3rd ed. John Wiley & Sons, Chichester, UK, 2011.
3. A.A. Beaujean. *Latent Variable Modeling Using R: A Step-by-Step Guide.* Taylor & Francis Group, New York, 2014.
4. J.C. Loehlin and A.A. Beaujean. *Latent Variable Models: An Introduction to Factor, Path, and Structural Equation Analysis*, 5th ed. Taylor & Francis Group, New York, 2016.
5. A.W. van der Vaart and J.A. Wellner. *Weak Convergence and Empirical Processes.* Springer Series in Statistics, New York, 1996.
6. A.W. van der Vaart. *Asymptotic Statistics.* Cambridge University Press, New York, 1998.
7. M. Kosorok. *Introduction to Empirical Processes and Semiparametric Inference.* Springer Series in Statistics, New York, 2006.
8. P.K. Andersen, O. Borgan, R.D. Gill, and N. Keiding. *Statistical Models Based on Counting Processes.* Springer Series in Statistics, New York, 1993.
9. G. McLachlan and D. Peel. *Finite Mixture Models.* John Wiley & Sons, New York, 2000.
10. W. Nelson. Theory and applications of hazard plotting for censored failure data. *Technometrics*, 14(4):945–966, 1972.
11. O. Aalen. Nonparametric inference for a family of counting processes. *The Annals of Statistics*, 6(4):701–726, 1978.
12. E.L. Kaplan and P. Meier. Nonparametric estimation from incomplete observations. *Journal of the American Statistical Association*, 282(53):457–481, 1958.
13. B.W. Turnbull. The empirical distribution function with arbitrary grouped, censored and truncated data. *Journal of the Royal Statistical Society, Series B*, 38:290–295, 1976.
14. J. Huang. Asymptotic properties of nonparametric estimation based on partly interval–censored data. *Statistica Sinica*, 9:501–519, 1999.
15. V. Patilea and J.M. Rolin. Product-limit estimators of the survival function for two modified forms of current-status data. *Bernoulli*, 12:801–819, 2006.
16. L. Bordes, J.Y. Dauxois, and P. Joly. Semiparametric inference of competing risks data with additive hazards and missing cause of failure under mcar or mar assumptions. *Electronic Journal of Statistics*, 8:41–95, 2014.
17. J.W. Vaupel, K.G. Manton, and E. Stallard. The impact of heterogeneity in individual frailty on the dynamics of mortality. *Demography*, 16(3):439–454, 1979.
18. P. Hougaard. *Analysis of Multivariate Survival Data.* Springer, New York, 2000.
19. L. Duchateau and P. Janssen. *The Frailty Model.* Statistics for Biology and Health, Springer, New York, 2008.
20. A. Wienke. *Frailty Models in Survival Analysis.* CRC Press, Boca Raton, FL, 2010.
21. P. Hougaard. Frailty models for survival data. *Lifetime Data Analysis*, 1:255–273, 1995.

22. S.A. Murphy. Consistency in a proportional hazard model incorporating a random effect. *Annals of Statistics*, 22:712–731, 1994.

23. S.A. Murphy. Asymptotic theory for the frailty model. *Annals of Statistics*, 23:182–198, 1994.

24. D. Commenges and H. Jacmin-Gadda. Generalized score test of homogeneity based on correlated random effects models. *Journal of the Royal Statistical Society B*, 59:157–171, 1997.

25. L. Bordes and D. Chauveau. Stochastic em algorithms for parametric and semiparametric mixture models for right-censored lifetime data. *Computational Statistics*, 31(4):1513–1538, 2016.

26. H. Teicher. Identifiability of mixtures of product measures. *Annals of Mathematical Statistics*, 38:1300–1302, 1967.

27. S.J. Yakowitz. On the identifiability of finite mixtures. *Annals Mathematical Statistics*, 39:209–214, 1968.

28. N. Atienza, J. Garcia-Heras, and J.M. Munoz Pichardo. A new condition for identifiability of finite mixture distributions. *Metrika*, 63(2):215–221, 2006.

29. N. Balakrishnan and D. Mitra. EM–based likelihood inference for some lifetime distributions based on left truncated and right censored data and associated model discrimination. *South African Statistical Journal*, 48:125–171, 2014.

30. T.H.K. Ng and Z. Ye. Comment: EM–based likelihood inference for some lifetime distributions based on left truncated and right censored data and associated model discrimination. *South African Statistical Journal*, 48:177–180, 2014.

31. L. Bordes and D. Chauveau. Comment: EM–based likelihood inference for some lifetime distributions based on left truncated and right censored data and associated model discrimination. *South African Statistical Journal*, 48:197–200, 2014.

32. S.F. Nielsen. The stochastic EM algorithm: Estimation and asymptotic results. *Bernoulli*, 6(3):457–489, 2000.

33. J.W. Boag. Maximum likelihood estimates of the proportion of patients cured by cancer therapy. *Journal of the Royal Statistical Society Series B*, 11:15–45, 1949.

34. V. Farewell. A model for binary variable with time-censored observations. *Biometrika*, 64:43–46, 1977.

35. Y. Peng and J.M.G. Taylor. Cure models. In J. Klein, H. van Houwelingen, J.G. Ibrahim, and T.H. Scheike, editors, *Handbook of Survival Analysis*, Chapter 6, pp. 113–134. Handbooks of Modern Statistical Methods Series, Chapman & Hall, Boca Raton, FL, 2014.

36. P. Sasieni. Proportional excess hazards. *Biometrika*, 83(1):127–141, 1996.

37. P. Sasieni and A.R. Brentnall. On standardized relative survival. *Biometrics*, 73(2):473–482, 2016.

38. M.P. Perme, J. Stare, and J. Estève. On estimation in relative survival. *Biometrics*, 68(1):113–120, 2012.

39. L. Bordes, S. Mercier, E. Remy, and E. Dautrême. Partially observed competing degradation processes: Modeling and inference. *Applied Stochastic Models in Business and Industry*, 32(5):677–696, 2016.

40. D. Cox and N. Reid. A note on pseudolikelihood constructed from marginal densities. *Biometrika*, 91:729–737, 2004.

41. M. Guida and F. Penta. A gamma process model for the analysis of fatigue crack growth data. *Engineering Fracture Mechanics*, 142:21–49, 2015.

42. S. Mercier, C. Meier-Hirmer, and M. Roussignol. Bivariate gamma wear processes for track geometry modelling, with application to intervention scheduling. *Structure and Infrastructure Engineering*, 8(4):357–366, 2012.

43. F.A. Buijs, J.W. Hall, J.M. van Noortwijk, and P.B. Sayers. Time-dependent reliability analysis of flood defences using gamma processes. In G. Augusti, G.I. Schueller, and M. Ciampoli, editors, *Safety and Reliability of Engineering Systems and Structures*, pp. 2209–2216; Proceedings of the Ninth International Conference on Structural Safety and Reliability (ICOSSAR), Rome, Italy, June 19–23, 2005, Mill-Press, Rotterdam, the Netherlands.

12 Expanded Failure Modes and Effects Analysis
A Different Approach for System Reliability Assessment

Perdomo Ojeda Manuel, Rivero Oliva Jesús, and Salomón Llanes Jesús

CONTENTS

12.1 BACKGROUND FOR DEVELOPING THE EXPANDED FAILURE MODES AND EFFECTS ANALYSIS

There is significant experience in the field of systems reliability analysis. Here the Fault Tree Analysis (FTA) technique has played an important role due to its power for emphasizing some aspects that exert an enormous influence on the reliability of redundant systems, specifically those designed to operate with high availability and safety requirements.

These analyses constitute a key support tool in decision-making process, which, in turn, are a crucial aspect when it concerns activities or processes in industrial facilities or services with significant hazards associated to the processes with which they deal [1].

To determine the dominant contributors of the risk or the reliability of a system, detailed information needs to be adequately processed so that the proposed objective can be accomplished.

However, the following difficulties are frequently present:

* Not all the necessary information for a correct decision is always available
* An important part of the information may be available, but not organized and processed in an appropriate manner

The solution of these key problems can be achieved by:

* Gathering the raw data of the facility, processing them adequately, and preparing a database oriented to reliability and safety, so that a specialized computer tool of reliability and risk analysis can use it in a proper manner
* Training and qualifying specialists and managers in the use of these databases and specialized computer programs so that the data can be used correctly and decisions can produce the expected results

Training and qualifying the staff of an industry in the use of specialized programs in this field is not a major problem, or at least its solution can be ready in the short term, because there is currently a significant amount of experience in that field.

Nevertheless, the collection of data, its handling, and developing of computerized databases, ready to be used in risk and reliability studies, are time-consuming tasks. On the other hand, the sample of available data should be sufficiently representative of the processes that are going to be modeled (e.g., failure rates of components-failure modes and average repair times).

Moreover, inaccuracies in the definition of the component boundaries and in the way the raw data are described, among other aspects, bring with them uncertainties in the data to be processed. The uncertainties degrees could be so high that, for example, the generic databases available for use in the Probabilistic Safety Assessment (PSA) indicate differences of up to 2 orders of magnitude in the values of the failure rates of the same failure mode and type of equipment [2,3].

In addition to this, it happens that current technological advances have in some fields such a dynamic that, when it begins to collect the data and organize them to be used, there may already be new designs somewhat different from those from which data has been collected.

Regarding these problems, there is a need to look for ways to reach useful results, even in the case of partial or almost total lack of data. Hence, qualitative analysis tools, such as Failure Modes and Effects Analysis (FMEA) and Hazard and Operability (HAZOP), need to become competitive with the powerful quantitative tools, such as FTA, whose results depend to a large extent on statistical data.

But, the system "analysis tool-analyst team" must be able to fulfill the objective of a detailed reliability or safety study with the lowest possible cost, the shortest execution time, and the least associated uncertainties. This result can be achieved by providing the analysis team with a tool that covers an exhaustive spectrum of safety-reliability aspects to be evaluated, the methods for their evaluation, and some useful analysis options.

The matrix shown in Table 12.1 presents a comparison among a set of reliability and risk analysis techniques widely used in industrial applications [4–9]. The compared attributes are based on important characteristics that an effective analysis tool should meet to support the decisions regarding safety and availability of facilities and services with potential risks associated with their operation.

It also includes, as a comparative pattern, the most frequently used techniques in the risk studies: the FTA and the Event Tree Analysis (ETA) [10–22], given their benefits and strengths in this subject.

In Table 12.1, the symbols used as qualificator of the technique characteristics mean:

TABLE 12.1

Comparative Matrix of Reliability and Risk Analysis Techniques

Techniques ▶ Items to Compare ▼	HAZOP	FMEA	Checklist	What If?	SR	PreHA	ETA	FTA
Completeness	++	++	–	–	–	–	+	++
Structured approach	++	++	–	–	–	–	++	++
Flexibility of application	–	++	++	++	++	++	+	+
Objectivity	+	+	–	–	–	–	+	+
Independence on quantitative data[a]	+	+	++	++	++	++	++	–
Capability of modeling dependences	–	–	–	–	–	–	++	++
Independence on the analysis team expertise[b]	–	+	++	+	+	–	––	––
Quickness in obtaining results	–	–	+	+	+	+	+	–

[a] In achieving quantitative results.

[b] Refers to the skill in using the technique.

++ High
+ Moderate
− Low
−− Very low or none

The acronyms used previously mean:

HAZOP: *HAZ*ard and *Op*erability *A*nalysis
FMEA: *F*ailure *M*ode and *E*ffect *A*nalysis
SR: *S*afety *R*eview
PreHA: *Pre*liminary *H*azard *A*nalysis
ETA: *E*vent *T*ree *A*nalysis
FTA: *F*ault *T*ree *A*nalysis

The characteristics to compare the different techniques in the matrix of Table 12.1 have been defined in a positive sense considering the benefits of the technique to achieve an exhaustive reliability or risk analysis of a system. This requirement means that the techniques with the greatest number of "++" and "+" results will be the best candidates to use in that kind of analysis.

Analyzing the previous matrix, the FMEA technique resulted one of the best candidates to improve for powering its characteristics to include some important analytical advantages of the FTA as, for example, the functional dependence and common cause failure (CCF) analyses.

On the other hand, reviewing some recently works published about the FMEA methodology [23–26], not one was found dedicated to treat the subject of the dependency analysis within FMEA/Failure Mode, Effects, and Criticality Analysis (FMECA).

Thus, an analytical tool has been developed that keeps the best characteristics of the qualitative techniques as FMEA does and that adds some of the greatest potentials of the quantitative ones. These strengths along with some other important features, which have been included in the expanded FMEA (FMEAe) methodology, are described in the next sections.

12.2 SOME DISTINCTIVE FEATURES OF THE FMEAe METHODOLOGY

The FMEA technique is recognized as a powerful analysis tool because it combines the structuring and the completeness of the method with descriptive capacities that improve its integrity, giving the analyst the flexibility to describe in a more complete way all the characteristics of the system from either the design or the operation standpoints. Thus, FMEA analyzes how these characteristics may influence the system reliability, or the risk they induce, and gives an order of the importance of system's postulated failure modes, allowing optimizing the corrective measures to reduce risk or increase reliability [4–10].

However, also recognized as an important limitation of the method is its inability for dependence analysis, an aspect well modeled by the FTA technique. Precisely one of the most frequent causes of accidents in complex industrial facilities has been the common cause failures and human errors, hence the importance of being able to treat them adequately in these studies.

An important part of the insufficiencies or limitations found in the qualitative techniques presented in Table 12.1 have been resolved in FMEAe. The most significant improvements in FMEAe methodology, in comparison with the traditional FMEA[1] were introduced through several procedures for:

- Determination of common cause basic events, estimation of their probabilities, and inclusion of them in the list of failure modes for the criticality analysis
- Analyzing the joint importance of components of the same type to determine those types of components with largest contribution to risk or unavailability and backing up, in that way, the standardization of the corrective measures
- Estimating the risk or reliability of the system under analysis by means of a global parameter called System Risk Index (IRS).

In FMEAe, these analyses are carried out through algorithms of identification and comparison of strings. These strings include the information in the fields of the traditional FMEA worksheet, together with some other that have been added to enlarge and complete the information about the design and functioning of the components involved in the analysis [27,28].

Figure 12.1 presents the work sheet of FMEAe in ASeC computer code, showing the CCF modes included at the final part of the list (those whose code begins with CM),

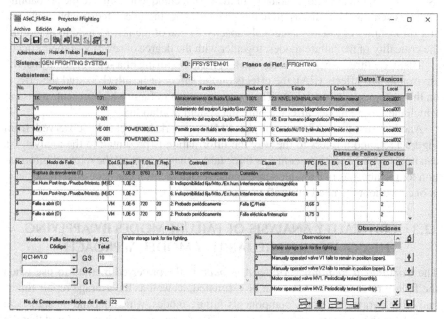

FIGURE 12.1 FMEAe worksheet in ASeC computer code.

---

[1] Not included in the computer codes considered in the state of the art of this methodology.

which form part of the criticality analysis. These events are generated automatically by algorithms that handle the information the analyst provides for the worksheet using the first two tables: Datos Técnicos (Engineering- related Data) and Datos de Fallos y Efectos (Failures and Effects Data). The third table of the worksheet, Observaciones (Remarks), serves to complete ideas or descriptions about the failure mode, mode of operation, mode of control of the components, or any consideration made for the analysis.

In the lower left corner of the worksheet appears a panel showing the quantity and code of the component-failure modes, which are the precursors of the CCF, classified by their degree of dependence (G1 to G3, in decreasing order of size). In this example, ten precursors having a G3 dependence degree were determined, where G3 represents the highest degree of dependence, G2 an intermediate degree, and G1 the lowest degree. As can be observed, for this example, there were no precursors of the G1 degree because all the five pairs of precursors share the attributes of the G3 failure. Later, in Section 12.4, the methodology developed for the automated determination of CCFs in FMEAe will be described in more detail.

After the worksheet fields had been filled, if there are data available it is convenient to start the criticality analysis by determining the CCFs so that their influence in the results is not missed. This latter issue is especially important in the case of redundant systems, which can be verified later (Section 12.6) through the example of application of the FMEAe methodology to a fire cooling system.

After determining the CCF, the criticality analysis is carried out by one of two approaches: the Component Reliability Model or the Risk Matrix. The first approach uses models similar to those included in the FTA technique to estimate the probability of basic events and it is discussed in this chapter. Here, the calculated reliability parameter (probability of loss of functional capacity) is one of the factors used to determine the criticality of the failure modes, together with the degree of severity of their effects.

There are three types of effects, each of them requiring of a separate analysis: the environmental effects (EA), the effects on the safety or health (ES), and the effects on the system availability (ED). The example shown in Figure 12.1 presents a case in which the failure modes affected only the system availability.

Another distinctive feature of the FMEAe worksheet can be observed from Figure 12.1, and it refers to the way the information is presented to the analyst. As can be seen, the worksheet contains three tables that present all the relevant information for the analysis in a unique screen page so that the user can access all the aspects at once without the need to scroll to another page.

12.3 CRITICALITY ANALYSIS OF FAILURE MODES BY APPLYING THE COMPONENT RELIABILITY MODEL APPROACH

The *Component Reliability Model Approach* is the preferred method to use when the failure rates are known or can be estimated, as well as the average repair times, and other parameters of the components-failure modes, which depend on statistical data. As in the analyses carried out by the FTA technique, here it is assumed that the failure rates and average repair times of the components-failure modes have a behavior described by the exponential distribution, in which they remain approximately constant in the time considering that the component is in its useful life.

To model the behavior of components reliability through the loss of their functional capacity, two parameters are considered. One of them is the probability of failure (p), which characterizes the reliability of the components that must operate during a given mission time; the other is the unavailability (q), for those in standby that must change their position or state at the time of the demand.

Together with the reliability parameter, the effects caused by the failure modes are considered to form a matrix (probability of occurrence vs. severity of the effects of each failure mode). This matrix is affected, in turn, by a weighting (quality) factor that considers the way the equipment is commanded, that is, auto-actuated or manual mode from either a remote panel or locally (at field).

The rest of the characteristics that influence the functional capacity of the component, such as degree of redundancy and the control mode (periodic testing, continuously monitoring, or non-controlled component), are already included implicitly in the reliability model of each component-failure mode and in the severity of the effect considering the information filled in the worksheet by the analyst.

There are five degrees of severity for the three kind of effects considered in FMEAe, which are described in Tables 12.2 through 12.4.

This approach assumes that once the failure mode has occurred, the effect will take place. In the case that more than one effect of the same kind (environment-related, safety/health-related, or facility's availability-related) can occur, the one with the highest severity is chosen.

TABLE 12.2
Severity of the Environmental Effect of a Failure Mode (EA)

Qualitative Classification of the Effect	Associated Value	Meaning
Low	1	There are no impacts on facility's site. It considers only internal minor effects. Corrective measures are not required.
Serious	2	There are minor impacts out of the facility boundaries, which demand some cleaning procedures, considering a recovery time of 1 week or less. There is presence of smoke, noise, and bad smells. The local traffic is affected by the evacuation.
Severe	3	There are minor impacts outside the facility boundaries, which require some cleaning processes with a recovery time of at least 1 month. Possible wounded or injured people.
Very severe	4	There are serious impacts outside the facility boundaries. Reversible damages are considered with a recovery time of up to 6 months. Moderated impacts on animal and vegetal life. Temporary disabilities of people.
Catastrophic	5	There are significant impacts outside the facility boundaries, with a recovery time of more than 6 months. Irreversible damages on animal and vegetal life are considered. Possible deaths or permanent disabilities of people are considered.

TABLE 12.3
Severity of the Effect of a Failure Mode on the Safety or Health (ES)

Qualitative Classification of the Effect	Associated Value	Meaning
Low	1	Local minor effects (including first aid procedures). There are no disabling damages.
Serious	2	Appreciable internal effects. Temporary injuries and disabilities.
Severe	3	Important internal effects. Some permanently injured and disabled people. The occurrence of up to 1 death is considered possible.
Very severe	4	Very important internal effects. Several permanently injured and disabled people. Up to 4 deaths are possible.
Catastrophic	5	Catastrophic internal effects. Multiple permanent affectations. Numerous deaths (5 cases or more).

TABLE 12.4
Severity of the Effect of a Failure Mode on the Facility's Availability (ED)

Qualitative Classification of the Effect	Associated Value	Meaning
Low	1	There is no effect on production/functioning. Additional maintenance tasks during shutdown could be required.
Serious	2	Loss of important redundancy/reserve. An unplanned shutdown within 72 hours could be required. Recovery time of up to 1 month is considered.
Severe	3	Immediate shutdown is required. Recovery time of 1–3 months is considered.
Very severe	4	Immediate shutdown is required. Recovery time of 3–6 months is considered.
Catastrophic	5	Immediate shutdown is required. Recovery time of more than 6 months.

Any necessary clarification in support of the analysis as, for example, some analysis hypothesis, basis of causes and effects, or assignment of certain parameters, whose certainty is not proven, is made in the Remarks table of the worksheet, for each failure mode analyzed. Finally, the corrective measures derived from all the information collected and the criticality analysis are incorporated in the *Recommendations* field of the *Results* sheet.

12.3.1 INDEXES USED IN THE CRITICALITY ANALYSIS OF COMPONENTS-FAILURE MODES

The criticality analysis in FMEAe uses some factors related to the following subjects:

- Probability of occurrence of the failure mode
- Severity of the induced effects
- Mode of control (periodic testing, continuously monitoring, etc.)
- Mode of command (auto-activated or manually-activated)
- Degree of redundancy
- Mechanisms of common cause failures

Next, a set of semi-quantitative indices is defined for the criticality analysis of the component-failure modes.

12.3.1.1 Component Risk Index

The Component Risk Index (IRC$_i$) gives a measure of the importance of each component-failure mode within the system function according to the three kind of effects (ED, ES, and EA) so that there can be three kind of risks related to the system function due to the occurrence of a failure mode i: IRCd$_i$, IRCs$_i$, and IRCa$_i$. Following the Component Reliability Model Approach, the expressions of these risk indexes are:

$$\text{IRCd}_i = (q_i)(\text{ED}_i)(\text{FPm}_i) \tag{12.1}$$

$$\text{IRCs}_i = (q_i)(\text{ES}_i)(\text{FPm}_i) \tag{12.2}$$

$$\text{IRCa}_i = (q_i)(\text{EA}_i)(\text{FPm}_i) \tag{12.3}$$

where:

q_i is the probability of failure or the unavailability of the component-failure mode i

ED$_i$, ES$_i$, and EA$_i$, are the severity degrees of the three kinds of effects (availability-related, safety-related, and environmental, respectively) induced by the component-failure mode i

FPm$_i$ is the weighting (quality) factor that considers the way the respective equipment that experiences a failure mode i is commanded when it is demanded for operation, that is, auto-actuated or in manual mode, from either a remote panel or locally at field

It takes the following values: 1 for components auto-activated; 3 for components commanded in remote manual mode (from a control room), and 5 for components commanded manually at field (by hand switch located near the equipment).

A qualitative scale to classify the criticality of each failure mode is established starting from a limiting quantitative goal for the IRC value defined by the following criteria:

1. A value of $q = 1.0$ E-3, which corresponds to systems with high requirements of safety and availability, as for the industries of good practices, such as, for example, the nuclear power plants. It means that the system may be unavailable 8 hours per year, considering that its availability is assessed for a typical year of operation, that is, 12 months (between planned shutdowns for maintenance).
2. Neither environmental effects nor effects on the health of people are present, and there is only a light effect on the system availability (EA = 0; ES = 0; ED = 1).
3. A weighting factor FPm = 1 (neutral) is considered, corresponding to the best engineering practice in which all the components are activated on an automatic signal.

In this way, the criticality scale starts with the target minimal value of IRC = 1.0E-03, and increases periodically by a factor of 5 until reaching the postulated upper limit of IRC = 1.2E-01, above which it considers that the criticality of the component-failure mode is extreme (extremely critical component-failure mode). The scale is as follows:

- IRC $\leq$ 1.0E-3: The risk index of the component-failure mode tends to excellence
- 1.0E-3 < IRC $\leq$ 5.0E-3: The risk index of the component-failure mode moves away from the target in a tolerable range
- 5.0E-3 < IRC $\leq$ 2.5E-2: The risk index of the component-failure mode has been degraded
- 2.5E-2 < IRC $\leq$ 1.2E-1: The risk index of the component-failure mode is critical
- 1.2E-1 < IRC: The risk index of the component-failure mode is extremely critical

12.3.1.2 System Risk Index

System Risk Index (*IRS*) is a measure of the average behavior of the functional capability of the system through the IRC index. It is directly related to the system's reliability. Then it considers the contribution of all its component-failure modes, through their IRC index, and it is estimated by expression (12.4):

$$\mathrm{IRS} = \frac{\sum_{i=1}^{n} \mathrm{IRC}_i}{n} \qquad (12.4)$$

where:
IRCi is the risk index of the component-failure mode i
n is the total of component-failure modes of the analyzed system

Following similar criteria to those defined for the IRC, it is postulated that the IRS target value is 1.0E-3. From this goal, a scale like that proposed for the IRC is established, but with more intermediate ranges for a finer classification of the system reliability. The scale is as follows:

- IRS ≤ 1.0E-3: The risk index of the system tends to excellence
- 1.0E-3 < IRS ≤ 2.25E-3: The risk index of the system moves away from excellence in a tolerable range
- 2.25E-3 < IRS ≤ 5.0E-3: The risk index of the system presents an incipient degradation
- 5.0E-3 < IRS ≤ 1.14E-2: The risk index of the system is degraded
- 1.14E-2 < IRS ≤ 2.56E-2: The risk index of the system approaches the critical zone
- 2.56E-2 < IRS ≤ 5.76E-2: The risk index of the system is in the critical zone
- 5.76E-2 < IRS ≤ 1.29E-1: The risk index of the system is very critical
- 1.29E-1 < IRS: The risk index of the system is extremely critical

12.3.1.3 Index of Relative Importance of the Component-Failure Mode i

The Index of Relative Importance of the Component-failure Mode i (IIR$_i$) gives the relative contribution or weight of the loss of the functional capacity of the component due to the failure mode i (IRC$_i$) to the IRS. It allows knowing how much the IRC value of the corresponding component-failure mode deviates from the IRS value either in excess or in defect. The greatest benefit of criticality analysis can be obtained when the results of both indices, (IRC$_i$ and IIR$_i$), are combined to make decisions. The IIR$_i$ is calculated according to the expression 12.5:

$$\text{IIR}_i = \frac{\text{IRC}_i}{\text{IRS}} \tag{12.5}$$

where:
IRC$_i$ is the risk index of the component-failure mode i
IRS is the risk index of the system

To classify the relative importance of the component-failure modes, the values of IIR$_i$ are ranked according to the following scale. It is recommended to use the following values for decision making together with the IRC values of the respective component-failure mode.

- IIR$_i$ > 10: Too important deviation in excess
- 5 < IIR$_i$ ≤ 10: Important deviation in excess
- 2.5 < IIR$_i$ ≤ 5: Appreciable deviation in excess
- 1 < IIR$_i$ ≤ 2.5: Light deviation in excess
- IIR$_i$ ≤ 1: No deviation or deviation in defect

In this way, those component-failure modes, very critical or critical (according with their IRC_i values) with too important or important deviations in excess, will receive the highest priority for proposing corrective measures to diminish their criticality.

12.3.2 Treatment of Redundant Components

The traditional FMEA technique does not make a global assessment of the system as such, but it is restricted to the individual analysis of each of its components-failure modes through their criticality (FMECA).

However, as it was previously mentioned, FMEAe defines an overall reliability/risk index at IRS that permits assessment of the system as a whole. However, for a system with redundancies, this way of assessing the reliability or risk will distort the value of IRS and the contribution of each redundant component to it. Thus, to avoid excessively conservative results of the IRC values of failure modes derived from redundant components to IRS, a procedure has been developed that considers the contribution of redundant components to the reliability or risk of the system as a function of its degree of redundancy.

To perform the weighting process within this procedure, first, the redundant components must be identified. To achieving this identification, two additional fields are added to the Datos Técnicos table in the FMEAe worksheet (see Figure 12.1):

- Degree of redundancy (*Reserv.* field)
- Redundancy coupling train (*C* field)

Thus, each group of redundant components is identified with a unique integer value in the "C" cell of the respective component-failure mode, and the following attributes must coincide for that group, which represent table fields in the worksheet of the FMEAe (see Figure 12.1):

- Component function (*Function* field).
- The mode of operation (*Estado* field).
- The failure mode (*Modo de Fallo* field).
- The mode of control (*Control* field).

After the group of redundant components has been identified, the unavailability of their values or probabilities of failure (represented by q_i) are weighted (penalized), as follows:

1. Case of groups with *n* identical redundant elements (lack of diversity). If there is lack of equipment diversity in a group of *n* redundant components, then all its components have the same generic code and the same model (*Cod. G.* and *Model* fields in the worksheet tables in Figure 12.1, respectively). Under this condition, the original q_i value is raised to a power equal to the degree of redundancy *n* and the result is divided by the latter. Finally, the result is assigned to the new unavailability value qp_i (weighted unavailability) of each redundant component of the group—this being the new value replaced by q_i in expressions 12.1, 12.2, or 12.3.

The general expression for this case is:

$$qp(n) = \frac{(q)^n}{n}$$ (12.6)

where:

$qp(n)$ is the weighted unavailability/probability of failure mode of redundant components with redundancy degree n

n is the degree of redundancy of the redundant component group

q is the original unavailability/probability of failure mode of the redundant component group degree of redundancy n

2. Case of groups with n redundant diverse components. This refers to those groups of redundant components, which have the same attributes of non-identical redundant elements, that is, they differ in the data of the *Cod. G.* and *Modelo* fields in the FMEAe worksheet tables (see Figure 12.1).

The general expression for this case is:

$$qp(n) = \frac{\displaystyle\prod_{i=1}^{n} q_i}{n}$$ (12.7)

The following example shows the usefulness of this weighting procedure.

Consider a system consisting of three identical components A, B and C, where A and B are arranged in parallel, and C is arranged in series with them. Each of them has the same value of unavailability, $q = 1E\text{-}3$. Assuming independence between components, the system unavailability (Qs) can be estimated as follows:

$$Qs = q(A \times B) + q(C)$$
$$= [q(A) \times q(B)] + [q(C)]$$
$$= 1E\text{-}6 + 1E\text{-}3$$
$$= 1.001E\text{-}3$$

If the reliability analysis of that system were performed applying FMEAe, without considering the weighting procedure and applying the expression 12.4, the IRS index would be (assuming no effects for simplicity):

$$IRS = (IRC[A] + IRC[B] + IRC[C])/N$$
$$= (3E\text{-}3)/3$$
$$= 1.0E\text{-}3$$

where:

IRC[A], IRC[B], and IRC[C] are the component risk indexes of components A, B, and C, respectively, assuming no effects

N is the total number of component-failure modes analyzed (assumed three for this example)

As previously established, the IIR index (relative importance of failure modes in FMEAe) for each component is calculated applying the expression 12.5:

$$IIR[A] = IRC[A]/IRS$$
$$= 1.0E\text{-}3/1.0E\text{-}3$$
$$= 1$$
$$IIR[B] = IRC[B]/IRS$$
$$= 1.0E\text{-}3/1.0E\text{-}3$$
$$= 1$$
$$IIR[C] = IRC[A]/IRS$$
$$= 1.0E\text{-}3/1.0E\text{-}3$$
$$= 1$$

From these results, the inconsistency of redundant components A and B can be concluded having the same contribution to the IRS of the component C. Different from C, the occurrence of failure mode A, or failure mode B, is not a sufficient condition for the system to fail. Then, this result means an excessively conservative contribution of the components A and B was caused by the previous procedure (without weighting of q values of redundant components).

The previous problem is solved by applying the weighting procedure in estimating the q values of the failure modes, in the case of redundant components, as established by expression 12.7. Thus, the contribution of the risk index of each of the failure modes of the previous example to the IRS, is estimated as follows:

$$qp\,(A,B) = (1E\text{-}3)^2/2$$
$$= 5E\text{-}7$$

where $qp\,(A,B)$ is the weighted unavailability of each component A and B, estimated by expression 12.7.

According to the procedure, the qp substitutes the original q value of the redundant components involved, in this case, A and B, so that the new values of q are $q(A) = 5E\text{-}7$, $q(B) = 5E\text{-}7$, while $q(C)$ keeps its value: $q(C) = 1E\text{-}3$, because the latter is not a redundant component.

Now the IRS can be estimated again, but using the new values of q for A and B, by expression 12.4:

$$IRS = \left(IRC\left[A\right] + IRC\left[B\right] + IRC\left[C\right]\right)/N$$
$$= \left(5E\text{-}7 + 5E\text{-}7 + 1E\text{-}3\right)/3$$
$$= 1.001E\text{-}3/3$$
$$= 3.34E\text{-}4$$

As can be observed, the new IRS value (3.34E-4) is smaller than the formerly estimated value without weighting the q values of the redundant components (1.0E-3), which is considered more realistic because it considers the expected effect of the redundancy on the system reliability.

The new IIR indexes for the component-failure modes A, B, and C are now estimated again, but substituting the modified values of q and IRS, applying expression 12.5:

$$IIR\left[A\right] = IRC\left[A\right]/IRS$$
$$= 5E\text{-}7/3.34E\text{-}4$$
$$= 0.001497$$
$$IIR\left[B\right] = IRC\left[B\right]/IRS$$
$$= 5E\text{-}7/3.34E\text{-}4$$
$$= 0.001497$$
$$IIR\left[C\right] = IRC\left[A\right]/IRS$$
$$= 1E\text{-}3/3.34E\text{-}4$$
$$= 2.99$$

Finally, it should be noted that the new values obtained for IRS and IIR[A], IIR[B], and IIR[C], are more representative of the system reliability, with a less value of IRS and more realistic relative importance or contribution to the IRS of the three components-failure modes (IRR[A] = IRR[B] << IIR[C]). This result proves the usefulness of the weighting procedure for treating redundant components included in FMEAe.

12.4 PROCEDURE FOR TREATING THE COMMON CAUSE FAILURES IN FMEAe

It is a fact that multiple failures due to common causes are less probable to occur than single failures due to independent causes, but when they occur, they tear down the design efforts to achieve high levels of reliability by means of redundancy. Because of that, is important for system designers and operators to be alert about those issues, even under lack of data for specific estimation of their failure rates or probabilities of occurrence.

In this way, the objective of treating the CCFs within a structured method like FMEAe is to avoid letting their effects on results go unnoticed in cases where their occurrence is possible despite the availability of quantitative specific data that reflect their occurrence. Thus, FMEAe employs generic data, as a first approximation, although later, the analysts can update them if the experience warrants it or depending on the existence of local defense measures.

The set of procedures developed to include the CCFs in an automated way within the worksheet of FMEAe summarizes criteria and steps of other procedures collected in the literature specialized in the subject [29–34], and the efforts to update and improve the methodology and general approaches of concern [34]. To achieving this, the typical structure of the traditional FMEA worksheet had to be modified and some new fields were included in the tables, as shown in Figure 12.1, to facilitate the analysis and comparison of the pertinent information.

The algorithms of these procedures include the following general tasks/steps:

- It starts with a list of generic components, based on engineering judgment, the operational experience, and on previous published studies [35–38] for these components for which knowledge justifies this type of analysis, given their functional characteristics and their failure mechanisms or effects. However, other components could be included in this list, if the specific operational data or the specific experience justifies it.
- From the list, some candidates are selected to form part of the CCFs analysis. Those components that match the condition of being active redundant components are, in principle, potential candidates. The FMEAe approach defines up to three degrees of dependency (depth of dependence) as a function of the quantity of coincident attributes among all the possible ways to share. In the group G1 are those components that only share internal attributes such as the mode of operation or command, the failure mode, the mode of control, etc. In the group G2 are comprised the components that share the same attributes of G1 plus one of the two external attributes (e.g., sharing the same room or the same working conditions). Finally, the G3 group includes the components that share all attributes, either internal or external. Then, depending on the degree of dependency, it is assigned a value for the generic β parameter, which is directly proportional to such dependency degree.
- The assignment of generic β parameters modifies the failure rate or the probability of the component-failure modes involved, which are candidates to CCF, whose value is adjusted as a function of the redundancy degree of the common cause events group, applying the next expression taken from [39]:

$$\beta_k = \prod_{i=2}^{k} \left[\frac{2^{i-2} - 1.0 + \beta_2}{2^{i-2}} \right]$$

(12.8)

where:

β_k is the beta factor to characterize the failure of k components from the same generic group of size m

β_2 is the generic beta factor to characterize the failure of two components from the same group (assumed here as $\beta_2 = 0.1$, which is the average of the values of β factors estimated for the component-failure modes involved in previous CCF studies [30,31])

- Finally, the resulting CCFs are added in the worksheet, so that they are part of the criticality analysis together with the rest of the single failure modes.

The next Sections 12.4.1 through 12.4.4 offer some details of these tasks.

12.4.1 List of Components with Potential to Generate Common Cause Failures

According to the operational experience in the industry—with the highest safety and availability requirements—and the recommendations resulting from important works within the field of dependent failures analysis for nuclear power facilities [12,29–38], a list of components with credible potential to generate CCFs has been established for using in FMEAe.

Most represent active-type components or, in some cases, passive ones with active failure modes (e.g., check valves). Other components, such as batteries, have been included based on the recommendations of the operational experience of the preceding references; although they do not have a macroscopic movement.

This preliminary list has a practical basis considering the generalized absence of data that allows for determining a complete CCF study. Hence, it does not mean that the occurrence of CCFs in other types of components is excluded; however, its inclusion would considerably complicate the models and the expected results would introduce high levels of unwarranted uncertainties.

A two-digit identifier accompanying each component of the following list is used by FMEAe to make the CCF analysis an easier task. That code is an excerpt of the 3-digit code used in references [2,3].

- BT: Battery
- DG: Emergency diesel generator
- KA: Circuit breaker general
- KB: Circuit breaker bus bar
- KG: Circuit breaker generator
- MA: Motor electrical
- PD: Diesel driven pump
- PM: Motor driven pump
- PT: Turbine driven pump
- QB: Blower fan
- QC: Compressor
- RT: Relay time delay

- RC: Relay control
- VA: Air operated valve
- VC: Check valve
- VD: Solenoid operated valve
- VM: Motor operated valve

12.4.2 Classification of Common Cause Failures into Groups by Their Degree of Dependency

The dependency degrees are used in FMEAe to qualify the depth with which the dependency mechanisms defined in [30,31,34] could act. Therefore, the ratio of failures due to common causes among the totality of causes is a value directly proportional to that dependency degree. Following is a general description of the procedure for classifying CCFs by their degree of dependency:

- *Degree of dependency G1*: The conditions that should be met by the components included in the previous list to be considered as precursors of potential CCF events of degree 1 (G1) is that they are identical redundant components. That is, they need to have an exact coincidence of their internal attributes. This analysis is carried out in FMEAe through an algorithm of identification and comparison of strings, which include the following fields of the worksheet.

  In the Datos Técnicos (Engineering-related data) table (see Figure 12.1), it should coincide the following fields for all the components involved:
  - *Redund*: It indicates the redundancy degree; and those component-failure modes with values equal to or greater than 200% are of interest (double degree of redundancy or higher)
  - *C*: It is the redundancy coupling train (those component-failure modes with the same value of C, means that they belong to the same redundant group, whose degree of redundancy can be checked at *Redund*.)
  - *Modelo*: Here appears the identification code of the component manufacturer's model
  - *Función*: It indicates the component function within the system
  - *Estado*: It indicates the component state and the mode in which the component is commanded (e.g., normally open, normally closed, auto-activated, or manually activated)

  In the Datos de Fallas y Efectos table (data related to failure modes, causes, and effects; see Figure 12.1), it should coincide with the strings filled in the following fields for all the components involved:
  - *Modo de Fallo*: It indicates the failure mode of the component (e.g., fail to open or fail to close)
  - *Cod. G*: Is the two-digits generic code which identifies the component type (e.g., manual operated valve, air operated valve, or motor driven pump)

- *Controles*: It indicates the mode of control of the component-failure modes (e.g., non-controlled, periodically tested, or continuously monitored)
- *Degree of dependency G2*: The conditions that should be met by the components included in the list of Section 12.3.1 to be considered as precursors of potential CCF events of degree 2 (G2) are the following:
  - Meet the same conditions of the CCFs of degree G1
  - Meet one of the following two external conditions (fields of the Datos Técnicos table; see Figure 12.1):
    - The components must operate under the same working conditions; that is, the strings in the field *Condiciones de trabajo* must coincide exactly
    - The components must be located inside the same room or very close to each other; that is, the information in the field *Local* must coincide exactly
- *Degree of dependency G3*: The conditions that should be met by the components included in the list of Section 12.3.1 to be considered as precursors of potential CCF events of degree 3 (G3) are the following:
  - Meet the same conditions of the CCFs of degree G1
  - Meet the following two external conditions (fields of the Datos Técnicos table; see Figure 12.1):
    - The components must operate under the same working conditions; that is, the strings in the field *Condiciones de trabajo* must coincide exactly
    - The components must be located inside the same room or very close to each other; that is, the information in the field *Local* must coincide exactly

12.4.3 ASSIGNMENT OF POSTULATED GENERIC β FACTORS

After the CCFs have been determined, the common-cause failure rates for the respective components are calculated, and then they are assigned to the corresponding cell of the field *Tasa F.* (failure rate) in the Datos de Fallas y Efectos table of the worksheet (see Figure 12.1).

The common-cause failure rate for each component involved is determined from the value of the failure rate of the precursor single component-failure mode and the generic β factor assigned. The latter refers to the β_2 parameter of the expression 12.9 and is chosen from the following list, depending of the corresponding degree of dependency (G1, G2, or G3):

- G1: $\beta_2 = 0.1$
- G2: $\beta_2 = 0.15$
- G3: $b_2 = 0.2$

The values of β_2 listed are based on the following criteria:

- The basic degree of dependency depth corresponds to the sharing of internal conditions (G1)
- The basic redundancy degree is the *double redundancy* ($2 \times 100\%$ of the component nominal functional capacity) from which the β factors were estimated in specialized studies as in [30,31] whose average value $\beta = 0.1$ is indicated in Table 1.1 of these references.
- Starting from this value of $\beta_2 = 0.1$, it is assumed that the addition of any other external attribute to a CCF of degree G1 produces a linear increase of its β factor in 0.05. This increase is postulated according to the range of values of β factors appearing in Table 1.1 of [30,31], so that the maximum postulated value does not exceed the maximum value of such a range.
- In this way, the CCF of components with double redundancy of degree G2 will have a $\beta_2 = 0.15$ and for those of G3 a $\beta_2 = 0.2$.

12.4.4 CORRECTION OF THE β FACTOR OF THE COMMON CAUSE FAILURE EVENTS, ACCORDING TO THE DEGREE OF REDUNDANCY

- The values of the β parameter listed in Section 12.4.3 are representative of CCFs generated by components with double redundancy.
- Since operational experience indicates that as redundancy level increases, so does the probability of survival to the CCF of the components involved, it is assumed that using these values for CCFs generated by higher redundancy components is a very conservative approach. Then, to fix that problem it is used the expression 12.8 in Section 12.4.
- Finally, these fixed values of CCF rates are added to the end of the list of tables in the FMEAe worksheet for revision and completion (e.g., to write the pertinent clarifications in the Remarks table or to modify some values according to pertinent engineering judgment or expert criteria). After the CCFs have been added to the FMEAe, the criticality analysis can be performed so that it may include the influence of this kind of events, as FTA normally does in a system reliability analysis.

12.5 ANALYSIS OF IMPORTANCE BY COMPONENT TYPE

This analysis depends on the results of the criticality analysis treated previously in Section 12.3 and it reveals the types of component or groups of components (e.g., motor driven pumps, diesel driven pumps, motor operated valves, check valves, etc.) with the highest criticality that support the decision making process in improving system reliability. The steps of the procedure are:

- After the IRC_i values have been calculated, they are grouped by component types, according to their generic code (field *Cod G.* in the Datos de fallas y efectos table of the FMEAe worksheet).

- Within each group k, the average values of IRC[k]$_i$ are calculated.
- Finally, the averaged IRC[k]$_i$ values are sorted in descending order and the IIC[k] indexes are computed as follows:

$$IIC[k] = \frac{\sum_{i=1}^{Nk} IRC_i^k}{Nk} \qquad (12.9)$$

where:

IIC[k] is the importance index of component type k (average IRC value within a given group k of generic components)

Nk is the total number of component-failure modes belonging to the group k

IRC_i^k is the risk index of the component-failure mode i belonging to group k

- Thus, the most important component types which engender the most critical failure modes are determined; that is, the types of components that most contribute to the risk can be known and, therefore, unique corrective actions for similar components can be typified or, otherwise, important design changes can be proposed.

12.6 RELIABILITY ASSESSMENT OF A GENERIC FIRE QUENCHING SYSTEM APPLYING FMEAe

Figure 12.2 shows the simplified drawing of a hypothetical fire quenching system whose reliability is assessed applying the FMEAe approach.

The design and operational information on which the reliability assessment is performed is summarized as follows:

1. The system stays in standby state and it is activated automatically through a signal of fire event from the instrumentation and control (I&C) circuits.
2. The standby positions/states of each component are indicated in Table 12.5.
3. The odd I&C circuit generates a signal to activate motor driven pump PM1 which is set in automatic position by its hand switch (HS), and this same signal closes motor operated valve MV3 and opens MV1.

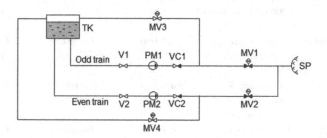

FIGURE 12.2   Simplified drawing of a fire quenching system.

TABLE 12.5
State/Position of Components System of Figure 12.3

No.	Component ID	Component Description	Standby State/Position	Demand State/Position	Control
1.	TK	Water storage tank	Full level	Empty after mission fulfilled	Continuously monitored
2.	V1	Manual operated valve for isolating the odd train	Normally open	Normally open	Periodically tested
3.	V2	Manual operated valve for isolating the even train	Normally open	Normally open	Periodically tested
4.	PM1	Motor driven pump. Odd train	Automatic	Running for 4 hours	Periodically tested
5.	PM2	Motor driven pump. Even train	Standby	Running for 4 hours if PM1 fails	Periodically tested
6.	VC1	Check valve. Odd train	Normally closed	Open while PM1 is running	Periodically tested
7.	VC2	Check valve. Even train	Normally closed	Open while PM2 is running	Periodically tested
8.	MV1	Motor operated valve at discharge of the odd train	Normally closed	Full open	Periodically tested
9.	MV2	Motor operated valve at discharge of the even train	Normally closed	Full open if train odd is failed	Periodically tested
10.	MV3	Motor operated valve for testing the odd train	Normally open	Full closed	Periodically tested
11.	MV4	Motor operated valve for testing the even train	Normally open	Full closed is PM2 is running	Periodically tested
12.	SP	Sprinkler	Empty	Cooling water flowing	Non-controlled
13.	Power6KV	Support system for power supply of both PM1 and PM2	Energized	Energized	Continuously monitored
14.	Power380V	Support system for power supply of all MOVs	Energized	Energized	Continuously monitored
15.	Odd-IC	I&C circuit for auto-activation of the odd train	Energized	Energized	Continuously monitored
16.	Even-IC	I&C circuit for auto-activation of the even train	Energized	Energized	Continuously monitored

4. Under signal of fire event, if the flow is not established, a signal for activation of the even train is produced, which starts PM2 (set in standby position of its HS), closes MV4, and opens MV2.

5. The odd train is tested every 720 hours through an MV3 valve flowing the cooling water in recirculation mode through MV3 to the water storage tank TK, and 15 days later, the even train is tested by starting PM2 and recirculating cooling water through VM4 to the water storage tank TK.

6. The motor operated valves (MOVs) MV3 and MV4 are tested monthly by closing them, following the same procedure used for testing each train. When the full closed position is verified, the valves are opened again and stay in that position.

7. In a way like the former case, the MOVs MV1 and MV2 are tested monthly by opening them. When the full open position is verified, the valves are closed again and stay in that position.

8. The motor driven pumps, PM1 and PM2, are powered from the same 6000 volts alternating current (6 kV AC) bus bar.

9. All MOVs, MV1, MV2, MV3 and MV4 are powered from the same 380 V AC bus bar.

12.6.1 GENERAL ASSUMPTIONS AND OTHER CONSIDERATIONS FOR THE ANALYSIS

Following is a set of general assumptions made for the analysis concerning data and modeling to gain simplicity for achieving the analysis purposes.

1. Only two types of support systems were considered: power supply and I&C circuits for auto-activating the fire quenching system.

2. The position of the HS of the active components in the case of the odd train is set to automatic. This means that under a real demand condition, the generated signal will act on the components of the odd train.

3. The position of HS for the active components of the even train is set in standby, which means that they will act only on the condition of coincidence of PM1 failed and fire alarm signal present.

4. The only human errors considered for this example refers to "V1 in wrong position on demand" (it fails to remain open on demand) due to human error after maintenance of the odd train; and "V2 in wrong position on demand" (it fails to remain open on demand) due to human error after maintenance of the even train. Both human actions are considered as independent events.

5. The pumps and valves are in the same room.

6. The boundary of pumps and MOVs includes the respective circuit breakers so that the interface for power supply is considered the BB-6KV bus bar for PM1 and PM2; and BB-380V bus bar for all MOVs from MV1 to MV4.

7. The interfaces for I&C circuit are assumed to be RC-101 for odd-IC circuit and RC-102 for even-IC circuit.

8. All quantitative data for component-failure modes were taken from generic databases starting from [2,3,39].

9. For simplicity, only one failure mode for each component was considered, except for the motor driven pumps. In the case of the manual operated

valves, V1 and V2, the hardware cause for "fail to remain in position" failure mode was neglected and only the human error was considered instead.

10. The sprinklers also were excluded from the assessment because they are passive components with very low failure rates.

12.6.2   PREPARING THE WORKSHEET FOR THE ANALYSIS AND RELIABILITY ASSESSMENT IN FMEAe

Figure 12.3 presents the FMEAe worksheet with all the essential information filled in the respective cells, before the CCF events are determined.

It can be observed from Figure 12.3 that 17 component-failure modes are included in the analysis, according to the system drawing of Figure 12.2, the assumptions, and other information of interest.

After the fields of the tables are filled, the next step is to proceed with the criticality analysis. To prove how the risk profile of the system is modified due to the inclusion of CCF events in the reliability assessment by means of FMEAe, the results of both cases are compared, which are presented in Figures 12.5 and 12.6. Figure 12.4 shows the modified FMEAe worksheet after the CCFs were determined.

The worksheet in Figure 12.4 shows an increase in the number of component-failure modes with respect to Figure 12.3. Thus, after determining CCF events the list of component-failure modes that participate in the criticality analysis encloses 22 elements, because five CCFs were added (those whose code begins with CM-). Since the degree of redundancy is two, five CCFs of double-failure were added as indicated at

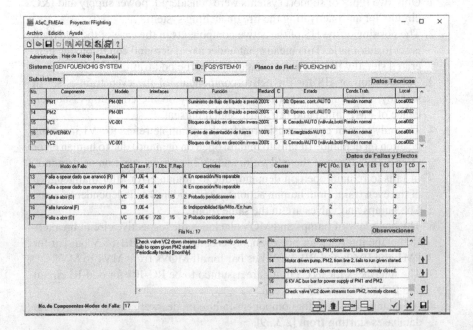

FIGURE 12.3   FMEAe worksheet showing the last five component-failure modes of the list (before determining the CCF events).

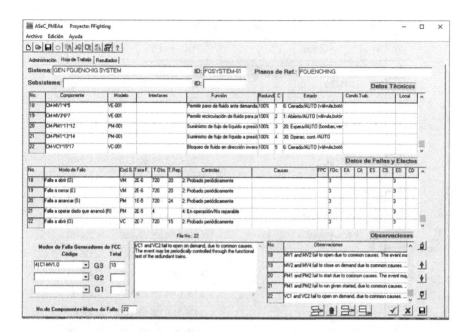

FIGURE 12.4 FMEAe worksheet modified after determining CCF events showing the last 5 out of 22 component-failure modes.

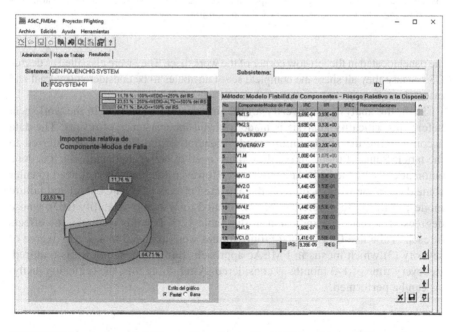

FIGURE 12.5 System reliability profile estimated by FMEAe without CCF contributions.

No.	Comp	Código	Prob	F-V	RRW	RAW	^
1	11	LF-PM1-S	1.91E-02	4.46E-01	1.81	2.39E+01	
2	20	LF-PM2-S	1.91E-02	4.46E-01	1.81	2.39E+01	
3	22	H1-V2-D	1.00E-02	2.33E-01	1.30	2.41E+01	
4	14	H1-V1-D	1.00E-02	2.33E-01	1.30	2.41E+01	
5	16	LF-MV2-O	3.78E-03	8.83E-02	1.10	2.42E+01	
6	9	LF-MV3-E	3.78E-03	8.83E-02	1.10	2.42E+01	
7	6	LF-MV1-O	3.78E-03	8.83E-02	1.10	2.42E+01	
8	18	LF-MV4-E	3.78E-03	8.83E-02	1.10	2.42E+01	
9	8	F-POWER380V	1.00E-04	6.22E-02	1.07	6.22E+02	
10	12	F-POWER6KV	1.00E-04	6.22E-02	1.07	6.22E+02	
11	21	LF-PM2-R	4.00E-04	9.33E-03	1.01	2.43E+01	
12	13	LF-PM1-R	4.00E-04	9.33E-03	1.01	2.43E+01	
13	10	LF-VC1-O	3.73E-04	8.72E-03	1.01	2.43E+01	
14	19	LF-VC2-O	3.73E-04	8.72E-03	1.01	2.43E+01	
15	7	F-ODDIC	5.00E-05	1.17E-03	1.00	2.43E+01	
16	17	F-EVENIC	5.00E-05	1.17E-03	1.00	2.43E+01	
17	15	LF-TK-T	1.00E-08	6.22E-06	1.00	6.22E+02	

Editor de importancia (RRW-Cociente)

FIGURE 12.6 List of component importance ranked by the F-V importance measure estimated by the ARCON code without CCF contributions.

the panel located in the left-low corner of the worksheet. They were classified as degree G3 because they all share the complete set of attributes to be considered (internal and external attributes).

Then, the analysts must verify all the information concerned in the worksheet before the criticality analysis is made to avoid inconsistence of results. The data accompanying the single failure modes generating the CCFs are transferred to the latter. Some of them, like failure rates, need to be recalculated, which is done automatically by the FMEAe algorithms. However, the analysts still need to enter some data in the worksheet, as is the expected effects of each CCF-related failure modes; in this case, it refers to effects related to system availability (ED), whose degrees of severity are indicated in the Table 12.3. For this example, the effects of each of the five CCF-related failure modes were assigned to a severe degree of severity (3) which means in FMEAe approach: Immediate shutdown is required. Recovery time of 1–3 months is considered. After doing this, the criticality analysis can be performed.

12.6.3 CRITICALITY ANALYSIS OF SYSTEM OF FIGURE 12.2 APPLYING THE FMEAe APPROACH

According to the data introduced by analysts, several indexes are estimated by algorithms of criticality analysis (through the corresponding expressions of Sections 12.3 and 12.4). Figure 12.5 shows the FMEAe Result sheet in which it can be seen the value of *IRS* and the ranking of criticality of the component-failure modes without CCF contribution.

Despite the low value of IRS (9.39E-5), which means that according to the data used the system presents high degrees of reliability/low degrees of risk, Figure 12.5 shows the component-failure modes which dominate the system reliability by means of the ranking made by the IIR_i values. These include, in decreasing order of importance, the single failure of both pumps to start under demand (PM1.S and PM2.S), the failure of both support systems of power supply, and the human errors on the valves V1 and V2 (V1.M and V2.M).

Figure 12.6 represents the results of the FTA for the system using the same set of data and assumptions made for FMEAe analysis by means of the ARCON code [40,41], which is used here as a way of comparison between the FTA and FMEAe approaches. The system failure probability estimated is $Ps = 1.61E-3$, which can be considered within the reliability target that should be established for safety systems at industrial facilities with high requirements for safety and availability. The reliability profile is shown in Figure 12.6 ranked by Fussell-Vesely (F-V) importance measure that represents the relative contribution of each component-failure mode to the system's probability of not fulfilling its safety function.

From Figures 12.5 and 12.6, it can be seen that the same group of component-failure modes dominate the reliability profile. The distinctive feature between both approaches lies in the fact that FMEAe adds the failure effects, which, in turn, considers the redundancy. Therefore, the single events Power380V and Power6KV are placed in a higher level of the ranking made by FMEAe. The same is made by F-V in the FTA performed with ARCON. In that sense, the approach of FMEAe can be used to follow the regulatory issues closer than that of FTA. Nevertheless, the list of the 12 more important failure modes coincides in both approaches.

When the CCFs are included in the analysis, whatever the method used, the reliability profiles change dramatically, as a function of the system's redundancy degree. In the case of the example used herein, the global values were not affected because of the relatively low values of unavailability used for the system's components, and the system redundancy itself, but importance profile of the component-failure modes has slight changes as shown in Figures 12.7 and 12.8.

The approach of minimal cut sets (MCSs) is responsible for the major difference between the results obtained by FMAEe and ARCON code and, once again, the inclusion of the effects in the former strengthen that difference even more.

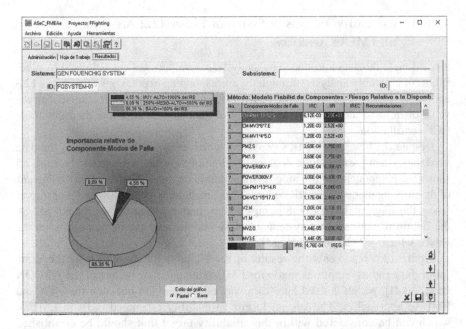

FIGURE 12.7 Reliability profile determined by FMEAe with CCF contributions.

The value of IRS increases five times when the CCFs are included in the analysis (see Figures 12.5 and 12.7), while the system failure probability estimated by ARCON after the inclusion of CCFs is $P = 4.55E-3$; that is, it increases 2.8 times. The reliability profile coincide in both cases with slight differences that are based on the same criteria explained previously. In this case, the most dominant failure modes in both approaches were the CCFs of motor-operated pumps to start, followed in the case of FMEAe by another two CCF events involving the failure to open on demand of both MV1 and MV2, and the failure to close on demand of MV3 and MV4 as shown in Figure 12.7.

On the other hand, the ranking of values estimated by FTA gives more priority to the single failures of the pumps PM1 and PM2 to start on demand over the CCFs of the MOVs to open and to close on demand, as shown in Figure 12.8, and as was stated before. This is due to the MCS approach, with respect to the failure mode and effect approach. Nevertheless, both approaches coincide in estimating the most important contributors to the system reliability, and therefore, they can be equally useful for decision making, despite their known major differences.

Finally, to complete the results from the FMEAe approach in evaluating the system reliability, an analysis of importance by component type can be done as was indicated in Section 12.5. The results of that kind of analysis for this example system is presented in Figure 12.9.

Figure 12.9 shows that the component type of greater importance was the PM type (motor driven pumps), which resulted in a medium value of importance according to the FMEAe postulated scale. This classification is quite logical because this type of component was the one with the highest values of criticality among all the system

No.	Comp	Código	Prob	F-V	RRW	RAW
1	3	CM-PM1_2-S	2.03E-03	4.46E-01	1.80	2.20E+0
2	11	LF-PM1-S	1.91E-02	1.58E-01	1.19	9.10
3	20	LF-PM2-S	1.91E-02	1.58E-01	1.19	9.10
4	5	CM-MV2_4-E	3.97E-04	8.73E-02	1.10	2.20E+02
5	1	CM-MV1_2-O	3.97E-04	8.73E-02	1.10	2.20E+02
6	22	H1-V2-D	1.00E-02	8.25E-02	1.09	9.17
7	14	H1-V1-D	1.00E-02	8.25E-02	1.09	9.17
8	6	LF-MV1-O	3.78E-03	3.12E-02	1.03	9.22
9	18	LF-MV4-E	3.78E-03	3.12E-02	1.03	9.22
10	9	LF-MV3-E	3.78E-03	3.12E-02	1.03	9.22
11	16	LF-MV2-O	3.78E-03	3.12E-02	1.03	9.22
12	12	F-POWER6KV	1.00E-04	2.20E-02	1.02	2.20E+02
13	8	F-POWER380V	1.00E-04	2.20E-02	1.02	2.20E+02
14	4	CM-PM1_2-R	8.00E-05	1.76E-02	1.02	2.20E+02
15	2	CM-VC1_2-O	3.87E-05	8.51E-03	1.01	2.20E+02
16	21	LF-PM2-R	4.00E-04	3.30E-03	1.00	9.25
17	13	LF-PM1-R	4.00E-04	3.30E-03	1.00	9.25
18	10	LF-VC1-O	3.73E-04	3.08E-03	1.00	9.25

FIGURE 12.8 List of component importance ranked by the F-V importance measure estimated by the ARCON code with CCF contributions.

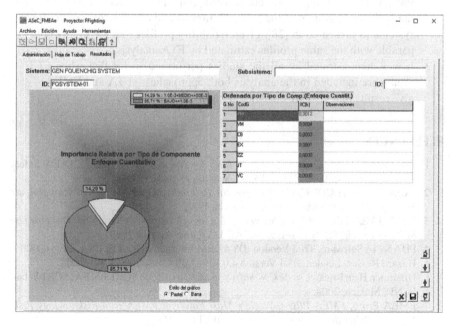

FIGURE 12.9 Results of the analysis of importance by component type using FMEAe.

components either due to their single or their common cause failures. This result supports the measures to be taken to improve the system reliability profile, even though the system reliability can be considered acceptable.

12.7 FINAL REMARKS

The new approach developed in FMEAe shows some important issues to consider for system reliability assessments. Following these are summarized:

- FMEAe do not substitute the reliability assessments made by other powerful quantitative techniques, such as FTA, but it can be used as an alternative method when the effects of failures need to be considered.
- FMEAe can be considered as an advanced variant of FMEA/FMECA, which solves some major recognized disadvantages of the latter regarding the dependency analysis.
- Unlike the traditional method of FMEA/FMECA, FMEAe can provide a global assessment of the system reliability by means of the new index IRS, which uses a postulated scale based on good practices criteria.
- Through the new index of relative importance of component-failure modes (IIR), the analysts can support their decisions on corrective measures to be proposed based on the analysis results.
- FMEAe keeps the strength of the qualitative methods regarding the descriptive potentialities and all the useful information that can be manage in the same analysis environment.
- FMEAe has been applied to reliability assessments of other systems such as the fuel supply system of the ATR-42 aircraft and other designs of cooling systems as presented in [28], given reliability profiles that are highly comparable with the same profiles estimated by FTA analysis.
- It was demonstrated by all the studies performed so far that when CCF events are included in the analysis both approaches—FTA and FMEAe—give results that also are considered highly comparable.

REFERENCES

1. IAEA. INSAG-12. *Basic Safety Principles for Nuclear Power Plants.* IAEA Safety Series No. 75-INSAG-3, Rev. 1. IAEA, Vienna, Austria, 1999.
2. IAEA. IAEA-TECDOC-478. *Component Reliability Data for Use in Probabilistic Safety Assessment.* IAEA, Vienna, Austria, 1988.
3. IAEA. IAEA-TECDOC-508. *Survey of Ranges of Component Reliability Data for Use in Probabilistic Safety Assessment.* IAEA, Vienna, Austria, 1989.
4. PHA5-Pro Software. Trial Version. DYADEM International LTD, USA, 1994–2000.
5. Hazard Review Leader. Trial Version 4.0.106. ABS Consulting, 2000–2003.
6. Dinámica Heurística, S.A. de C.V. *Software SCRI-HAZOP.* SCRI-FMEA, SCRI-What/If. NL, México, 2004.
7. FMEA Pro 6. *FMEA-PRO 6 World's Most Powerful FMEA Tool Risknowlogy Risk, Safety & Reliability.* Dyadem International LTD, 2003.
8. Relex FMECA. Relex Software Corporation, 2003.

9. MIL-STD-1629A. Military standard. *Procedures for Performing a Failure Mode, Effects and Criticality Analysis*. DOE, Washington, DC, 1980.
10. US NRC. NUREG/CR-2300. *Probabilistic Risk Analysis Procedures Guide*. USNRC, Washington, DC, 1983.
11. IAEA. IAEA Safety Series No. 50-P-4. *Procedures for Conducting PSA of Nuclear Power Plants*. IAEA, Vienna, Austria, 1992.
12. US NRC. NUREG/CR-4550. Vol. 1, Rev. 1. *Analysis of Core Damage Frequency: Internal Events Methodology*. US NRC, Washington, DC, 1991.
13. US NRC. NUREG/CR-2815. *Probabilistic Safety Analysis Procedures Guide*. US NRC, Rev. 1, Washington, DC, 1985.
14. US NRC. NUREG/CR-2728. *Interim Reliability Evaluation Programme (IREP) Procedures Guide*. US NRC, Washington, DC, 1983.
15. CNE.APS.IF.103. *Análisis Probabilístico de Seguridad*. Informe Final de la Fase I. Volúmenes I y II. Rev. 1, Central Nuclear Embalse, Córdoba, Argentina, 2003.
16. Análisis Probabilista de Seguridad de la Central Electro Nuclear de Juraguá. Reporte Preliminar del estudio de 15 sucesos iniciadores, GDA/APS, ISCTN-CNSN, Habana, Cuba, 1995.
17. US NRC. NUREG/CR-2787. *Interim Reliability Evaluation Programme (IREP)*. Analysis of the Arkansas Nuclear One -Unit 1- Nuclear Power Plant (NPP). US NRC, Washington, DC, 1982.
18. US NRC. NUREG/CR-1659. Vol. 1. *Reactor Safety Study Methodology Applications Program*. Study of Sequoyah PWR Unit 1. US NRC, Washington, DC, 1981.
19. US NRC. NUREG/CR-1659. Vol. 3. *Reactor Safety Study Methodology Applications Program*. Study of Calvert Cliffs PWR Unit 2. US NRC, Washington, DC, 1982.
20. Análisis Probabilístico de Seguridad de la Central Nucleoeléctrica de Laguna Verde. Unidad 1. Informe Ejecutivo. México, 5 de Diciembre de 1989.
21. EPRI NP-1804-SR. *The German Risk Study for Nuclear Power Plants*. Ministry for Research & Technology, TUV Rheinland, Koeln, Germany, 1979.
22. WASH-1400-MR (NUREG-75/014). *Reactor Safety Study: An Assessment of Accidents Risks in US Commercial Nuclear Power Plants*. United States Nuclear Regulatory Commission, Washington, DC, 1975.
23. Estorilo, C., and Posso, R. 2010. The reduction of irregularities in the use of process FMEA. *International Journal of Quality and Reliability Management* Vol. 27, No. 6, pp. 721–733.
24. Arabian Hoseynabadi, H., Oraee, H., and Tavner, P. J. 2010. Failure mode and effects analysis (FMEA) for wind turbines. *Electrical Power and Energy Systems* Vol. 32, pp. 817–824.
25. Sawant, A., Dietrich, S., Svatos, M., and Keal, P. 2010. Failure mode and effect analysis-based quality assurance for dynamic MLC tracking systems. *Medical Physics* Vol. 37, No. 12, p. 6466.
26. Yang, F., Cao, N., Young, L. et al. 2015. Validating FMEA output against incident learning data: A study in stereotactic body radiation therapy. *Medical Physics* Vol. 42, p. 2777.
27. Perdomo, M., Salomon, J., Rivero, J. et al. 2010. *ASeC*, An advanced system for operational safety and risk assessment of industrial facilities with high reliability requirements. *IBP3090_10*. *Rio Oil & Gas 2010 Expo and Conference*, September 13–16, 2010.
28. Perdomo, M., and Salomón, J. 2016. Análisis de modos y efectos de falla expandido: Enfoque avanzado de evaluación de fiabilidad. *Revista Cubana de Ingeniería* Vol. VII, No. 2, pp. 45–54.
29. US NRC. NUREG/CR-4780. *Procedures for Treating CCF in Safety and Reliability Studies*. US NRC, Washington, DC, Vol. 1, 1988, Vol. 2, 1989.

30. EPRI NP-3967. *Classification and Analysis of Reactor Operating Experience Involving Dependent Events*. EPRI, Palo Alto, CA, 1985.
31. EPRI TR-100382. *A Data Base of Common-Cause Events for Risk and Reliability Applications*. EPRI, Palo Alto, CA, 1992.
32. US NRC. NUREG/CR-5460. *A Cause-Defense Approach to the Understanding and Analysis of Common Cause Failures*. SNLs, JBF Associates, NUS Corporation, Washington, DC, 1990.
33. US NRC. NUREG/CR-5801. *Procedure for Analysis of Common-Cause Failures in Probabilistic Safety Analysis*. USNRC, Washington, DC, 1993.
34. US NRC. NUREG/CR-6268, Rev. 1. *Common-Cause Failure Database and Analysis System: Event Data Collection, Classification, and Coding*. INL, Washington, DC, 2007.
35. US NRC. NUREG/CR-6819, Vol. 1. *Common-Cause Failure Event Insights: Diesel Generators*. INEEL, Washington, DC, 2003.
36. US NRC. NUREG/CR-6819, Vol. 2. *Common-Cause Failure Event Insights: Motor Operated Valves*. INEEL, Washington, DC, 2003.
37. US NRC. NUREG/CR-6819, Vol. 3. *Common-Cause Failure Event Insights: Pumps*. INEEL, Washington, DC, 2003.
38. US NRC. NUREG/CR-6819, Vol. 4. *Common-Cause Failure Event Insights: Circuit Breakers*. INEEL, Washington, DC, 2003.
39. OREDA. *Offshore Reliability Data*, 4th ed. Det Norske Veritas, Høvik, Norway, 2002.
40. Mosquera, G., Rivero, J., Salomón, J. et al. 1995. *Disponibilidad y Confiabilidad de Sistemas Industriales. El sistema ARCON*. Anexo B, pp. 137–140. Ediciones Universitarias UGMA, Barcelona, Venezuela, Mayo.
41. Salomón, J. *Manual de Usuario Práctico del Código ARCONWIN Ver 7.2*. Registro de autor, CENDA, La Habana, Cuba, 2015.

13 Reliability Assessment and Probabilistic Data Analysis of Vehicle Components and Systems

Zhigang Wei

CONTENTS

13.1 INTRODUCTION

Fatigue-related durability and reliability performance is a major concern for the design of vehicle components and systems [1]. Durability describes the ability of a product to sustain required performance over time or cycles without undesirable failure. Reliability is defined as the ability of a system or component to perform its required functions under stated conditions for a specified period. Both load/stress, as experienced by a vehicle component or a system, and the strength of the components or systems being studied are random variables and normally follow stochastic and probabilistic processes. Eventually, probability distribution functions can

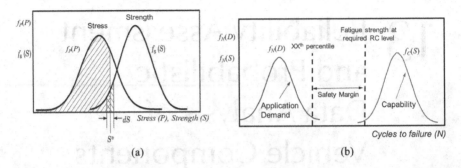

(a)   (b)

FIGURE 13.1 Reliability assessment based on (a) stress-strength interference model and (b) demand-capability in terms of fatigue cycle. (Adapted from Wei, Z. et al., Reliability analysis based on stress-strength interface model, *Wiley Encyclopedia of Electrical and Electronics Engineering*, Chichester, UK, Wiley, 2018.)

be used to characterize the load/stress and strength of the vehicle components and systems for a given cycle. The stress–strength interference model is a fundamental probability-based method for reliability analysis [2] (Figure 13.1a) and it can be applied to fatigue-related reliability assessment if both stress/load distribution and fatigue strength distribution at a given common cycle are known. Another approach, which is like the stress-strength interference model but is more commonly used in practical reliability assessment, is the life-based demand-capability model (Figure 13.1b). In contrast to the stress-strength interference model, the life distributions of demand and capability at a given certain stress or load level must be known in advance.

To make the stress-strength interference model applicable, the stress distribution—probability density function (PDF) $f_P(P)$ and the strength distribution $f_S(S)$—must be available. Similarly, to make the life-based demand-capability model applicable, the demand distribution $f_N(D)$ and the capability distribution $f_N(C)$ must be provided in advance. How to obtain a representative stress distribution and a life demand distribution is a challenging topic. A simplified method often is used in practice. For example, instead of using the whole set of life demand information, a single life demand point is set as a target, which represents XXth (e.g., 95%) percentile usage. Corresponding to the life demand point, a single capability point, which represents a certain reliability and confidence (RC) levels, for example, R90C90 (90% in reliability and 90% in confidence), as obtained from the life capability is identified to compare it with the demand point [1]. A safety factor then can be defined as the ratio of life capability over the life demand. The stress-strength interference model can be simplified in a similar way. How to obtain a fatigue life distribution and a fatigue strength distribution from a given set of stress-cycle (S-N) fatigue data is one of the main focuses in this chapter. The relationship between these two distributions for a given set of fatigue data is a key to accomplishing reliability assessments; however, the relationship between them is often unclear. To reveal the relationship, a new fatigue S-N curve transformation technique, which is based on the fundamental statistics definition and some reasonable assumptions, is specifically introduced in this chapter.

Numerous testing methods are available for product durability validation and reliability demonstration, and such methods include life testing (test-to-failure), binomial testing (pass or fail), and degradation testing [1,3]. The test-to-failure method tests a component to the occurrence of failure under a specified loading. The binomial (Bogey) testing method is used often in reliability demonstration in which the customers' specifications must be met for acceptance into service. The degradation testing is used to test a product to a certain damage level, which is often at a level far below complete failure. Additionally, the associated accelerated testing methods [3,4] (i.e., accelerated life testing, accelerated binomial testing, and accelerated degradation testing) are used often to shorten the development time and reduce the associated cost while not significantly sacrificing the accuracy of the assessment. All these methods are treated separately, and their relationships are not clear, which impedes the wide and proper applications of these methods and their combinations. In this chapter, a unified framework of the reliability assessment method is presented in a damage-cycle (D-N) diagram [5], which consists of the following major constituents: (1) test data, either test-to-failure, binomial, degradation, or combined, for estimating the continuous probabilistic distribution function, (2) damage accumulation rules, such as the linear or nonlinear damage accumulation rules, for data interpolation and extrapolation, and (3) a variable transformation technique, which converts a probabilistic distribution of a variable into a probabilistic distribution of another variable.

In addition to these two transformation techniques, the probabilistic analysis on data with large sample size with two- and three-parameter Weibull distribution functions, the uncertainty for data with small sample size, and the sample size reduction approaches based on the Bayesian statistic also are investigated. Furthermore, the basic assumptions and theories in assessing the reliability of systems are provided to complement these two basic transformation techniques. It should be noted that software reliability of the modern vehicle components and systems is very important [3] and it is especially true when vehicle-to-vehicle (V2V), vehicle-to-infrastructure (V2I), and autonomous vehicle are the mainstream topics in the automotive industry. However, only fatigue-related reliability is considered in this chapter because of space limitations.

This chapter is organized as follows:

Section 13.2 provides a brief and general background about the reliability assessments of vehicle components and systems with an emphasis on vehicle exhaust components and systems. Section 13.3 presents a fatigue S-N curve transformation technique in which distributions of load/stress and life can be properly selected based on data pattern and converted to each other when necessary. Section 13.4 introduces a variable transformation technique in a damage-cycle (D-N) diagram, which is a tool that can effectively interpret the commonly used fatigue-testing methods and seamlessly reveal the interrelationship among these testing methods. Section 13.5 provides the basic concepts on reliability assessment of systems. Section 13.6 provides some basic methods for processing data with probabilistic distributions with a special attention to the differences between the two-parameter and three-parameter Weibull distribution functions in terms of predictability and applicability. Uncertainty analysis on data with small sample size and the potential capability of the Bayesian statistic in sample size reduction also are discussed in Section 13.6. Pertinent examples are provided in each section to demonstrate the concepts and techniques developed. Finally, Section 13.7 summarizes this chapter with several key observations.

13.2   RELIABILITY OF VEHICLE COMPONENTS AND SYSTEMS

A vehicle usually consists of several systems, such as powertrain, chassis, body, electrical, and exhaust systems. Each system can be further divided into subsystems and their constituent components. During vehicle operation, vehicle components and systems are subjected almost invariably to road load and engine vibration. With the increased mileage demand of the vehicle life (e.g., 10 years/150,000 miles), the durability and reliability performance of the vehicles is an important factor in vehicle design and development. Some vehicle systems may be subject to various other operating environments and conditions. For example, vehicle engine and systems are constantly exposed to high temperature and corrosive environments [6].

Based on the temperature level, the associated failure mechanisms, and related analysis approaches, the failure type can be categorized into three groups: (1) isothermal fatigue, (2) anisothermal fatigue, and (3) high-temperature thermal-mechanical fatigue (TMF) [7]. Temperature remains relatively low and constant in isothermal fatigue. Temperature varies and does not have a single fixed temperature in anisothermal fatigue. The applied temperature in isothermal fatigue and anisothermal fatigue should be low enough to avoid triggering other failure mechanisms such as creep and oxidation, which are time-dependent failure mechanisms. Corrosion in vehicle exhaust systems is usually caused by salt, condensate, urea, and other corrosive agents. Creep begins at a temperature of approximately half the absolute temperature (degrees Kelvin or Rankine) of the metal melting point [6]. By contrast, fatigue is essentially a cycle-dependent failure mechanism. The temperature in high-temperature TMF is high enough to trigger creep and oxidation.

Product durability and reliability validation testing and associated life assessment are becoming routine processes for the development of exhaust components and systems. In product validation, how to handle the temperature effects is still a controversial issue. Generally, there are two approaches: (1) cold-testing and (2) hot-testing. To reduce cost and shorten product development cycle, the hot gas in a vehicle often is bypassed during the road load data acquisition (RLDA) process in cold-testing; hence, room or near-room temperature information is collected during RLDA. For consistent performance evaluation, subsequent calibration testing and component bench testing also are conducted in the same cold conditions [8]. With the cold-testing information, the performance of the component or system at high operating temperatures can be estimated by introducing a temperature factor, which is used to correct and compensate the temperature effects. With the introduced temperature factor, the product designed in the cold-testing condition could be reliable if the dominating factors are properly considered in the temperature factor. After these factors are identified, quantified, and applied to the RLDA load data, the rainflow cycle counting can be performed with the help of the linear Miner's damage rule. Miner's rule predicts that failure occurs when damage is greater than or equal to 1 [8].

As the name implies, in hot-testing all parts of the RLDA, calibration, and bench testing are conducted in service or equivalent high temperature conditions. The fatigue life can be assessed in service condition and no temperature correction factor is required in the fatigue life assessment. The hot-testing method is still evolving [8] and, without losing generality, only the cold-testing related topics will be addressed in this chapter.

13.3   FATIGUE S-N CURVE TRANSFORMATION TECHNIQUE

The fatigue S-N data in a (2D) fatigue S-N plot characterize the capability of the material in fatigue failure resistance. The higher the location of the data in the plot, the higher the resistance of the material to fatigue failure. Fatigue data often show large scatters in life as well as in load/stress due to a wide variety of intrinsic uncertainties, such as material, loading, and manufacturing uncertainties. Figure 13.2 schematically shows the major characteristics of a fatigue S-N mean curve, its lower and upper bounds, and the life distributions around the mean curve. A fatigue S-N curve can be roughly divided into three regimes: Regime-I, Regime-II, and Regime-III, which represent low-cycle fatigue, medium-cycle fatigue, and high-cycle fatigue, respectively. In many engineering applications, the Regime-II for medium-cycle fatigue is of significant interest, and the mean curve in Regime-II often can be treated using a linear approximation in an appropriate plot, such as log-log plot. Mean curve is used often to characterize the general trend of a material in fatigue failure resistance. However, in many applications, such as product validation, quality control, and life management, the scatter of the fatigue life around the mean is also of significant importance.

There are two basic ways to describe the statistical variability of the fatigue S-N data in the linear Regime-II:

1. Life distribution as a function of load or stress, $f[N_f(S)]$ (Figure 13.3)
2. Load/stress (strength) distribution as a function of fatigue cycle, $f[S(N_f)]$ (Figure 13.3)

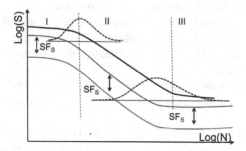

FIGURE 13.2   A schematic of a general fatigue S-N curve.

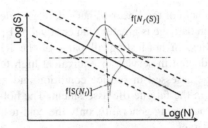

FIGURE 13.3 Cycle and load based probabilistic distributions for the same set of fatigue S-N data.

The life distribution is used much more commonly than the strength distribution in fatigue data analysis. However, the strength distribution has many unique characteristics and important applications, such as:

1. Relatively invariant to the levels of load/stress for some engineering materials [9]
2. The load/stress-based safety factor is more reasonable to assess the margin of safety of a product as a unifier across the whole range of the fatigue regimes
3. The probabilistic distributions of stress/load (strength) at a given cycle to failure is an essential part of the stress-strength interference model based reliability analysis
4. The strength distribution makes the stress-strength interference model possible at the system level, Although the life distribution and strength distribution can be obtained directly from, respectively, the horizontal offset method and the vertical offset method, the recommended standard fatigue life data analysis is the horizontal offset method [10]. How to transform the life distribution to load/stress distribution is an open challenge. In addition, the vertical offset method is not always feasible, but the load/stress distribution is often desirable. For example:
5. The raw fatigue S-N data, such as the data plotted in literature and reports, are not always available. However, the values of the fit parameters based on the horizontal offset approach are often provided.
6. The patterns of some fatigue S-N data lead to inaccurate fitting results if the vertical offset approach is used, whereas the data patterns match the horizontal offset method well [10]. This situation often is the case for two-stress level fatigue data.
7. Fatigue S-N data are available only at one stress level while the slope of the fatigue S-N curve is known already based on the historical data.

In all these cases, a new technique is required to transform the distribution of life to the distribution of strength or vice versa. The following is such a technique to accomplish this goal.

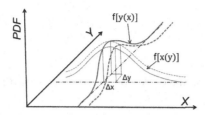

FIGURE 13.4   Schematic of probabilistic distributions of x and y.

The only assumption of this new technique is mathematically expressed in Equation 13.1, which indicates that the amplitude of the PDF of y at a given x level is proportional to the amplitude of the PDF, $f\left[x(y)\right]$, of x at that point (Figure 13.4).

$$f\left[y(x)\right] = Kf\left[x(y)\right] \tag{13.1}$$

where K is a constant, which is determined by satisfying the basic probability law (Equation 13.2):

$$\int_{-\infty}^{+\infty} f(y)\,dy = 1 \tag{13.2}$$

Equation 13.1 indicates implicitly that the peak of the PDF of the strength distribution corresponds to that of the PDF of life distribution, and the valley of the PDF of the strength distribution corresponds to that of the PDF of life distribution for single-mode probabilistic distributions (Figure 13.4). This assumption makes sense intuitively based on the observations of a wide variety of fatigue data. The following lognormal (normal) distributions is provided to demonstrate the transformation technique.

The selection of probabilistic distribution functions is a critical issue in reliability assessment. The real distribution of a fatigue life given stress level is essentially unknown. However, the two-parameter Weibull and log-normal distribution functions are commonly used in probabilistic fatigue life assessments [11]. In the automotive industry, the two-parameter Weibull often is preferred in fatigue life assessments because of its simplicity and seemly meaningful interpretation of the shape parameter. Years of experience and data collection show that both functions empirically fit the fatigue data equally well as far as the mean behavior is concerned [11]. The pairs of the two fit parameters for the two distribution functions are, respectively, μ(mean)/σ (standard deviation) and η(scale)/β(shape). The bell-shaped normal PDF and the corresponding cumulative distribution function (CDF) are expressed in Equations 13.3a and 13.3b, respectively:

$$f(x) = \frac{1}{\sigma\sqrt{2\pi}} \exp\left[-\frac{1}{2}\left(\frac{x-\mu}{\sigma}\right)^2\right] \tag{13.3a}$$

$$F(x) = \frac{1}{2}\left[1 + erf\left(\frac{x-\mu}{\sigma\sqrt{2}}\right)\right] \tag{13.3b}$$

where $erf()$ is the error function.

The Weibull PDF and the corresponding CDF are listed in Equations 13.4a and 13.4b, respectively:

$$f(x) = \frac{\beta}{\eta}\left(\frac{x}{\eta}\right)^{\beta-1}\exp\left[-\left(\frac{x}{\eta}\right)^{\beta}\right] \tag{13.4a}$$

$$F(x) = 1 - \exp\left[-\left(\frac{x}{\eta}\right)^{\beta}\right] \tag{13.4b}$$

With an added threshold parameter, a, Equation 13.4 can be generalized to the three-parameter Weibull function in Equation 13.5:

$$f(x) = \left(\frac{\beta}{\eta}\right)\left(\frac{x-a}{\eta}\right)^{\beta-1}\exp\left[-\left(\frac{x-a}{\eta}\right)^{\beta}\right] \tag{13.5a}$$

$$F(x) = 1 - \exp\left\{-\left[\frac{(x-a)}{\eta}\right]^{\beta}\right\}; 0 < a \leq x < \infty, \eta, \beta > 0 \tag{13.5b}$$

The normal distribution is used in this section to show the fatigue S-N curve transformation technique. Based on Equations 13.1 through 13.3a, the PDF of the normal distribution function $f(y)$ as a function of y can be written as Equation 13.6:

$$\int_{-\infty}^{+\infty}Kf[x(y)]dy = K\int_{-\infty}^{+\infty}\frac{1}{\sigma_x(y)\sqrt{2\pi}}\exp\left\{-\frac{1}{2}\left[\frac{x-\mu_x(y)}{\sigma_x(y)}\right]^2\right\}dy = 1 \tag{13.6}$$

where both mean μ_x and standard deviation σ_x in Equation 13.6 are assumed to be functions of y. When σ_x is assumed to be a constant and the linear relationship is held for the mean curve in Equation 13.7:

$$\mu_x = a + by \tag{13.7}$$

Equation 13.6 can be much simplified and after rearrangement it can be transformed to Equation 13.8:

$$K\int_{-\infty}^{+\infty}\frac{1}{\sigma_x\sqrt{2\pi}}\exp\left\{-\frac{1}{2}\left[\frac{y-(a-x)/(-b)}{(-\sigma_x/b)}\right]^2\right\}dy = 1 \tag{13.8}$$

Based on the linear assumption in Equation 13.7, the term $(a-x)/(-b)$ in Equation 13.8 is actually the mean of y and can be expressed as $\mu_y = (a-x)/(-b)$, which is essentially Equation 13.7 but in a different format. The unknown K can be solved from Equation 13.8 and the result is expressed in Equation 13.9:

$$K = -b \tag{13.9}$$

For a linear fatigue S-N curve in a log-log plot with $x = \log(N)$ and $y = \log(S)$, assume that the distribution of the cycles to failure at a stress level follows a normal distribution in a log-log plot, then Equation 13.8 simply becomes Equation 13.10:

$$f\left[S(N)\right] = \frac{1}{\sqrt{2\pi}\,(-\sigma/b)} \exp\left\{ -\frac{\left[\log(S) - \left(\dfrac{a - \overline{\log(N)}}{-b}\right)\right]^2}{2(-\sigma/b)^2} \right\} \tag{13.10}$$

Clearly, the probabilistic distribution as a function of strength is still a normal distribution with a new mean (see Equation 13.11a), which is essentially Equation 13.7, and a new standard deviation in Equation 13.11b:

$$\mu_y = \left[a - \overline{\log(N)}\right]/(-b) \tag{13.11a}$$

$$\sigma_y = -\sigma_x/b \tag{13.11b}$$

Example 13.1 Two-stress level fatigue data

Table 13.1 lists a set of fatigue S-N data of welded exhaust components made of a steel. Tests are conducted by controlling the applied load and only two load levels are tested with six data points at each load level. The fatigue data show a wide scatter band because many factors, such as material inhomogeneity and welding quality, are involved in the failure of the exhaust components. Since the data in Figure 13.5 belong to the "standard horizontal pattern" [12], the horizontal offsets method, which is the ASTM standard recommended method [10], should provide a reasonable fit curve. The fit curves with the horizontal offset method as well as the vertical offset method are plotted in Figure 13.5, and the fit parameters are listed in Table 13.2. The results of the horizontal offset methods are very different from those of the vertical offset methods, which provide a poor fit to the set of data. Based on the linear assumption in the log-log plot and the estimated mean curve from the horizontal offset method, the standard deviation of the strength can be calculated, and the results are listed in Table 13.2. This example belongs to type (f) listed in Section 13.3. To accurately assess the reliability of a system, the reliability of each constituent component must be accurately assessed as well. However, the reliability assessment of each component often is conducted with limited sample size and under certain testing conditions because of budget and testing constraints, which brings significant uncertainty in test results and their interpretations. Example 13.1 indicates that the obtained results (the mean and the standard deviation) could be inaccurate and even misleading if the load/stress distribution is obtained directly from fitting the data with the vertical offsets method. By contrast, the load/stress distribution as obtained by transforming the life distribution, which is obtained by fitting the data with the horizontal offsets method, is logically sound and meaningful; therefore, surely it will lead to a more accurate system reliability assessment.

TABLE 13.1
Fatigue Cycles to Failure at the Two-Stress Levels

Load, lbs	No. 1	No. 2	No. 3
520	86188	130708	153282
620	45823	55775	73715
Load, lbs	No. 4	No. 5	No. 6
520	168718	177465	304998
620	89524	108583	135140

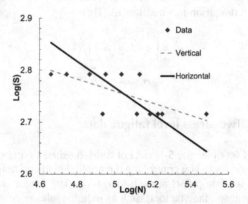

FIGURE 13.5 Vertical and horizontal offsets methods for fatigue data of an automotive exhaust component.

TABLE 13.2
Calculated Fit Parameters with $a=log(C)$ and $b = 1/h$ for the Power Law $S = CN^h$

	$C(a)$	$h(b)$	STD_N	STD_S (Equation 13.11b)
Vert.	2213.1 (28.6)	−0.117 (−8.547)	0.262	—
Hori.	10889.3 (15,9)	−0.254 (−3.937)	0.178	**0.045**

13.4 REPRESENTATION OF RELIABILITY TESTING METHODS IN THE DAMAGE-CYCLE DIAGRAM

The reliability testing methods (life testing, Bogey testing, and degradation testing) are treated as three different methods in practice. To better understand and fully use these testing methods, a general framework in which the three testing methods can be evaluated in a consistent way is required. The damage-cycle (D-N) diagram [5] is a tool to bring all these three reliability testing methods together in the same framework. Figure 13.6 schematically shows the three reliability testing methods in the (D-N) diagram. In Figure 13.6, the horizontal axis represents the applied fatigue cycle, N, while the vertical axis represents damage, D. The intersection of the two axes, where the applied cycle $N = 0$ and $D = 0$, represents the beginning of the testing process. The D is always an increasing function of N because a damage process is often assumed to be an irreversible process. When the applied cycle $N = N_f$, $D = 1$ indicating a complete failure of the product. The dashed lines shown in Figure 13.6 represent the evolution trajectory of the damage bounds (lower and upper), which can be linear or nonlinear depending on the assumption of the damage process.

The PDF and CDF as obtained from fitting the test-to-failure data can be described as $f\lfloor N_f(D=1)\rfloor$ and $F\lfloor N_f(D=1)\rfloor$. In Figure 13.6 the probabilistic distributions are assumed to be representative of the population so that the uncertainty caused by the sample size can be ignored. The obtained probabilistic distribution of failure can be compared against the established reliability criterion to assess the reliability performance of the product. The most appropriate reliability criterion is the reliability function $R = 1 - F$. For example, a product specification R99 states that 99% of the product is expected to pass a specified target. The uncertainty of the fatigue behavior of a population also can be described by $f\lfloor D(N)\rfloor$ and $F\lfloor D(N)\rfloor$ which are the PDF and CDF of the damage at a specific applied cycle, respectively, and also shown in Figure 13.6. As compared to the life data from the life testing, obtaining the damage distribution below $D < 1$ is more difficult. Eventually in practice, instead of the detailed continuous distribution, a discrete assessment (i.e., pass or fail), which is exact the measure used in the binomial testing, often is used. Mathematically, all products with $D < 1$ are characterized as "pass" and all others are characterized as "fail."

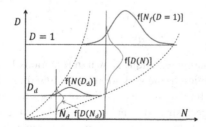

FIGURE 13.6 The representation of population as revealed in life testing, binomial testing, and degradation testing in the (D-N) diagram.

The distribution of the cycles at any given damage level below $D = 1$ (i.e., $f\lfloor N(D_d)\rfloor$ shown in Figure 13.6 and the damage distribution at any applied cycle below N_F, i.e., $f\lfloor D(N_d)\rfloor$ also shown in Figure 13.6 can be represented in the (D-N) diagram. The lower case d stands for degradation.

Intuitively, close interrelationships among the distributions as shown in Figure 13.6 should exist. The key to reveal the interrelationships among the distributions is to find the relationship between two different variables for a given damage evolution equation. Mathematically, the problem is equivalent to seeking a target distribution function, $F_Y(y)$, for a given initial distribution function, $F_X(x)$, and transformation functions, $y = \varphi(x)$ and $x = \psi(y)$. In fact, closed-form solutions can be obtained by the following procedure [13] that is well developed and described briefly herein.

The target distribution function, $F_Y(y)$, can be expressed as:

$$F_Y(y) = P(Y \le y) = P\big[\varphi(X) \le y\big] \tag{13.12}$$

First, consider the case where $y = \varphi(x)$ is a strictly monotone increasing function. $x = \psi(y)$ is then a unique inverse function and:

$$F_Y(y) = \int_{-\infty}^{\psi(y)} f_X(x)dx = F_X\big[\psi(y)\big] \tag{13.13}$$

The target PDF $f_Y(y)$ of Y is obtained as:

$$f_Y(y) = \frac{dF_Y(y)}{dy} = f_X\big[\psi(y)\big]\frac{d\psi(y)}{dy} \tag{13.14}$$

For cases where $\varphi(x)$ is a strictly monotone decreasing function:

$$F_Y(y) = \int_{\psi(y)}^{\infty} f_X(x)dx = 1 - F_X\big[\psi(y)\big] \tag{13.15}$$

$$f_Y(y) = \frac{dF_Y(y)}{dy} = -f_X\big[\psi(y)\big]\frac{d\psi(y)}{dy} \tag{13.16}$$

The two cases in Equations 13.14 and 13.17 can be combined as:

$$f_Y(y) = \frac{dF_Y(y)}{dy} = f_X\big[\psi(y)\big]\left|\frac{d\psi(y)}{dy}\right| \tag{13.17}$$

The variable transformation technique shown in Equations 13.14 through 13.17 is the essential part of the unified framework for representing these three reliability testing methods. With this technique, the distribution of cycles to failure can be calculated easily from the damage distribution at a given cycle or the cycle distribution at a given damage with the help of a damage evolution equation, which can be either linear or nonlinear. In reverse, if the final life distribution is known, then the distribution of damage at any given cycle and the cycle distribution at any given damage level can be calculated in the same manner. In practice, the PDF, $f\lfloor N_f(D=1)\rfloor$ and CDF $F\lfloor N_f(D=1)\rfloor$, can be estimated by fitting the life testing data. It is noted that Equation 13.17 is obtained by assuming that the transformation functions are either monotone increasing or decreasing, which is the case for the fatigue-based reliability analysis. For complex cases where the assumptions of monotone increasing or decreasing are not valid, a more general theoretical framework as provided in [13] can be followed.

Corresponding to the three commonly used testing methods, there are three corresponding accelerated testing methods: accelerated life testing, accelerated binomial testing, and accelerated degradation testing methods. For example, accelerated fatigue life testing can be achieved through increasing stress/load levels. At least two higher stress levels (lower and upper levels) often are introduced to conduct accelerated fatigue life testing. Then the design parameters at service stress level are estimated from the accelerated testing through extrapolation. With probabilistic distribution functions (i.e., $f\lfloor N_F(S_U)\rfloor$ and $f\lfloor N_F(S_L)\rfloor$) at the two higher stress levels, S_U and S_L, the probabilistic distribution $f\lfloor N_F(S_s)\rfloor$ of the life at the service stress level, S_S, can be obtained appropriately by extrapolating data obtained from the higher stress levels. It should be noted that in accelerated testing data analysis, the farther the accelerated stress level is from the normal stress level, the larger the uncertainty in the extrapolation [4]. All these testing methods can be interpreted in a single D-N diagram for one-stress level testing (Figure 13.6) and the S-N curve for multiple-stress level testing.

Example 13.2 The Damage Distribution at a Specific Given Applied Cycle N Obtained from the Weibull Life Distribution with the Linear Damage Rule, $D = N/N_f$

For the linear damage rule, Equation 13.17 results in $D = \varphi(N_f) = N/N_f$, $N_f = \psi(D) = N/D$, and $d\psi(D)/dD = -N/D^2$. The damage distribution at a given applied cycle, N, can be obtained from the variable transformation technique. For example, the two-parameter Weibull distribution, $f\lfloor N_f(D=1)\rfloor$ and $F\lfloor N_f(D=1)\rfloor$, as shown Equation 13.4 can be expressed in Equation 13.18:

$$f(D) = \left(\frac{\beta}{\eta}\right)\left[\frac{N}{D^2}\right]\left[\frac{N}{D\eta}\right]^{(\beta-1)} \exp\left\{-\left[\frac{N}{D\eta}\right]^{\beta}\right\}, D \geq 0 \qquad (13.18a)$$

$$F(D) = \exp\left[-\left(\frac{N}{D\eta}\right)^{\beta}\right], D \geq 0 \qquad (13.18b)$$

Clearly, when $D = 1$, the complementary part (i.e., $1-F[D(1)]$) of Equation (13.18) is exactly the same as Equation 13.4, $F\lfloor N_f\rfloor$.

**Example 13.3 The Cycle Distribution at Damage D Obtained
             from the Weibull Life Distribution with
             the Linear Damage Rule, $D = N/N_f$**

For the same two-parameter Weibull distribution of cycles to failure (Equation 13.4), the transformed distribution of cycle at a given damage D is Equation 13.19:

$$f(N) = \frac{1}{D}\left(\frac{\beta}{n}\right)\left(\frac{N}{D\eta}\right)^{\beta-1} \exp\left[-\left(\frac{N}{D\eta}\right)^{\beta}\right], N \geq 0 \qquad (13.19a)$$

$$F(N) = 1 - \exp\left[-\left(\frac{N}{D\eta}\right)^{\beta}\right], D \geq 0 \qquad (13.19b)$$

The distribution function is still a two-parameter Weibull function with a shape parameter β and a scale parameter of $D\eta$, which is a proportional to the scale parameter η with a factor D.

13.5   PROBABILISTIC DATA ANALYSIS

In practice it is difficult and often impossible to get the distribution of the population because of limited sample size. A schematic of reliability testing methods with limited sample size based on the corresponding population distributions (Figure 13.6) is illustrated in Figure 13.7. The dashed distribution lines indicate the true population distributions are essentially unknown. The circles on the horizontal line, $D = 1$, represent the cycles to failure of the samples. The solid circles stand for the samples below a specified cycle N, and the hollow circles stand for the samples above the cycle N. The diamonds on the vertical line at the specific cycle N represent the pass/fail status of the samples. The solid diamonds stand for pass (i.e., $D < 1$) and the hollow diamonds stand for fail (i.e., $D > 1$). Following the binomial testing and life testing based on the damage-cycle diagram in Figure 13.7 are addressed.

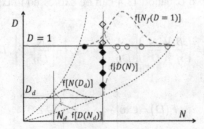

FIGURE 13.7   The representation of unknown population and samples in the (D-N) diagram.

13.5.1 BINOMIAL RELIABILITY DEMONSTRATION

Bernoulli (or binomial) trials can be used to describe an independent random event that has only two possible outcomes: success or failure. The discrete probability distribution generated from the Bernoulli trial is binomial distribution. Suppose an experiment is repeated n times, where p is the probability of success (reliability $R = p$), the probability of a product to survive (based on the binomial CDF) can be presented in the form of [3]:

$$C = 1 - \sum_{i=0}^{r} \frac{n!}{i!(n-i)!} R^{n-i} (1-R)^{i} \qquad (13.20)$$

where:
  R is reliability
  C is confidence level
  r is the number of failed items

When $r = 0$ (no failure), Equation 13.20 is a simple equation for a successful run testing (Equation 13.21):

$$C = 1 - R^{n} \qquad (13.21)$$

The binomial test methods have been used widely in the automotive industry. However, the sample size required for achieving high confidence and reliability are significant. Based on the assumption that the probabilistic distribution follows the two-parameter Weibull distribution (Equation 13.4), a general accelerated testing procedure (Equation 13.22) can be developed by following the Lipson equality [14]:

$$C = 1 - R^{n(L_1 L_2)^{\beta}} \qquad (13.22a)$$

$$R = (1-C)^{1/\left[n(L_1 L_2)^{\beta}\right]} \qquad (13.22b)$$

$$n = \frac{\ln(1-C)}{(L_1 L_2)^{\beta} \ln R} \qquad (13.22c)$$

where:
  $L_1 = t_2 / t_1$ is life test ratio
  $L_2 = \eta_1/\eta_2$ is load test ratio indicating that the change in the characteristic life is caused by the change in load

In Equation 13.22, the shape parameter β is assumed to be a constant for simplicity even though a formula similar to Equation 13.22 can be derived when $\beta_1 \neq \beta_2$. Based on Equation 13.22c, $(L_1 L_2)^\beta$ times fewer test units are needed than would be required by using the conventional successful run approach. In addition, the larger the value of the shape parameter β, the greater the ratio effects on sample size reduction. The ratio η_1/η_2 can be estimated from historical data or expert opinions. Equation 13.22 can be reduced to the extended test method when the effect of L_2 is ignored (i.e., $\eta_1 = \eta_2$ [3]).

From Equation 13.22, with the same confidence and reliability, the sample size reduction can be achieved in three ways:

Way-1: extend or increase the test time at the same stress/load level [3]
Way-2: increase the load/stress level and eventually reduce the characteristic life η
Way-3: combine Way-1 and Way-2

Figure 13.8a and b illustrate Way-1 and Way-2, respectively.

13.5.2 LIFE TESTING

The two-parameter Weibull distribution function often is used in the life testing and almost exclusively used in the extended time testing, which can be considered as an accelerated testing method by appropriately extending the testing time but with significantly reduced testing samples as shown in Equation 13.22c in Section 13.5.1. However, the fatigue data from a wide variety of sources indicate that the three-parameter Weibull distribution function with a threshold parameter at the left tail is more appropriate for fatigue life data with large sample sizes [14]. The uncertainties introduced from the assumptions about the underlying probabilistic distribution would significantly affect the interpretation of the test data and the assessment of the performance of the accelerated binomial testing methods; therefore, the selection of a probabilistic model is critically important. Product validation and reliability demonstration, designs targeting the low percentiles of the fatigue life at the left tail, are required [11]. Therefore, the characteristics of the left tail of a selected model needs to be thoroughly examined test data with a large sample size against the physical mechanisms when the left tail of a distribution is a concern. For test data with

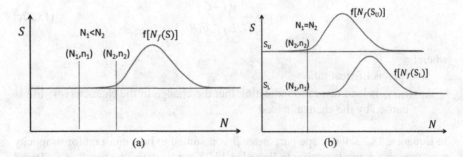

FIGURE 13.8 Schematic of accelerated binomial (Bogey) testing procedure through (a) extended testing time and (b) increased load/stress level as represented in the (S-N) diagram with $D = 1$.

a small sample size, the benefit of using the three-parameter Weibull distribution is not clear because the third fit parameter (threshold) brings significant uncertainty in data analysis and often results in abnormal values of the fit parameters. However, meaningful results can be obtained for data with even very small sample sizes if Bayesian statistics are used and the historical data are available. Three examples following demonstrate these three respective aspects.

Example 13.4 Fatigue Data of 2024-T4 with Relative Large Sample Size

A set of high-cycle fatigue data at room temperature [9] with sample size of 30 for 2024-T4, which is a commonly used aluminum alloy, is selected for fitting the Weibull distribution functions. The probability plots estimated using Minitab for the two- and three-parameter Weibull functions are shown in Figure 13.9a and b. The values of fit parameters for the set of test data also are listed in Figure 13.9. The three-parameter Weibull distribution has a much better fit in terms of visual examination and the Anderson-Darling (AD) statistic value. The AD values for the two- and the three-parameter Weibull distribution functions are, respectively, 1.246 and 0.526. A lower value of the AD statistic indicates a better data fit.

An important observation from the data shown in Figure 13.9 is that the values of the shape parameters for the two- and three-parameter Weibull functions are, respectively, 1.74758 and 0.908975, a change from $\beta > 1$ to $\beta < 1$. The values of the scale parameters are, respectively, 2092213 and 1510218 cycles. It is noted that one of the advantages of the Weibull function over other distribution functions is supposed to be its capability to distinguish among several possible failure mechanisms [3], $\beta < 1$ for infantile or early-life failure, $\beta = 1$ for constant failure rate, and $\beta > 1$ wear-out failure. Clearly, the characterization of data based on β can be significantly compromised by the fact that two- and three-parameter Weibull functions can lead to very different conclusions for the same set of data. All the fitting parameters are listed in Table 13.3. The results indicate that the three-parameter Weibull distribution function with a threshold parameter at the left tail is more appropriate for fatigue life data with large sample sizes. By contrast, the two-parameter Weibull with a long left tail (zero at the left end) does not reflect the intrinsic incubation time caused by the fatigue crack initiation and propagation.

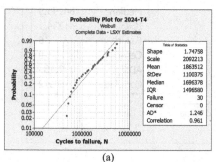
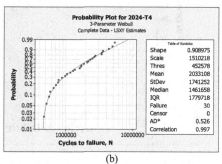

FIGURE 13.9 Probability plots of (a) two-parameter Weibull distribution and (b) three-parameter Weibull distribution for a set of 2024-T4.

TABLE 13.3

The Values of the Fit Parameters of Two-Parameter (2P) and Three-Parameter (3P) Weibull Distribution Functions for the High-Cycle Fatigue Data of 2024-t4

Distribution Functions	Parameters	
2P-Weibull	Shape	1.74758
	Scale	2092213
	AD statistic	1.246
3P-Weibull	β	0.908975
	η	1510218
	δ	452578
	AD statistic	0.526

Example 13.5 Fatigue Data with Small Sample Size

The probabilistic plot and the corresponding estimated parameters obtained using Minitab from six fatigue failure data is shown in Figure 13.10. The obtained shape parameter $\beta = 2.37219$ and a scale parameter of $\eta = 750355$ can be calculated directly from the probability plot. In Figure 13.10, the 90% (confidence)

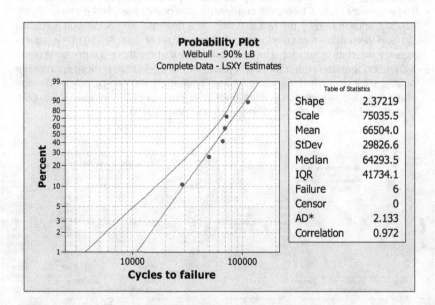

FIGURE 13.10 Probability plots obtained from cycles to failure.

lower bound is shown with a large scatter band indicating the uncertainty nature of the calculated values of the fit parameters caused by the small sample size. The value of the test data at any given reliability and confidence levels (RxxCyy) can be obtained readily. The smaller the sample size, the wider the scatter band and the larger the uncertainty. Clearly, how to obtain accurate estimated parameters from small sample size is a big challenge.

It should be noted that even though the suitability of the three-parameter Weibull distribution in fatigue life testing and associated product validation is obvious from Figure 13.9, for the data with a relatively large sample size, its application to data with small sample size is not recommended because of the high possibility of unstable solutions with the introduced third fit parameter in the three-parameter Weibull distribution. Instead, a two-parameter Weibull distribution is recommended because, although it cannot provide accurate information about the tails of the distribution, it does provide reliable information of the mean, which is often useful. To obtain accurate parameter estimation of three- or multiple Weibull distributions with limited sample sizes, the Bayesian statistics, which uses historical data, can be considered.

13.5.3 BAYESIAN STATISTICS FOR SAMPLE SIZE REDUCTION

When sample size is extremely small, the disadvantage of the traditional Frequentist (based on the current test data only) method is obvious: (1) significant loss in certainty and confidence and (2) high sensitivity to the specific pattern of the test data. However, reliability assessment based on extremely small sample size, such as 3, 2, and 1, often is desired. To overcome these drawbacks of the Frequentist method, a Bayesian statistics-based approach has been developed [11].

The Bayes's rule in the modern version can be expressed as:

$$p(\theta \mid x) = \frac{l(\theta; x) p(\theta)}{\int_0^1 l(\theta; x) p(\theta) d(\theta)} \tag{13.23}$$

where $p(\theta \mid x)$ is posterior PDF, for the parameter θ given the data x. $p(\theta)$ is prior PDF for the parameter θ. $l(\theta; x)$ is the likelihood function, which is defined as $l(\theta; x) = \prod_{k-1}^{n} f(\theta; x_k)$, where x_k is kth experimental observation and $f(\theta; x_k)$ is the PDF of cycles to failure. The denominator in Equation 13.27 is simply a normalizing factor which ensures that the posterior PDF integrates to one. The Bayesian process is schematically shown in Figure 13.11. The posterior distribution usually is narrower than the prior distribution, and results with improved confidence and accuracy can be obtained by analyzing the posterior data.

Two key steps to realize the Bayesian statistics in constructing a reliability RxxCyy are (1) posterior distributions from the historical data and (2) efficient numerical algorithms to implement Equation 13.23.

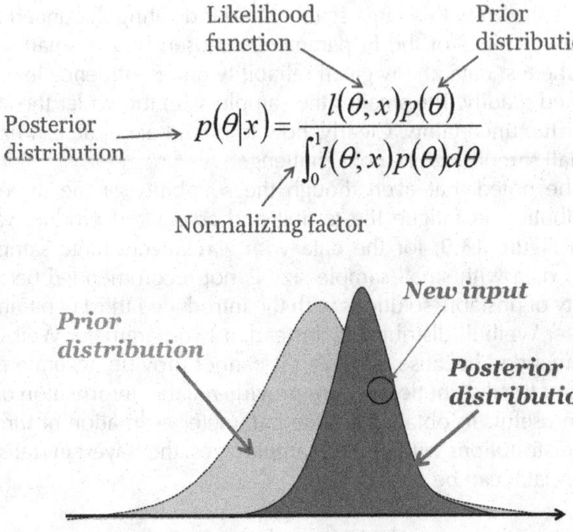

Likelihood function

Prior distribution

Posterior distribution $\longrightarrow$ $$p(\theta|x) = \frac{l(\theta;x)p(\theta)}{\int_0^1 l(\theta;x)p(\theta)d\theta}$$

Normalizing factor

New input

Prior distribution

Posterior distribution

FIGURE 13.11   Schematic of the basic concept of Bayesian statistics procedure.

Example 13.6 Bayesian Statistic for Design Curve Construction

A large amount of reliable historical fatigue test data for welded structures has been systematically collected and analyzed, and the associated probabilistic distributions of the mean and standard deviation of the failure cycles have been successfully obtained [15]. An advanced acceptance-rejection resampling algorithm and a Monte Carlo simulation procedure have been implemented.

Figure 13.12a shows a mean life-standard deviation (Mean-STD) plot (in log-log) of 110 sets of fatigue failure data of a type of welded exhaust component. Based on the data pattern shown in Figure 13.12a, a probabilistic distribution and the values of the corresponding fit parameters have been obtained. With the probabilistic

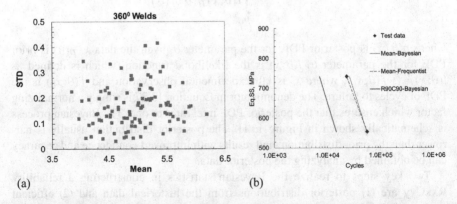

FIGURE 13.12   (a) The STD-Mean plot based on historical weld fatigue data and (b) the R90C90 design curve constructed with only one data point at each of the two stress levels by using a Bayesian statistics procedure.

distribution of the historical data, the Bayes's rule (Equation 13.23) and an advanced numerical algorithm, a design curve obtained with only one data point at each of the two stress levels, can be constructed and is shown Figure 13.12b. It should be noted that a design curve cannot be constructed with the traditional probability plot even with two data points at each stress level. The advantage of the Bayesian statistic is clearly demonstrated from this example.

13.6   SYSTEM RELIABILITY

The reliability of a system can be cascaded into the reliability of its components. A system is often complex and the reliability of a system often can be idealized with the following simple models and their combinations [3].

13.6.1   SERIES SYSTEM MODEL

A system is called series system if its life is the smallest of all those potential times (or cycles) to failure. Such a system fails when the first failure mode occurs. Mathematically:

$$F(x) = 1 - \prod_{i=1}^{n}\left[1 - F_i(x)\right] \tag{13.24a}$$

or, equivalently:

$$R(x) = \prod_{i=1}^{n} R_i(x) \tag{13.24b}$$

Equation 13.24a is referred to as the product rule of reliability since it establishes that the reliability of a series system is the product of the individual component reliabilities. A series system model also is called a competing risks model if bi-modal or multiple failure mechanisms are involved.

13.6.2   PARALLEL SYSTEM MODEL

A parallel system is the system that fails when all components fail. Mathematically:

$$F(x) = \prod_{i=1}^{n} F_i(x) \tag{13.25a}$$

or, equivalently:

$$R(x) = 1 - \prod_{i=1}^{n}\left[1 - R_i(x)\right] \tag{13.25b}$$

The parallel system model represents statistically the polar opposite from the series system model but with $F(x)$ and $R(x)$ interchanged. Like the product rule of reliability, Equation 13.25a can be referred to as the product rule of unreliability since it establishes that the unreliability of a parallel system is the product of the individual component unreliabilities. A parallel system model also is called a dominant modal model [4] if bi-modal or multiple failure mechanisms are involved.

13.6.3 Mixtures Model

$$F(x) = \sum_{i=1}^{n} p_i F_i(x); \qquad \sum_{i=1}^{n} p_i = 1; \qquad 0 \le p_i \le 1 \qquad (13.26a)$$

or, equivalently:

$$R(x) = \sum_{i=1}^{n} p_i R_i(x); \qquad \sum_{i=1}^{n} p_i = 1; \qquad 0 \le p_i \le 1 \qquad (13.26b)$$

where $F(x)$ is CDF and, again, $R(x)$ is reliability or survival function. p_i is proportion or probability of occurrence for failure mechanism i. $F(x)$ and $R(x)$ have the same mathematical structure.

With these system models, the reliability analysis can be conducted like that used for component analysis. An example of using Equation 13.26 to calculate system reliability from the known component reliability is provided as follows. Suppose a series system consisting of five identical components is subjected to constant amplitude vibrational loading. What would be the system reliability if the reliability of each of the five identical components is 0.98 as calculated from the stress-strength interference model? The system reliability as calculated from Equation 13.26b is simply $R = 0.98^5 = 0.90$.

13.7 CONCLUSIONS

This chapter introduces several recently developed new methodologies for fatigue associated reliability assessment of vehicle components and systems. The most important two of these methodologies are the fatigue S-N curve transformation technique and a variable transformation technique. In principle, these methodologies can be applied to the reliability assessment of other similar engineering components and systems. With these new methodologies, the current S-N data analysis and reliability testing methods can be interpreted in a new unified probabilistic framework. The importance of selecting two-parameter and three-parameter Weibull distributions in a probabilistic analysis of data with large sample size has been illustrated with examples. The uncertainty introduced in test data with small sample size and the benefits of using Bayesian statistics approach in cost reduction also has been demonstrated with examples.

REFERENCES

1. Lee YL, Pan J, Hathaway R, Barkey M. *Fatigue Testing and Analysis: Theory and Practice*. Boston, MA: Elsevier Butterworth-Heinemann; 2005.
2. Wei Z, Hamilton J, Ling J, Pan J. Reliability analysis based on stress-strength interface model, *Wiley Encyclopedia of Electrical and Electronics Engineering*, Chichester, UK: Wiley; 2018.
3. O'Connor PDT, Kleyner A. *Practical Reliability Engineering*, 5th ed. Chichester, UK: Wiley; 2012.
4. Nelson WB. *Accelerated Testing: Statistical Models, Test Plans, and Data analysis*, Hoboken, NJ: John Wiley & Sons; 2004.
5. Wei Z, Start M, Hamilton J, Luo L. A unified framework for representing product validation testing methods and conducting reliability analysis. SAE Technical Paper 2016-01-0269.
6. Wei Z, Kotrba A, Goehring T, Mioduszewski M, Luo L, Rybarz M, Ellinghaus K, Pieszkalla M. Chapter 18: Failure mechanisms and modes analysis of automotive exhaust components and systems, pp. 392–432, *Handbook of Materials Failure Analysis*, Abdel Salam Hamdy Makhlouf (Ed.). Amsterdam, the Netherland: Elsevier; 2015.
7. Wei Z, Luo L, Voltenburg R, Seitz M, Hamilton J, Rebandt R. Consideration of temperature effects in thermal-fatigue performance assessment of components with stress raisers, SAE Technical Paper 2017-01-0352.
8. Seitz M, Hamilton J, Voltenburg R, Luo L, Wei Z, Rebandt R. Practical and technical challenges of the exhaust system fatigue life assessment process at elevated temperature, *ASTM Selected Technical Papers (STP) 1598*, Zhigang Wei, Kamran Nikbin, D. Gary Harlow, Peter C. McKeighan (Eds.). ASTM International; 2016.
9. Shen CL. The statistical analysis of fatigue data. PhD dissertations, Tucson, AZ: The University of Arizona; 1994.
10. Standard practice for statistical analysis of linear or linearized stress-life (S-N) and strain-life $(\varepsilon - N)$ fatigue data, ASTM Designation: E739-10.
11. Wei Z, Luo L, Yang F, Lin B, Konson D. Product durability/reliability design and validation based on test data analysis, pp. 379–413, *Quality and Reliability Management and Its Applications*, Hoang Pham (Ed.). Springer; 2016.
12. Wei Z, Yang F, Cheng H, Maleki S, Nikbin K. Engineering failure data analysis: Revisiting the standard linear approach. *Engineering Failure Analysis*, 2013; 30: 27–42.
13. Elishakoff I. *Probabilistic Theory of Structures*, 2nd ed. Mineola, NY: Dover Publications; 1999.
14. Wei Z, Mandapati R, Nayaki R, Hamilton J. Accelerated reliability demonstration methods based on three-parameter Weibull distribution. SAE Technical Paper 2017-01-0202.
15. Wei Z, Zhu G, Gao L, Luo L. Failure modes effect and fatigue data analysis of welded components and its applications in product validation. SAE Technical Paper, 2016-01-0374.

14 Maintenance Policy Analysis of a Marine Power Generating Multi-state System

Thomas Markopoulos and Agapios N. Platis

CONTENTS

14.1　INTRODUCTION

This study is an attempt to analyze the reliability performance of a marine power generation system with the auxiliary systems attached and to develop an alternative for maintenance policy. The main scope of this study is to analyze the methodology and to conduct reliability analysis of the marine electric power system, focusing rather on the mathematical modeling than on the field of the research on pure electric and mechanical systems and their technical details. This aspect leads to generic inferences that are applicable in most systems providing the big picture of the problem and its solution. Nevertheless, authors use references to certain technical issues to help the reader to understand the basic principles of a marine electrical power generating system with the attached auxiliary systems.

This chapter is organized as follows. In this section, there is a short description and general information concerning the marine power generating system as a part of the ship and some related references. Section 14.2 is a presentation of the reliability assessment and multi-state systems in brief. In Section 14.3, there is a description of a typical electric power generation system and reliability characteristics. Section 14.4 presents the development of the semi Markov model. Section 14.5 is a description of the auxiliary diesel engines system driving the electric power generators and a reliability analysis of the multi-state system including the probabilities related with its operation. In Section 14.6 the basic outlines on maintenance policy and maintenance implications and ideas on how stochastic analysis and its inferences could contribute in real world management issues are presented. In addition, there are empirical results concerning the availability of the power generating system under different system configurations. Finally, in Section 14.7, the conclusion sums up maintenance policy and suggests some ideas for further research.

The design of a vessel follows certain basic principles given as guidelines by organizations such as International Maritime Organization (IMO) and Marine Technology Society (MTS), covering all possible sectors of a ship building project and all systems of the vessel. Consequently, such guidelines (MTS DP Technical Committee; MSC/Circular 1994) as a design philosophy and for all essential calculations (IMO MEPC 1-CIRC 866 2014) exist for the electrical power generating system as a part of the whole vessel. Currently, and due to issues related to environment and modern economics, major challenges arise concerning the ship's technology (MUNIN D6.7 2014). There is an increasing pressure for more efficiency in energy, environmental effect, and safety. IMO has developed certain regulations (IMO 2016) concerning a ship's efficiency quantification providing guidelines for all essential calculations (MEPC 61/inf.18 2010; MEPC.1-Circ.681–2 2009; MEPC.1-Circ.684 2009). One major problem designers have in ship technology and design is systems efficiency. Especially, the ship's energy is a sector where a lot of challenges arise continuously. Climate change and the problem of the greenhouse gas emissions lead research to more efficient energy systems on ships and intensifying the demand for improved safety levels and environmental protection to be competitive. The quantitative analysis of this effort could be summarized using certain indices such as the Energy Efficiency Design Index (EEDI) and the Energy Efficiency Operational Indicator (EEOI). Presumably, diesel engine driven electric power generating systems depend on these regulations. Previous research (Prousalidis et al. 2011) has shown that the evolution of a ship's technology leads to new trends. Concerning the energy efficiency, use of vessel and energy management means research on optimization of routes and vessel's speed, which implies optimization of power systems and management and finally presenting advantages through an extensive electrification of ship systems. All those challenges and trends could lead to increased complexity of the systems and requirements concerning the technical background and skills of crewmembers. Unfortunately, all improvements mentioned do not assure full ship safety and there is always the probability that unpredictable incidents will happen (Mindykowski and Tarasiuk 2015). Since electric power is a basic and essential factor of the normal operation of a ship, the electric power system of a ship is dedicated to meet its electric load requirements according to the type of mission during the different phases of its operation, such as overseas voyage, charging and discharging, berthing, etc. According to international regulations in the case of an electrical system failure, the usual and anticipated consequence is a blackout (Brocken 2016), which leads to a deadship condition initiating event. The meaning of the term "deadship" is a condition under which the main propulsion plant, boilers, and auxiliaries are not in operation and in restoring the propulsion, no stored energy for starting the propulsion plant, the main source of electrical power, and other essential auxiliaries should be assumed available. It is assumed that the means are available to start the emergency generator at all times (IMO 2005).

The research about blackout incidents shows that there are many different factors causing a blackout in a ship, such as human error, control equipment failure, automation failure, electrical failure, lack of fuel, mechanical failure, and other causes (Miller 2012) leading to certain questions such as:

- Do the available electric power generators meet the ship's power requirements?
- What is the probability of a total system failure?
- What would be the financial cost of the system failure?

All these questions are closely related with the issue of electric power system reliability, which in the case of a vessel is manageable by following strategies on its architecture such as the use of multiple power sources, sectioning of the distribution grid (Stevens et al. 2015), and use of auxiliary safety subsystems, such as earthing and protection systems (Maes 2013). More specifically, the primary and standby generators are driven by diesel engines with different technical characteristics and attributes related to the requirements and the mission of the vessel such as:

- Load acceptance and rejection
- Starting time
- Load up time and emergency loading ramp
- Time on hot standby
- Minimum load and part load ratings
- Black start requirements

The subsystems of generators are:

- Excitation system
- Lubrication system
- Cooling system
- Facilities for alarms, monitoring, and protection
- Neutral earthing

The importance of the marine electric power system and its components could be understood easily if electric failures are considered that led to marine accidents, such as that of RMS Queen Mary 2 (MAIB 2011), which is obvious since its main tasks could be summarized as follows (Patel 2012):

- The optimal system configuration
- Load analysis and selection of the necessary equipment (e.g., generators and electric motors)
- The power distribution system
- Optimization of the routing cables
- Fault current analysis and the necessary safety devices
- Optimization of the power monitoring system

Since a ship operates in an autonomous mode at sea and usually when moored, the design of the power system faces major challenges to meet the established standards and other requirements. The ship designers must consider the electrical power requirements during each phase of the ship operation. A major concept affecting the use of electric energy on a ship is the quality of power. According to the established standards, by the term "quality of power," we mean "the term of power quality referring to a wide variety of electromagnetic phenomena that characterize the voltage and current at a given time and at a given location on the power system" (IEEE 1159–1995). There are several direct and indirect consequences of a poor electric power supply quality on a ship, which leads to several problems and distortions that could take place resulting in systems failures and a reduced level of reliability.

These problems could be summarized as follows (Prousalidis et al. 2008):

- Harmonics
- Short duration voltage events
- Voltage unbalance

According to other research, the operation of the electric power generating system could be summarized by two major groups of parameters:

- Parameters of voltage and currents in all the points of the analyzed system
- Parameters describing a risk of loss of power supply continuity

Attempting to evaluate the levels of quality and to deal with these problems, researchers have developed certain quality indices concerning voltage and frequency deviations (Prousalidis et al. 2011). The importance of those indices is obvious if their limit values and the standards established (Table 14.1) concerning the issues

TABLE 14.1
Standards Concerning Power Quality of a Ship

#	Standard	Range
1	IEEE Std. 45:2002	IEEE Recommended Practice for Electrical Installations on Shipboard
2	IEC 60092-101:2002	Electrical installations in ships. Definitions and general requirements
3	STANAG 1008:2004	Characteristics of Shipboard Electrical Power Systems in Warships of the North Atlantic Treaty Navies, NATO, Edition 9, 2004
4	American Bureau of Shipping, ABS, 2008	Rules of building and classing, steel vessels
5	Rules of international ship classification societies, e.g., PRS/25/P/2006	Technical Requirements for Shipboard Power Electronic Systems

Source: Mindykowski, J., Power quality on ships: Today and tomorrow's challenges, *International Conference and Exposition on Electrical and Power Engineering (EPE 2014)*, Iasi, Romania, 2014.

of electric power quality assessment in ship networks are considered. The usual causes of the power quality problems on ships are human factors, the assigned loads, overloading, and technical failures (Mindykowski 2014). It should be noted that the quality of electric power passes through two stages: assessment and improvement. The improvement stage is possible through the technical solutions and the investment in the staff and human capital (Mindykowski 2016). Technical solutions refer to new distribution systems such as Zonal Electrical Distribution System (ZEDS) or hybrid technology solutions (Shagar et al. 2017). The needs for electrical power differ from phase to phase of operation depending on the devices and systems that are necessary for the normal operation of the ship. According to expert opinions, the phase of charging and discharging are the most demanding and stressful for the electrical power generating system of a ship. Thus, a reliability modeling and analysis of the system related to these phases provides valuable inferences about the safety of a ship.

14.2 RELIABILITY ASSESSMENT AND MULTI-STATE SYSTEMS

The term "reliability" refers generally to the capability of any system or element to perform its assigned task. The analysis of reliability of a marine electric power system starts from the elements of the system, continues to the subsystems, and finally examines the whole system (Wu et al. 2013). Multi-state systems (MSS) theory covers a wide range of applications in reliability analysis with significant theoretical advances as well (Lisnianski et al. 2010). An MSS can operate passing through a finite number of states that are called state spaces (Lisnianski et al. 2010), describing different states (Eryilmaz 2015), and consequently working in different rates of output. This finite number of states indicates the difference between the MSS and the binary systems that operate in two states only (on-off) (Levitin and Xing 2018). The complexity of MSS depends on the number the subsystems, whereas its availability depends on the availability of these subsystems (Markopoulos and Platis 2018). Based on the requirements set, the structure of a MSS provides flexibility to the research of reliability to manage both theoretical problems and applications. It is well known that reliability is the capability of an element or a system to operate normally without failures or interruptions. In the case of an MSS, reliability could be the system capability to operate among specific states related to acceptable limits of operation according to the requirements established. The general mathematical form of an MSS operating among several states depicts the set of them such as:

$$S_j = \left\{ S_{j1}, S_{j2}, ..., S_{ji}, ..., S_{jk} \right\} \tag{14.1}$$

where:

s_{ji} is the state representing a specific level of performance of the subsystem j
$i \in \left\{ 1, 2, ..., k \right\}$ is the set of the states of each subsystem

Introducing the factor of time in the model, the state of the MSS over time is a random variable representing a stochastic process (Lisnianski et al. 2010) with its major parameters such as mean and variance. The function describing the reliability of the MSS can be defined as:

$$R(t,w) = P\{S(t) \geq w\} \tag{14.2}$$

Based on literature findings, one of the major research fields of MSS is reliability assessment and more specifically the electric—electronic systems such as power generation and communication systems (Lisnianski et al. 2012). To assess the expected performance of a complex or composite system, it is necessary to determine the states of the system and the sojourn time of each state (Barbu and Karagrigoriou 2018). This aspect implies the use of the semi-Markov methodology to take advantage of the flexibility it provides compared with the ordinary two-state Markov binary systems (operation or failure). The trade-off of the flexibility is the complexity of the system and the implied difficulties for understanding and performance evaluation (Yingkui and Jing 2012). There are more advantages concerning the flexibility of MSS. Since the focus is on the acceptable and non-acceptable sides, the analysis is closer to real world problems (Liu and Kapur 2006) than the ordinary simple systems that focus on "time to failure." This advantage leads to better accuracy assessments (Lisnianski et al. 2012) and improving the time needed to analyze the model (Billinton and Li 2007).

14.3 DESCRIPTION OF SHIP'S ELECTRIC POWER GENERATION SYSTEM

All these characteristics mentioned that are related to flexibility imply the capability of MSS to describe a lot of systems either technical or not. It is known that major research fields for MSS analysis are the electrical power generation and distribution systems (Markopoulos and Platis 2018) and telecommunications as well. In this current analysis, the attempt is to expand the reliability analysis using the theoretical tool of MSS in a marine electrical power generating system, considering its specific particularities against the terrestrial ordinary power generating systems. Depending on the phase of the operation, they should meet all minimum energy and power requirements without remaining in "out of order" status, even if some of their elements fail during the repair process. Thus, this chapter considers failure of the system as all those levels of output that do not meet minimum requirements for the normal operation of each operational phase of the ship.

According to the existing standards (DNV 2011), three general assumptions on the ship's systems structure are necessary to meet the established requirements. First, an electric power generation station should be arranged (DNV 2011, B301). The next requirement is that depending on the ship and operational phase, there are a minimum required number of independent electric power sources capable to meet the load requirements for normal operation of the ship without use of emergency power generators (DNV 2011, B302). The third main requirement is that the electric power generation system should be able to be restored within 45 seconds (DNV 2011, B303) using the existing automatic control switching (ACS) system. A typical example of an electric power system in a ship consists of operating components for power generation, energy transmission, and energy distribution for all energy consuming devices. Usually, there are ships with a configuration of three main generators and one emergency unit (Wärtsilä 2014) where the main system consists of two primary and one secondary and the switchboard (Mindykowski 2016), or a set of four, consisting of two main generators (primary) and two standby ones (Mennis and Platis 2013).

Considering the standards of IEEE (IEEE 45-2002) as shown in Figure 14.1, a typical example of the electrical power generation system of a large cargo ship consists of four generator units dedicated to serve the ordinary loads during different phases of the ship's use. An emergency generator unit exists in case of a total failure (blackout) of all four main generators. In this case, the capacity of the emergency generator is lower than that of the main ones, since it serves only the basic loads such as emergency lighting and basic instruments and devices of the ship such as internal communication and basic electronic systems (Patel 2012). In addition, many

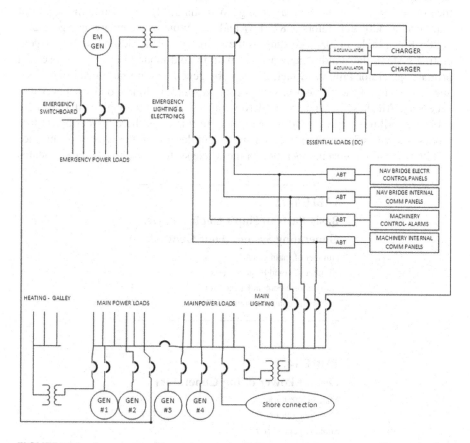

FIGURE 14.1   Large cargo ship power system with emergency generator and battery backup based on Standard IEEE 45-2002. (Based on Patel, M.R., *Shipboard Electrical Power Systems*, CRC Press, Boca Raton, FL, 2012.)

batteries exist to serve the ship in case of a total blackout. The case of four generators is the generic one covering more complex systems.

Starting the description of the system, we examine the case of the four-generator system assuming that it consists of two primary generators and two standby generators as shown in Figure 14.1 and in Table 14.2 (Patel 2012). When generators are in automatic startup, they need specific time to acquire their operational parameters such as the voltage and the frequency of their output current. All generators are controlled by an automatic control system which activates the standby generator or generators when necessary.

According to the same standard (IEEE 45-2002), we assume that when a generator startup failure takes place there are two ways of activation: automatic switching by the automatic control system and manual switching by the crew. The automatic switching time to activate the standby generators is 45 seconds.

The switching time for the manual activation depends on the current position of the crew members in the ship and for the current analysis we assume it is 5 minutes as the time to proceed to the machinery room from anywhere in the ship. Concerning the nominal power of the generators (e.g., Wärtsilä 2014), we assume the output of main and standby generators is 875 KW and the output of the emergency generator is 200 KW (Table 14.3). According to the ordinary use of the marine power generating system, the standby generators remain in cold mode to operate in case of a primary generator failure. In fact, the standby generators are not in running mode and only some of their essential subsystems are running to respond whenever it is necessary. All these generators are driven by auxiliary diesel engines that also are subject to failures, repairs, and maintenance. In this case, there are certain failure modes (shown in Table 14.4) for each subsystem describing the type of occurrence and its effect to the normal operation of the whole system. Due to the standby status

TABLE 14.2

Basic Parts of Ship's Electric Power Generating System (Four Generators)

Number of main generators	2
Number of stand by generators	2
Number of emergency generator	1
Automatic control system (ACS)	1

TABLE 14.3

Output Power of the Generators

Main generator #1	875 KW
Main generator #2	875 KW
Standby generator #1	875 KW
Standby generator #2	875 KW
Emergency generator	200 KW

TABLE 14.4

Failure Modes of Standby Generator

	Effect	
Occurrence Type	Prevents the Operation	Does Not Prevent the Operation
Monitored	Monitored Critical	Monitored Non-critical
Latent	Latent Critical	Latent Non-critical

Source: Alzbutas, R., *Energetika*, 4, 27–33, 2003.

of the secondary systems, their failures are probable to remain latent and they would be realized during a simultaneous failure of a main generator, whereas the time of this failure combined with the status of the whole system would be critical, especially when the specific generator is the last one available, since all main and standby ones have failed.

We should notice when the ship is in anchorage without additional electric systems in operation, one generator meets all load requirements. During additional operations such as cargo charging and discharging, one more generator is considered necessary (Mennis and Platis 2013). A general block diagram of the whole system is shown in Figure 14.2, where in case of a primary generator failure the automatic control system will switch normally to one of the two standby generators or the emergency one in case of failure of all main generators. We assume that according to the switching sequence, the ACS activates the first available secondary generator anticipating a failure with the probability (γ).

Considering the block diagram of Figure 14.2, the next step is to construct the Markov model diagrams for each phase of the vessel's operation. Since it is an ordinary electric power structure, we can use the same guidelines from previous research (Mennis and Platis 2013) adapting to the requirements of the current analysis.

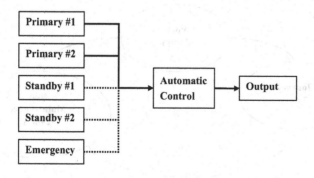

FIGURE 14.2   Block diagram of the power generating system. Use of primary #2 generator depends on the phase of the operation (e.g., it is necessary only during port phase).

Although the operational combinations of generators remain the same, we add some additional assumptions concerning the repair of generators and the automatic control system, which is supposed to present a certain level of availability. Specifically, when all primary generators fail, the priority is to switch to a secondary one first if it is available and after that to repair one or to complete the maintenance that is in progress. Considering other research on electric power systems (Wu et al. 2013), the reliability of ACS exceeds 5,500 hours. Since it is a pure electronic system, we suppose that only replacement of modules or rearrangements of cables are possible or necessary on board. The engine crew assigned to operate and maintain all the mechanical and electric power systems assure that the systems will be activated even manually; thus, we could consider the probability to fail a start-up of a generator as close to zero. The time of manual switching is 5 minutes during the manual activation of a standby generator. Major maintenance works on generators take place when the ship is in shipyard. In this current analysis, we should notice that there are differences in the requirements for electrical power between the phases of the operation. A model describing the states of the electric power generating system concerning the operational sequence is shown in Figure 14.3.

Because of the complexity of the system and for better understanding, a short description of the model is necessary. The meaning of the term "phases" is closely related to the minimum available power to meet the operational requirements.

There are three phases. The first phase is the *journey* phase that lasts on average 7 days when the vessel departs a port following a route to another one. During this phase, one primary generator provides the necessary level of power covering the needs of all operational systems. The *port* phase follows and lasts 3 days and takes place when the ship is in the port. The necessary machinery operates to charge or discharge cargo and to complete loading and unloading cargo processes according to the type of the ship and its cargo. Therefore, this phase is the most demanding one for power requirements and at least two primary generators are necessary for cargo handling. Finally, the *maintenance* phase follows the port phase and lasts two days. During this phase the engine crew conducts all necessary maintenance works, including both corrective (repairing generators that fail) and preventive

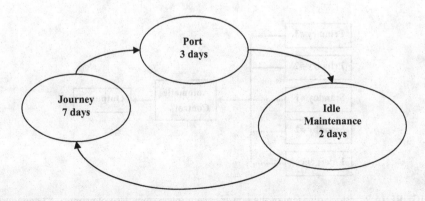

FIGURE 14.3 Phases of a ship's operation.

maintenance (conducting inspections or overhaul maintenance). The maintenance refers to all four (or three) primary generators covering routine inspections with ad hoc repairs and overhaul maintenance according to the manufacturer's maintenance plan.

14.4   SEMI-MARKOV MODEL DEVELOPMENT

The development of a semi-Markov model aims to develop a probabilistic study of the system and to assess the probabilities of the system to run in a specific output mode (state) and the time to remain in this state as well. To develop the semi-Markov model, it is necessary to determine and solve the system of steady-state equations. We will assume that the sojourn time in each state depends on the rate that will lead the system to another state through either failure or repair, and, in addition, each rate is distributed randomly following the exponential distribution. The process of transition from one state to another one consists of two components. The first one is the probability of transition of the system between two different states and the other component is the time spent in each specific state. The parameter that rules the jump from one state to another is the failure or repair rate for the affected generators or subsystems. The transition matrix of the probabilities will have the following form:

$$P = \begin{bmatrix} p_{1,1} & p_{1,2} & \cdots & p_{1,13} \\ p_{2,1} & p_{2,2} & \cdots & p_{2,13} \\ \cdots & \cdots & \cdots & \cdots \\ p_{13,1} & p_{13,2} & \cdots & p_{13,13} \end{bmatrix} \tag{14.3}$$

To calculate the steady-state probabilities, it is necessary to solve the following equation system (Trivedi 2002):

$$v = vP \tag{14.4}$$

where P is the matrix of the steady-state probabilities of Equation 14.3 and v is the vector of the discrete time Markov chain:

$$v = [v_1, v_2, ..., v_{13}] \tag{14.5}$$

The solution of Equation 14.4 is feasible under the restriction (Trivedi 2002):

$$\sum_{i=1}^{13} v_i = 1 \tag{14.6}$$

The general formula of mean sojourn time is (Trivedi 2002):

$$h_i = \int_0^\infty [1 - H_i(t)]dt \tag{14.7}$$

which after the integration of exponential distributions is:

$$h_i = \frac{1}{\sum_i \lambda_i + \sum_j \mu_j} \tag{14.8}$$

where λ and μ are the failure and repair rates, respectively. Since the manual time and automatic repair time are considered constant, the mean sojourn time is:

$$h_i = t_{man} \text{ and } h_i = t_{aut} \tag{14.9}$$

The expression of formula (14.8) is a general one that implies that the transition of the system from one state to another depends on the combination of all probable failures and repairs between the two states. The state probabilities of the semi-Markov model are given by the following formula:

$$\pi_i = \frac{v_i h_i}{\sum_j v_j h_j} \tag{14.10}$$

The matrix equation is:

$$V \cdot P_{semi} = U \Leftrightarrow V = U \cdot P_{semi}^{-1} \tag{14.11}$$

where U is the vector:

$$U = \left[1_{(1)}, 0_{(2)}, ..., 0_{(13)} \right] \tag{14.12}$$

and V is the matrix that will be combined with the set of mean sojourn times to calculate the final steady-state probabilities. Considering the general model of Figure 14.3, the one-step transition probability matrix is given by Table A14.4 of the Appendix to this chapter. A typical scenario of the operation cycle of a ship as previously mentioned consists of three phases: the system runs for 7 days in the journey phase, for 3 days in the port phase, and 2 days in the maintenance phase for a total of 12 days and a total of approximately 30 cycles on an annual basis. Proceeding to further analysis, the failures on the operating components of the system are events that take place in a random order; thus, they could be assumed to follow the Poisson distribution with a mean rate of failure (λ), whereas the mean time to repair $1/\mu$ follows the exponential distribution and, consequently, the rate of repair is (μ). A series of state diagrams could describe the system. The number of possible states of the system depends on its complexity. The model of the main electric power generation system as mentioned previously consists of two primary generators and two secondary (or standby) ones. Their output is identical, providing 875 KW. There is also a fifth generator (emergency generator) for providing a lower power level at 200 KW and its mission is to provide power for auxiliary loads (Wärtsilä 2014), thus providing a certain level of reliability in the case of a total blackout of all main generators.

TABLE 14.5

Failure and Repair Rates

System	MTTF (Hrs)	Failure Rate (per 10⁶ Hrs)	MTTR (man-hours)	Failure Rate (per hour)	Repair Rate (per hour)
Prim. Gen. 1	2,208.04	452.89	58.00	0.000453	0.017241
Prim. Gen. 2	2,208.04	452.89	58.00	0.000453	0.017241
Standby Gen. 1	2,208.04	452.89	58.00	0.000453	0.017241
Standby Gen. 2	2,208.04	452.89	58.00	0.000453	0.017241
Autom. Control	2,828.97	353.49	0.0833	0.000353	N/A

According to the available data (OREDA 2002), the failure rates and the repair times are given in three forms: min, mean, and max. Examining the worst case scenario, we assume the highest rate of failure and the longest repair time expressed in man-hours for each case. The failure rates and time to repair for all five generators are shown in Table 14.5. The failure rates are expressed in failures per 10^6 hours and the repair rates in hours considering a basic crew of six in the engine room.

The automatic control system is responsible for the activation process of a standby generator when a primary generator fails. In this study, we assume an automatic system (Wu et al. 2013) that is connected to the marine generators provides reliability parameters and characteristics as shown in Table A14.1 of the Appendix. The systems that are used in our study consist of three serial subsystems (Figure 14.4): Sys1—the main switch with a failure rate $\lambda_{SYS1} = 59.9998 \times 10^{-6}$/hr, Sys2—the excitation system with a failure rate $f_{SYS2} = 18.7 \times 10^{-6}$/hr, and Sys3—the main switching system with a failure rate $f_{SYS3} = 361.4859 \times 10^{-6}$/hr. The system is serial; thus, its failure rate is the sum of its components failure rates and totally $f_{SYSTEM} = 432.1857 \times 10^{-6}$/hr.

Since it is an electronic system, in the case of a failure its repair includes replacement of a module or rearrangement of the cables and contacts start-up a secondary generator when a primary one fails. The time the crew needs to repair the system manually is considered mean time to repair (MTTR) = 5 minutes or 0.0833 hours. Considering the structure of the whole power system, the automatic control system is vital for its normal operation. Consequently, the calculation of the probability (even if it is close to zero) to switch from a failed generator to a standby one is necessary. This probability is identical with the availability (A) of the control system and is expressed by the following formula:

$$A = \frac{MTBF\text{-}MTTR}{MTBF} \Leftrightarrow A = 1 - \gamma = 0.999964 \tag{14.13}$$

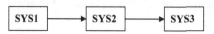

FIGURE 14.4 Layout of ACS blocks.

where $1 - \gamma$ is the probability to work normally when a primary generator fails. The probabilities of ACS normal operation impose certain difficulties in the development of the models and to the calculations process. To deal with this issue, the calculations are based on the expected time (t_{switch}) of the system response and the rate of response is:

$$t_{switch} = \gamma \cdot t_{man} + (1 - \gamma) \cdot t_{aut} \qquad (14.14)$$

Equation (14.14) represents the weighted average of the switching time either manually or automatically.

14.5   MULTI-STATE SYSTEM ANALYSIS

The electric power generating machine as an MSS can be assumed in operating mode (acceptable level of operation) when it operates in specific states which meet the operational requirements of each phase with the assumption that the power output for all other states is lower than the required one. In this case all latter states are considered failure states (Levitin and Lisnianski 1999). As mentioned in previous sections, the electric power generating system is driven by auxiliary diesel engines. Based on the available information[1] about the maintenance schedule and manufacturer's instructions for the auxiliary systems, there are certain restrictions and/ or limitations concerning the operation and handling of the whole system. Thus, the typical crew in an engine room consists of three engineers, two assistants, and one electric expert. These persons are responsible for the normal operation of all electromechanical systems during the ship's operation. In addition, they are responsible for the minor routine inspections and maintenance and the major ones such as overhauls whenever necessary. According to the available information, some typical rates of failure and repair times for auxiliary diesel engines are provided in Table A14.2 of the Appendix and a typical maintenance routine is summarized in Table A14.3 of the Appendix as well. During overhaul maintenance, when the parts are new, they are checked for good condition. Most parts are replaced with new ones at 8,000, 16,000, 24,000, and 32,000 hours, or confirmed whether they are in good condition. Exceptions to this rule are the fuel system check (fuel injection pump) in 2,000 hours, the lubricating system and cooling water system (thermostatic valves) in 12,000 hours, and the supercharging system (clean charge air cooler) in 4,000 hours. Obviously, the overhaul maintenance is scheduled according to the manufacturer's instructions but there is always the probability of discrepancies due to the quality of fuel and lubricating oil. The planned maintenance assures that no major failures will take place during the time between overhaul maintenances. Except the maintenance previously described, there are additional minor maintenance steps in shorter intervals such as a daily pressure check of the air filters and supercharging the compressor, the weekly check for the functionality of the control system and the compressed air system. Furthermore, the monthly check

---

[1] Information given by the marine engineer expert based on major engine manufacturer's data.

is adopted for other elements such as centrifugal oil filter and the compressed air system. There is a major factor affecting the maintenance of certain subsystems in the auxiliary engines. Due to crew and other restrictions, overhaul maintenance and all minor inspections (daily, weekly, and monthly) take place during the maintenance phase. All inspections or maintenance cover the respective ones of lower levels, for example, when the monthly inspection takes place then the respective weekly or daily inspections are omitted and engine crew members repair failures of auxiliary engines when they appear. Concerning the detailed analysis of the Markov model for each operational phase, each state presents the specific conditions of the system's operation. The code of label in each state describes the operational state (1st character), the number of active primary generators (2nd character), the number of the active secondary generators (3rd character), and whether one generator primary or secondary is in the maintenance process (4th character). Transitions and their rates for all states of the model are provided in Table A14.4 of the Appendix to this chapter. Starting with the maintenance phase as shown in Figure 14.5, the system enters the maintenance phase and leaves the port phase. The possible states are all those with one primary generator active (M,1,3,0 – M,1,2,0 – M,1,1,0, – M,1,0,0). These states represent the preparation of the maintenance process. The rate of maintenance is four generators per 48 hours (2 days of maintenance). During this phase, if a primary generator fails, then a secondary generator is activated either automatically (by ACS) or manually by the crew. This situation refers to the states M,0,3,0 – M,0,2,0 – M,0,1,0 – M,0,2,1, and M,0,1,1.

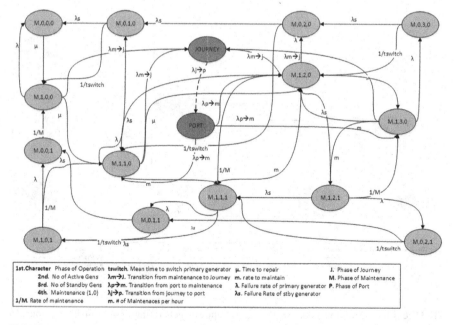

1st.Character Phase of Operation	tswitch. Mean time to switch primary generator	μ. Time to repair	J. Phase of Journey
2nd. No of Active Gens	λm→J. Transition from maintenance to Journey	m. rate to maintain	M. Phase of Maintenance
3rd. No of Standby Gens	λp→m. Transition from port to maintenance	λ. Failure rate of primary generator	P. Phase of Port
4th. Maintenance (1,0)	λJ→p. Transition from journey to port	λs. Failure Rate of stby generator	
1/M. Rate of maintenance	m. # of Maintenaces per hour		

FIGURE 14.5 Markov model of the electric power generating system (4-Gen)—maintenance phase.

If the failure happens while scheduled maintenance is in progress, then the crew continues to complete the maintenance because this time is shorter than that of a repair. If a secondary generator and a primary one operate normally, then the crew starts the process to repair it. Concerning the failure of a secondary generator while maintenance is in progress, the crew follows the same steps as in primary's failure. When all generators fail, the crew repairs one to recover normal power for maintenance. In this phase, the system is considered in normal operation when at least one primary generator is in normal operation, including states (M,1,3,0 – M,1,2,0 – M,1,1,0 – M,1,0,0 – M,1,2,1 – M,1,1,1 – M,1,0,1) and fails when it falls in any of the other states.

Next is the journey phase shown in Figure 14.6 when the system enters the journey phase leaving the maintenance phase. The strategy of the crew to repair or maintain the generators is the same as that of the maintenance phase with the difference that there is no generator under maintenance process. The possible states are all those with one primary generator active (J,1,3,0 – J,1,2,0 – J,1,1,0, – J,1,0,0). The activation of a secondary generator after a primary one's failure follows the same steps through the ACS and the normal operation includes the states J,1,3,0 – J,1,2,0 – J,1,1,0 – J,1,0,0. In this phase, there is an additional characteristic. Whereas the journey phase requires at least one primary generator, the transition to the next phase, the port phase, requires at least two primary generators. Thus, there are three

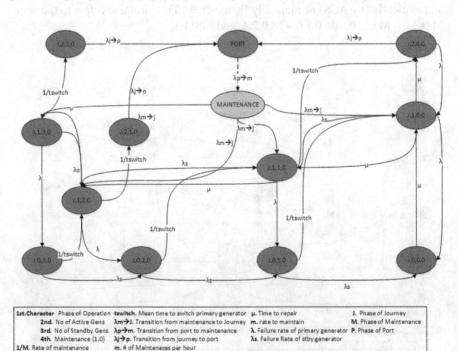

1st.Character	Phase of Operation	tswitch. Mean time to switch primary generator	μ. Time to repair	J. Phase of Journey
2nd.	No of Active Gens	λm→J. Transition from maintenance to Journey	m. rate to maintain	M. Phase of Maintenance
3rd.	No of Standby Gens	λp→m. Transition from port to maintenance	λ. Failure rate of primary generator	P. Phase of Port
4th.	Maintenance (1,0)	λj→p. Transition from journey to port	λs. Failure Rate of stby generator	
1/M. Rate of maintenance		m. # of Maintenaces per hour		

FIGURE 14.6 Markov model of the electric power generating system (4-Gen)—journey phase.

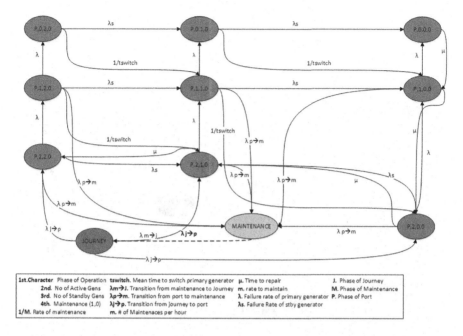

1st.Character	Phase of Operation	tswitch.	Mean time to switch primary generator	μ. Time to repair	J. Phase of Journey
2nd.	No of Active Gens	λm→J.	Transition from maintenance to Journey	m. rate to maintain	M. Phase of Maintenance
3rd.	No of Standby Gens	λp→m.	Transition from port to maintenance	λ. Failure rate of primary generator	P. Phase of Port
4th.	Maintenance (1,0)	λJ→p.	Transition from journey to port	λs. Failure Rate of stby generator	
1/M.	Rate of maintenance	m.	# of Maintenaces per hour		

FIGURE 14.7   Markov model of the electric power generating system (4-Gen)—port phase.

additional states in the journey phase (J,2,2,0 – J,2,1,0 – J,2,0,0) aiming to assure the activation of the second primary generator to prepare the system for the next phase requiring increased power.

The next and last phase, the port phase, shown in Figure 14.7 is when the system enters this phase after the journey phase. The possible states are P,2,2,0 – P,2,1,0 – P,2,0,0. The repair strategy for activation of secondary generators using ACS is the same with that of the journey phase.

Following the semi-Markov methodology as described in formulas (14.3) through (14.13), we can construct the transition matrix easily using Table A14.4 of the Appendix to this chapter followed by the one step probability matrix and the matrix of mean sojourn times. The V vector after calculations and the mean sojourn times are shown in Table A14.6 of the Appendix to this chapter and the final matrix of the steady-state probabilities in Table A14.8. As previously mentioned, systems with four generators are usual in large vessels. The analysis of the model with four generators shows that the level of availability of the system is high and there is no serious variation when the number of crew changes proving that investment in backup systems can reduce the need for crewmembers.

At this point it would worthwhile to investigate the sensitivity of the system's structure concerning the crewmembers and the backup systems. One test is to assume fewer generators for the system (e.g., three generators) as shown in Figures 14.8 through 14.10.

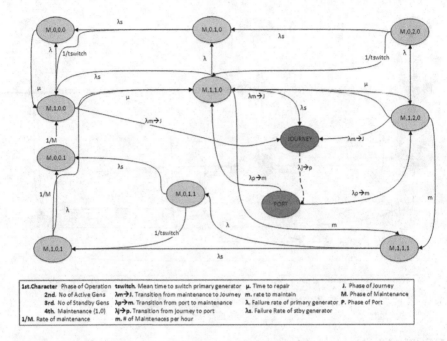

FIGURE 14.8 Markov model of the electric power generating system (3-Gen)—maintenance phase.

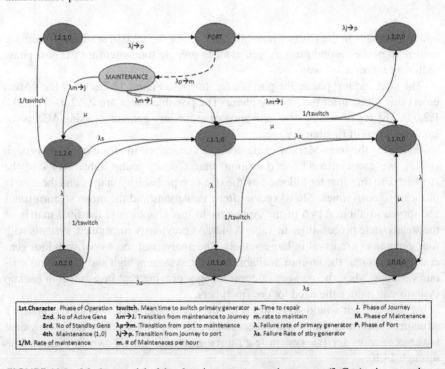

FIGURE 14.9 Markov model of the electric power generating system (3-Gen)—journey phase.

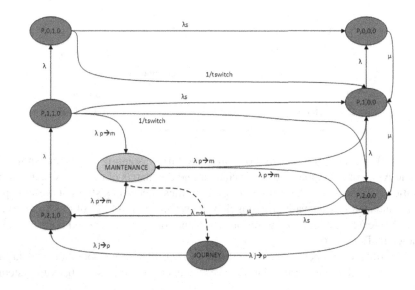

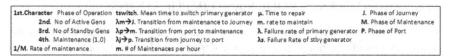

FIGURE 14.10 Markov model of the electric power generating system (3-Gen)—port phase.

Given that the needs for power are the same for each phase in both configurations (three and four generators), the systems differ only in the number of secondary generators. Following the same methodology of semi-Markov modeling as in four generator configurations, we can see simpler diagrams. The Markov models of the phases for system with three generators are shown in Figures 14.7 through 14.9 and all transitions are shown in Table A14.5 of the Appendix to this chapter. Concerning the maintenance phase (Figure 14.8), there are five out of ten states in normal operation $(M,1,2,0 - M,1,1,0 - M,1,0,0 - M,1,1,1 - M,1,0,1)$, while all others are considered failure. The next phase, the journey phase, as shown in Figure 14.9, is where the system enters the phase and leaves the maintenance phase. The strategy of crew to repair or maintain the generators is the same as that of the maintenance phase with the difference that there is no generator under a maintenance process. The possible states are all those with one primary generator active $(J,1,2,0 - J, 1,1,0, - J,1,0,0)$.

The activation of a secondary generator after a primary one's failure follows the same steps through the ACS and the normal operation includes the states $J,1,2,0 - J,1,1,0 - J,1,0,0$. In this phase, there is an additional characteristic. Following the same transition states of preparation (states $J,2,1,0$ and $J,2,0,0$), the system passes to the port phase (Figure 14.10). Implementing the semi-Markov methodology, we can construct the transition matrix using transitions of Table A14.5 in the Appendix to this chapter followed by the one step probability matrix and the matrix of mean sojourn times.

TABLE 14.6

Steady-State Probabilities of the Semi-Markov Model

	Configuration	
State of	4 Gen	3 Gen
Unavailability	2.893424E-06	1.024447E-05

The V vector after calculations and the mean sojourn times are shown in Table A14.6 of the Appendix to this chapter and the matrix of the steady-state probabilities in Table A14.9. A summary of probabilities concerning states of normal operation and states of unavailability (as they shown in Table A14.14 of the Appendix to this chapter) for system configuration (three and four generators) and crew of six are shown in Table 14.6.

The differences between probabilities in state of normal operation and failure are in line with the general reliability theory concerning the use of backup systems.

14.6 MAINTENANCE POLICY AND IMPLICATIONS

The maintenance policy of each operation system is a wide concept. The form and characteristics of the maintenance policy depend on the nature of the system, its complexity, and the requirements it is called to meet. It is also well known that maintenance aims to keep the system at a sufficient level of availability through the management of its parts and subsystems, according to the established requirements along with the reduction or minimization of the required time and cost. Concerning the general maintenance methodologies, they are classified in three basic groups which are (Chowdhury 1988):

1. Replacement and/or repair on failure
2. Planned maintenance (repair or replacement)
3. Condition-based maintenance

Considering the first group, this method is applied once a failure takes place. Following this strategy, it can be handled as a stochastic renewal process and the implied cost can be expressed by the following formula:

$$C_{RR} = C_R + C_D \tag{14.15}$$

where:

C_{RR} is the total cost of maintenance

C_R is the repair cost

C_D is the indirect cost while the system is not operative

Compared with the alternative of planned maintenance, the repair on failure policy is preferred when:

$$C_{RR} \leq C_{PM} \tag{14.16}$$

where C_{PM} is the total cost of planned maintenance in predetermined intervals, which refers to the second group. The replacement as a concept is included in the wider concept of maintenance. The adequate actions could refer to age, periodic, or block replacement (Nakagawa 2006). All above parameters of maintenance summarize the preventive maintenance which in the case of a ship and due to its particularities, plays a crucial role. According to planned maintenance policy and depending on the conditions, the replacement takes place in blocks, when the parts have reached their age or operational limit, using the optimal number of failures or using cycle time. Implementing the block replacement/repair policy, each part or subsystem is replaced at times kT with $k = 1, 2 \ldots$ or at failure, whichever comes first (Barlow and Proschan 1996). In this case, the interval that minimizes the total cost is (Chowdhury 1988):

$$T_0 m(T_0) - M(T_0) = \frac{C_2}{C_1} \qquad (14.17)$$

where:
   T_0 is the optimum time interval
   $M(T_0)$ is the renewal function
   $m(T_0)$ is the renewal density
   C_1 is the expected cost of failure
   C_2 is the expected cost for exchanging non-failed item

The study of the maintenance problem also is related to the MSS methodology. In general, two major categories of maintenance are followed (Liu and Huang 2010). The corrective one is conducted when a system failure takes place and the preventive one is conducted when the user's intention is to keep its performance within the desired limits during specific periods of operation. Concerning the ships and shipping industry, the application of the corrective maintenance refers to onboard repairing activities whereas the preventive one refers to repairs in shipyards during major overhauls. Due to the existing restrictions to repair failed systems on board, the corrective maintenance presents inherent difficulties. One major challenge is the optimization of maintenance policy through a combination of maintenance policy to achieve the ship's unobstructed operation. Depending on the management policy of the ship's owner, the maintenance policy (consisting of corrective and preventive maintenance) of the ship's subsystems should consider the expected time between failures organize the transportation assignments of each ship aiming to minimize the cost. Thus, minor failures subjected to repair by the crew members would not affect the ship's transportation capability. It is obvious that repair of major failures should be scheduled during the overhaul inspections and repair in the shipyard.

One important parameter that affects the development of the maintenance policy is the time horizon. This horizon determines the strategy of the maintenance policy management. In the case of a long or medium term, the maintenance policy could focus more on planning and spare parts' inventory management, while the short-term horizon focuses more on monitoring and control (Ben-Daya et al. 2000).

If the complexity of the system is high, the replacement policy should focus rather on block replacement than on other policies such as age replacement, proving that the first one is preferred to the second one (Barlow and Proschan 1996). The basic analysis of the maintenance policy refers to the failures and their distributions of the parts, subsystems, and systems. Nevertheless, it does not provide universal answers concerning maintenance, because further questions could arise, such as what the optimal maintenance policy is, considering the specific conditions of the system. The term "optimal" refers not only to a cost minimization, but also to maximization of the availability (Barlow and Proschan 1996). These two terms follow the same principles concerning the optimization of problems with more than two decision variables under optimization.

Considering the findings from previous sections, we could propose certain basic principles concerning a ship's operation. One example is the maintenance of the electrical power system and the results of this study as a way of thinking could be expanded to other subsystems and finally to the whole ship. The electric system of a ship as a complex system consists of many different parts that are subject to deterioration and a possible gradual degradation. Since it is a complicated system, it could be an MSS operating in different output levels. As described in previous sections, a marine electric power generating system consists of many different main and emergency generators. The system follows a typical configuration of four main generators and one emergency, whereas the set of the four main consists of two primary and two standby generators. Considering the generating system as an MSS, in general, whenever a system's performance falls under the threshold of acceptance it is assumed failed, thus maintenance actions should take place (Liu and Huang 2010). Since there is always the probability of transition of the MSS from one state to another, the restoration of the system is subject to a factor of randomness, because one subsystem could fail during the restoration of a previously failed one.

As maintenance policy depends on the strategic goals of the decision makers, it is closely related to the minimization of the maintenance cost. Although cost is a single concept, there are different aspects to describe it in the same result. Concerning a ship, the sufficient level of maintenance implies direct and indirect cost savings. Direct cost savings refers to reduced needs for repairs, reduced man-hours, and the losses due to not using the equipment. The indirect cost savings refer to meeting the requirements of contracts and penalties due to delays. Another valid assessment of this cost is the reliability associated cost (RAC), which is expressed by (Lisnianski et al. 2010) by the formula:

$$RAC = OC + RC + PC \tag{14.18}$$

where:

OC is the operational cost and the fuel cost, when it comes to power systems driven by auxiliary diesel engines

RC the repair cost including the repair and maintenance cost in man-hours and spare parts

PC the penalty cost when the system's failure leads to delays of the operation

TABLE 14.7

Probabilities of Operation at Acceptable Level (min power requirements)

Crew Members	4 Generators Unavailability	3 Generators Unavailability
1	1.400095E-05	2.204020E-04
2	4.306027E-06	5.963794E-05
3	3.288975E-06	2.927190E-05
4	3.031658E-06	1.834814E-05
5	2.936747E-06	1.314772E-05
6	2.893424E-06	1.024447E-05
7	2.870731E-06	8.447592E-06
8	2.857641E-06	7.252148E-06
9	2.849532E-06	6.413197E-06
10	2.844226E-06	5.799750E-06
11	2.840598E-06	5.336362E-06
12	2.838030E-06	4.976965E-06

According to our findings, the probability of the system to reach a non-acceptable level of operation depending on the crew varies from 1.400095E-05 to 2.838030E-06 in a four generator configuration and from 2.204020E-04 to 4.976965E-06 in a three generators configuration (Table 14.7). Attempting to understand the sensitivity of availability subjected to changes of crew, this chapter developed all models of previous sections for different crews, from 1 up to 12 members. The final probabilities for each phase of the auxiliary engines' operation and for different crews are shown in Table 14.7.

Alternatively, the availabilities of both configurations are shown in Figure 14.11. There is an obvious difference between the availability of the two systems showing the possible interaction between systems and manpower.

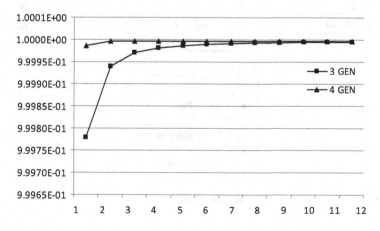

FIGURE 14.11 System availability according to the number of crew members.

Concerning all models previously presented, certain implications referring to the maintenance policy and its cost minimization arise. More specifically, there are different ways to improve the availability of the auxiliary engines and the reduction of their reliability cost. Compared to any proposed solutions, presumably the lower the cost and the simpler the action, the more preferable the solution is.

According to formula (14.18), the structure of reliability cost for the marine generator system is shown in Figure 14.12. Considering this cost and its minimization, there are two possible options for action concerning the ship. Either, the ship is in a *planning* and *construction* phase or it is already *in operation*. If it is in planning or construction phase, adjustment to the equipment such as changes of the number or the power level of generators and the attached auxiliary engines and systems are possible. On the other hand, if it is in phase of operation, the management team should search for alternative options to reduce/optimize maintenance cost. This study aiming to search for alternative maintenance policy options focuses on the potential adjustment of parameters based on Figure 14.12.

Supplies, transportation, and fuel cost are related to geographical and financial factors which are subject to changes beyond planning and construction. Other costs (e.g., spare parts cost) are related with planning and construction decisions. For a ready-made ship (in which case this study refers), only any cost related to its use is adjustable, such as penalty cost and man-hours cost that could be adjusted to deal with the problem of the maintenance cost optimization. The penalty cost due to delays is related to the reliability of the ship and its capability to operate on time and to increase its utility and efficiency. The labor cost is related to the number of the crew members in the engine room. Obviously, the more crew members, the higher maintainability and the less time to repair subjected to the restrictions between the cost and benefits of each personnel change.

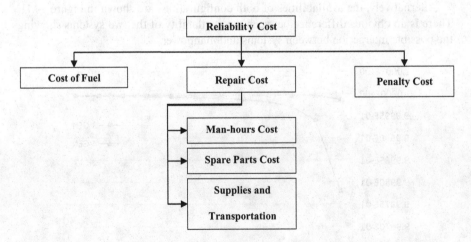

FIGURE 14.12   Structure of dependability cost.

14.7 CONCLUSIONS

This study evaluated the reliability performance using MSS theory of a marine electric power generation unit that consists of four generators (one or two primary and two or three secondary) driven by auxiliary diesel engines. An additional alternative analysis of a system including three generators (one or two primary and one or two secondary) was conducted to investigate the differences between two system configurations and to identify possible alternatives concerning the decision making.

One main characteristic of the system is that it uses a single type of generators. This strategy implies a managerial aspect that is the simplification of the system's management concerning, in general, the schedule of supplies and the maintenance. The analysis of the generators as a MSS shows that the probability of operation in non-acceptable level of output is low implying high availability, which is along with the operational requirements of a ship. This fact is possible through the increased number of generators (primary, secondary, and emergency) and the expected rates of failure of the subsystems (generators) that tend to reach low levels following the technological progress. Due to technological limitations, the continuous lowering of failure rates could be considered difficult. To achieve the goal of continuous improvement of the systems there are two options. The first one is the additional backup systems (configuration with four generators) and the second one is to increase repair rates using the parameter of maintainability (configuration with three generators) using additional highly qualified personnel in the engine room.

The analysis of the system shows that except the ordinary aspect of maintenance focusing on materials, another aspect focusing on manpower and human capital also exists. Considering the reliability parameters of the systems and the modeling process, the empirical results show that the efficiency of the systems is very high. Comparison of both system configurations shows that increasing the engine crew, the probability of normal operation tends to be almost the same for both options. This fact implies the existence of an interaction between system and crew. The findings that the level of reliability increases along with the increase of crew members are valid within limits that depend on the cost of penalty to cost of wage ratio. Obviously, these limits differ depending on the specifications and operation requirements of each system. The analysis conducted in this chapter covers a theoretical approach of the maintenance problem optimization in a specific sector of the ship. This model could focus to specific devices and subsystems or could be expanded in an appropriate way to include more systems or integrated blocks. All the above findings would be a starting point for further research, combining additional mathematical methods such as Monte Carlo Simulation, leading to uncertainty reduction of the models.

ACKNOWLEDGMENT

We would like to thank Dr. J. Dagkinis, whose expert knowledge on marine engineering issues was most helpful.

APPENDIX

TABLE A14.1
Failure Rates for Automatic Control System EEA-22

Type of Failure	MTBF (Days)	Failure Rate (per 10^{-6} Hrs)
5C147	1,818.38	22.9142
Input 5C15	8,998.89	4.6302
Power 5C3	301.02	138.4184
Start 5C21	2,324.97	17.9214
Stop 5C27	3,929.78	10.6028
General control 6C109	2,270.22	18.3536
Stop cascade 6C103	10,659.16	3.9090
Voltage monitor 5C153	671.80	62.0226
V/f monitor 6C39	4,030.83	10.3370
Additional reference value 6C67	5,374.96	7.7520
Closing pulse EVG23	8,726.73	4.7746
Power SNT23	19,007.65	2.1921
Relay output RAG23	2,102.36	19.8190
Frequency presetting FAG23	9,780.68	4.2601
Active power measuring WMG23	3,499.05	11.9080
Frequency controller FRG23	9,063.09	4.5974
Load distribution LAG23	4,592.13	9.0735
Total Rate	**117.87**	**353.4859**

TABLE A14.2
Failure and Repair Rates for Engine Drivers of Power Generators

Type of Failure	MTBF (Days)	Failure Rate (per 10^{-6} Hrs)	Time to Repair (Hrs)	Repair Rate (per 10^{-6} Hrs)
Fuel oil filters	20	2,083.33	1.5	27,777.78
Oil filters	20	2,083.33	1.5	27,777.78
Air filters	20	2,083.33	1.5	27,777.78
Water filters	30	1,388.89	1.5	27,777.78
Fuel injector	60	694.44	1.5	27,777.78
Leaking of gasket	90	462.96	1.5	27,777.78
Piping system	60	694.44	1.5	27,777.78
Water pump	60	694.44	3.0	13,888.89
Fuel pump	30	1,388.89	3.0	13,888.89
Fuel injector	180	231.48	3.0	13,888.89
Dirty water cooler	60	694.44	3.0	13,888.89
Dirty oil cooler	60	694.44	3.0	13,888.89
Cover gasket damage	365	114.16	3.0	13,888.89
Exhaust inlet valve	365	114.16	7.0	5,952.38
Cracking of cyl. heads	240	173.61	7.0	5,952.38
Piston ring damage	365	114.16	7.0	5,952.38
Turbo charger	365	114.16	7.0	5,952.38

TABLE A14.3
General Plan of Maintenance

Action—Description	Daily	Weekly	Monthly	Overhaul Maintenance Interval (Hours of Operation)									
				1,000	2,000	4,000	8,000	12,000	16,000	20,000	24,000	28,000	30,000
Major fasteners—retightening	X	X	X										
Major bearing—inspection							X				X		X
Resilient mounts—inspect-retighten							X						
Cylinder and rod—inspection		X							X				
Crankshaft—gears—inspection							X		X		X		
Valve mechanism									X				
Control system	X	X					X						
Fuel system					X		X						
Lubricating oil system			X					X	X				
Cooling water system			X					X	X				
Compressed air system	X	X	X										
Supercharging system	X	X				X							

TABLE A14.4
States of the Marine Power Generating System (4 Generators)[a]

State	FR	TO	Rate	State	FR	TO	Rate	State	FR	TO	Rate
M,1,3,0	S1	S2	λs	J,1,3,0	S15	S16	λs	P,2,2,0	S26	S1	$\lambda p \rightarrow m$
		S5	m		S15	S18	λ		S26	S27	λs
		S8	λ		S15	S23	$1/tswitch$		S26	S29	λ
		S15	$\lambda m \rightarrow j$	J,1,2,0	S16	S15	μ	P,2,1,0	S27	S2	$\lambda p \rightarrow m$
M,1,2,0	S2	S1	μ		S16	S17	λs		S27	S26	μ
		S3	λs		S16	S19	λ		S27	S28	λs
		S5	m		S16	S24	$1/tswitch$		S27	S30	λ
		S9	λ	J,1,1,0	S17	S16	μ	P,2,0,0	S28	S3	$\lambda p \rightarrow m$
		S16	$\lambda m \rightarrow j$		S17	S20	λ		S28	S27	μ
M,1,1,0	S3	S2	μ		S17	S21	λs		S28	S31	λ
		S6	m		S17	S25	$1/tswitch$	P,1,2,0	S29	S2	$\lambda p \rightarrow m$
		S10	λ	J,0,3,0	S18	S16	$1/tswitch$		S29	S27	$1/tswitch$
		S17	$\lambda m \rightarrow j$		S18	S19	λs		S29	S30	λs
M,1,0,0	S4	S3	μ	J,0,2,0	S19	S17	$1/tswitch$		S29	S32	λ
		S11	λ		S19	S20	λs	P,1,1,0	S30	S3	$\lambda p \rightarrow m$
		S21	$\lambda m \rightarrow j$	J,0,1,0	S20	S21	$1/tswitch$		S30	S31	λs
M,1,2,1	S5	S1	$1/M$		S20	S22	λs		S30	S28	$1/tswitch$
		S6	λs	J,1,0,0	S21	S17	μ		S30	S33	λ
		S12	λ		S21	S22	λ	P,1,0,0	S31	S4	$\lambda p \rightarrow m$
M,1,1,1	S6	S2	$1/M$	J,0,0,0	S22	S21	μ		S31	S28	μ
	S6	S7	λs	J,2,2,0	S23	S26	$\lambda j \rightarrow p$		S31	S34	λ
	S6	S13	λ	J,2,1,0	S24	S27	$\lambda j \rightarrow p$	P,0,2,0	S32	S30	$1/tswitch$
M,1,0,1	S7	S3	$1/M$	J,2,0,0	S25	S28	$\lambda j \rightarrow p$		S32	S33	λs
		S14	λ					P,0,1,0	S33	S30	$1/tswitch$
M,0,3,0	S8	S2	$1/tswitch$						S33	S34	λs
	S8	S9	λs					P,0,0,0	S34	S31	μ
M,0,2,0	S9	S3	$1/tswitch$								
		S10	λs								
M,0,1,0	S10	S4	$1/tswitch$								
		S11	λs								
M,0,0,0	S11	S4	μ								
M,0,2,1	S12	S6	$1/tswitch$								
	S12	S13	λs								
M,0,1,1	S13	S7	$1/tswitch$								
	S13	S14	λs								
M,0,0,1	S14	S4	$1/M$								

[a] The transition to another state is either through failure or repair of a generator or automatic control system.

TABLE A14.5
States of the Marine Power Generating System (3 Generators)[a]

State	FR	TO	Rate	State	FR	TO	Rate	State	FR	TO	Rate
M,1,2,0	S1	S2	λs	J,1,2,0	S11	S12	λs	P,2,1,0	S19	S1	$\lambda p \to m$
		S4	m			S13	λ			S20	λs
		S6	λ			S17	$1/tswitch$			S21	λ
		S11	$\lambda m \to j$	J,1,1,0	S12	S11	μ	P,2,0,0	S20	S2	$\lambda p \to m$
M,1,1,0	S2	S1	μ			S14	λ			S19	μ
		S3	λs			S15	λs			S22	λ
		S4	m			S18	$1/tswitch$	P,1,1,0	S21	S2	$\lambda p \to m$
		S7	λ	J,0,2,0	S13	S12	$1/tswitch$			S20	$1/tswitch$
		S12	$\lambda m \to j$			S14	λs			S22	λs
M,1,0,0	S3	S2	μ	J,0,1,0	S14	S15	$1/tswitch$			S23	λ
		S8	λ			S16	λs	P,1,0,0	S22	S3	$\lambda p \to m$
		S15	$\lambda m \to j$	J,1,0,0	S15	S12	μ			S20	μ
M,1,1,1	S4	S1	$1/M$			S16	λ			S24	λ
		S5	λs	J,0,0,0	S16	S15	μ	P,0,1,0	S33	S22	$1/tswitch$
		S9	λ	J,2,1,0	S17	S19	$\lambda j \to p$			S24	λs
M,1,0,1	S5	S2	$1/M$	J,2,0,0	S18	S20	$\lambda j \to p$	P,0,0,0	S34	S22	μ
		S10	λ								
M,0,2,0	S6	S2	$1/tswitch$								
		S7	λs								
M,0,1,0	S7	S3	$1/tswitch$								
		S3	μ								
M,0,0,0	S8	S5	$1/tswitch$								
		S10	λs								
M,0,1,1	S9	S3	$1/M$								
		S2	λs								
M,0,0,1	S10	S4	m								

[a] The transition to another state is either through failure or repair of a generator or automatic control system.

TABLE A14.6
V-Vector and Mean Sojourn Times (4 Generators)

State	vi	hi	State	vi	hi
1	2.9110E-01	1.4128E+00	18	1.1644E-06	1.2502E-02
2	1.6476E-03	1.0933E+00	19	5.1062E-09	1.2502E-02
3	7.2131E-06	1.0938E+00	20	2.2407E-11	1.2502E-02
4	1.0464E-08	1.4137E+00	21	7.4578E-09	4.8228E+00
5	8.5464E-02	4.8123E+00	22	1.6289E-11	4.8333E+00
6	3.7416E-04	4.8123E+00	23	2.0564E-01	7.0000E+00
7	1.6319E-06	4.8228E+00	24	9.0063E-04	7.0000E+00
8	1.8626E-04	1.2502E-02	25	3.9522E-06	7.0000E+00
9	8.1683E-07	1.2502E-02	26	2.0620E-01	2.9919E+00
10	3.5778E-09	1.2502E-02	27	1.4607E-03	1.8480E+00
11	6.7201E-12	4.8333E+00	28	6.3973E-06	1.8495E+00
12	1.8626E-04	1.2502E-02	29	2.7939E-04	1.2451E-02
13	8.1650E-07	1.2502E-02	30	1.2257E-06	1.2451E-02
14	3.5691E-09	4.8333E+00	31	5.3700E-09	1.8495E+00
15	2.0564E-01	1.2502E-02	32	1.5754E-09	1.2502E-02
16	9.0297E-04	1.2470E-02	33	6.9200E-12	1.2502E-02
17	3.9625E-06	1.2470E-02	34	4.4978E-12	4.8333E+00

TABLE A14.7
V-Vector and Mean Sojourn Times (3 Generators)

State	vi	hi	State	vi	hi
1	2.9111E-01	1.4128E+00	13	1.1644E-06	1.2502E-02
2	1.6468E-03	1.0933E+00	14	5.1164E-09	1.2502E-02
3	3.2053E-06	1.4137E+00	15	2.2809E-06	4.8228E+00
4	8.5465E-02	4.8123E+00	16	4.9820E-09	4.8333E+00
5	3.7253E-04	4.8228E+00	17	2.0564E-01	7.0000E+00
6	1.8626E-04	1.2502E-02	18	9.0243E-04	7.0000E+00
7	8.1641E-07	1.2503E-02	19	2.0620E-01	2.9919E+00
8	2.0523E-09	4.8333E+00	20	1.4605E-03	1.8495E+00
9	1.8626E-04	1.2502E-02	21	2.7940E-04	1.2451E-02
10	8.1473E-07	4.8333E+00	22	1.2276E-06	1.8495E+00
11	2.0564E-01	1.2502E-02	23	1.5754E-09	1.2502E-02
12	9.0477E-04	1.2470E-02	24	1.0282E-09	1.0000E+00

TABLE A14.8
Steady-State Probabilities of Electric Power Generating System (4 Generators)

Phase		State	Probability Crew of 6
Maintenance	S1	M,1,3,0	1.4147949E-01
	S2	M,1,2,0	1.2390664E-03
	S3	M,1,1,0	1.0848216E-05
	S4	M,1,0,0	2.5590879E-08
	S5	M,1,2,1	1.4147978E-01
	S6	M,1,1,1	1.2387772E-03
	S7	M,1,0,1	1.0802759E-05
	S8	M,0,3,0	8.0109202E-07
	S9	M,0,2,0	7.0204378E-09
	S10	M,0,1,0	6.1465043E-11
	S11	M,0,0,0	1.1230434E-10
	S12	M,0,2,1	8.0109365E-07
	S13	M,0,1,1	7.0188000E-09
	S14	M,0,0,1	4.7324521E-08
Journey	S15	J,1,3,0	8.8442730E-04
	S16	J,1,2,0	7.7457435E-06
	S17	J,1,1,0	6.7974990E-08
	S18	J,0,3,0	5.0078470E-09
	S19	J,0,2,0	4.3886679E-11
	S20	J,0,1,0	3.8513977E-13
	S21	J,1,0,0	1.2428462E-07
	S22	J,0,0,0	5.4411190E-10
	S23	J,2,2,0	4.9517823E-01
	S24	J,2,1,0	4.3367313E-03
	S25	J,2,0,0	3.8058227E-05
Port	S26	P,2,2,0	2.1221906E-01
	S27	P,2,1,0	1.8575841E-03
	S28	P,2,0,0	1.6281384E-05
	S29	P,1,2,0	1.1966432E-06
	S30	P,1,1,0	1.0487945E-08
	S31	P,1,0,0	1.6892712E-08
	S32	P,0,2,0	6.7756912E-12
	S33	P,0,1,0	5.9423717E-14
	S34	P,0,0,0	7.3956051E-11

TABLE A14.9
Steady-State probabilities of Electric Power Generating System (3 Generators)

Phase	State		Probability Crew of 6
Maintenance	S1	M,1,2,0	1.4148529E-01
	S2	M,1,1,0	1.2381783E-03
	S3	M,1,0,0	3.8518151E-06
	S4	M,1,1,1	1.4148464E-01
	S5	M,1,0,1	1.2334183E-03
	S6	M,0,2,0	8.0112483E-07
	S7	M,0,1,0	7.0154487E-09
	S8	M,0,0,0	1.6863003E-08
	S9	M,0,1,1	8.0112119E-07
	S10	M,0,0,1	5.4033344E-06
Journey	S11	J,1,2,0	8.8446355E-04
	S12	J,1,1,0	7.7642464E-06
	S13	J,0,2,0	5.0080522E-09
	S14	J,0,1,0	4.3991448E-11
	S15	J,1,0,0	1.8685111E-05
	S16	J,0,0,0	8.1802426E-08
	S17	J,2,1,0	4.9519852E-01
	S18	J,2,0,0	4.3470908E-03
Port	S19	P,2,1,0	2.1222833E-01
	S20	P,2,0,0	1.8595168E-03
	S21	P,1,1,0	1.1966955E-06
	S22	P,1,0,0	1.9305769E-06
	S23	P,0,1,0	6.7759872E-12
	S24	P,0,0,0	8.7434201E-10

TABLE A14.10
Steady-State Probabilities for Different Crews (4 Generators)

State		Crew 1	2	3	4	5	6
				Probabilities (1 of 2)			
M,1,3,0	1	1.35245E-01	1.38982E-01	1.40230E-01	1.40855E-01	1.41230E-01	1.41479E-01
M,1,2,0	2	7.10546E-03	3.65104E-03	2.45599E-03	1.85026E-03	1.48420E-03	1.23907E-03
M,1,1,0	3	3.73045E-04	9.58536E-05	4.29911E-05	2.42938E-05	1.55917E-05	1.08482E-05
M,1,0,0	4	1.11179E-06	2.72044E-07	1.16200E-07	6.26382E-08	3.84186E-08	2.55909E-08
M,1,2,1	5	1.35246E-01	1.38982E-01	1.40231E-01	1.40855E-01	1.41230E-01	1.41480E-01
M,1,1,1	6	7.10492E-03	3.65068E-03	2.45566E-03	1.84996E-03	1.48391E-03	1.23878E-03
M,1,0,1	7	3.63724E-04	9.46614E-05	4.26364E-05	2.41436E-05	1.55141E-05	1.08028E-05

(Continued)

TABLE A14.10 (*Continued*)
Steady-State Probabilities for Different Crews (4 Generators)

Probabilities (1 of 2)

State		Crew 1	2	3	4	5	6
M,0,3,0	8	7.65792E-07	7.86949E-07	7.94018E-07	7.97555E-07	7.99677E-07	8.01092E-07
M,0,2,0	9	4.02372E-08	2.06775E-08	1.39109E-08	1.04812E-08	8.40845E-09	7.02044E-09
M,0,1,0	10	2.11250E-09	5.42864E-10	2.43505E-10	1.37617E-10	8.83315E-11	6.14650E-11
M,0,0,0	11	2.92597E-08	3.58011E-09	1.01957E-09	4.12243E-10	2.02297E-10	1.12304E-10
M,0,2,1	12	7.65794E-07	7.86951E-07	7.94020E-07	7.97556E-07	7.99679E-07	8.01094E-07
M,0,1,1	13	4.02342E-08	2.06755E-08	1.39091E-08	1.04794E-08	8.40677E-09	7.01880E-09
M,0,0,1	14	9.55521E-06	1.24354E-06	3.73441E-07	1.58617E-07	8.15479E-08	4.73245E-08
J,1,3,0	15	8.45455E-04	8.68813E-04	8.76618E-04	8.80522E-04	8.82865E-04	8.84427E-04
J,1,2,0	16	4.44182E-05	2.28236E-05	1.53530E-05	1.15665E-05	9.27816E-06	7.74574E-06
J,1,1,0	17	2.33895E-06	6.00907E-07	2.69475E-07	1.52259E-07	9.77078E-08	6.79750E-08
J,0,3,0	18	4.78718E-09	4.91944E-09	4.96363E-09	4.98573E-09	4.99900E-09	5.00785E-09
J,0,2,0	19	2.51534E-10	1.29261E-10	8.69609E-11	6.55207E-11	5.25635E-11	4.38867E-11
J,0,1,0	20	1.32452E-11	3.40322E-12	1.52633E-12	8.62498E-13	5.53544E-13	3.85140E-13
J,1,0,0	21	3.23649E-05	3.96043E-06	1.12799E-06	4.56127E-07	2.23855E-07	1.24285E-07
J,0,0,0	22	8.50149E-07	5.20156E-08	9.87656E-09	2.99535E-09	1.17603E-09	5.44112E-10
J,2,2,0	23	4.73358E-01	4.86436E-01	4.90806E-01	4.92992E-01	4.94304E-01	4.95178E-01
J,2,1,0	24	2.48691E-02	1.27786E-02	8.59595E-03	6.47592E-03	5.19471E-03	4.33673E-03
J,2,0,0	25	1.30955E-03	3.36440E-04	1.50875E-04	8.52475E-05	5.47052E-05	3.80582E-05
P,2,2,0	26	2.02868E-01	2.08473E-01	2.10345E-01	2.11282E-01	2.11844E-01	2.12219E-01
P,2,1,0	27	1.06571E-02	5.47545E-03	3.68290E-03	2.77433E-03	2.22527E-03	1.85758E-03
P,2,0,0	28	5.60425E-04	1.43969E-04	6.45574E-05	3.64737E-05	2.34044E-05	1.62814E-05
P,1,2,0	29	1.14391E-06	1.17552E-06	1.18608E-06	1.19136E-06	1.19453E-06	1.19664E-06
P,1,1,0	30	6.01056E-08	3.08880E-08	2.07803E-08	1.56572E-08	1.25612E-08	1.04879E-08
P,1,0,0	31	7.24062E-07	1.77306E-07	7.59545E-08	4.10781E-08	2.52784E-08	1.68927E-08
P,0,2,0	32	6.47713E-12	6.65607E-12	6.71586E-12	6.74577E-12	6.76372E-12	6.77569E-12
P,0,1,0	33	3.40369E-13	1.74933E-13	1.17701E-13	8.86932E-14	7.11628E-14	5.94237E-14
P,0,0,0	34	1.90194E-08	2.32870E-09	6.65049E-10	2.69756E-10	1.32802E-10	7.39561E-11

TABLE A14.11
Steady-State Probabilities for Different Crews (4 Generators)

Probabilities (2 of 2)

State		Crew 7	8	9	10	11	12
M,1,3,0	1	1.41658E-01	1.41792E-01	1.41896E-01	1.41979E-01	1.42048E-01	1.42104E-01
M,1,2,0	2	1.06343E-03	9.31411E-04	8.28553E-04	7.46157E-04	6.78668E-04	6.22377E-04
M,1,1,0	3	7.98124E-06	6.11720E-06	4.83750E-06	3.92115E-06	3.24256E-06	2.72606E-06
M,1,0,0	4	1.80554E-08	1.32918E-08	1.01108E-08	7.89429E-09	6.29624E-09	5.11149E-09
M,1,2,1	5	1.41658E-01	1.41792E-01	1.41896E-01	1.41980E-01	1.42048E-01	1.42105E-01

(Continued)

TABLE A14.11 (*Continued*)
Steady-State Probabilities for Different Crews (4 Generators)

				Probabilities (2 of 2)			
State		Crew 7	8	9	10	11	12
M,1,1,1	6	1.06315E-03	9.31136E-04	8.28284E-04	7.45894E-04	6.78411E-04	6.22126E-04
M,1,0,1	7	7.95212E-06	6.09728E-06	4.82316E-06	3.91040E-06	3.23423E-06	2.71943E-06
M,0,3,0	8	8.02103E-07	8.02861E-07	8.03451E-07	8.03922E-07	8.04308E-07	8.04630E-07
M,0,2,0	9	6.02595E-09	5.27842E-09	4.69602E-09	4.22948E-09	3.84734E-09	3.52861E-09
M,0,1,0	10	4.52259E-11	3.46670E-11	2.74177E-11	2.22265E-11	1.83820E-11	1.54556E-11
M,0,0,0	11	6.79228E-11	4.37567E-11	2.95897E-11	2.07948E-11	1.50791E-11	1.12227E-11
M,0,2,1	12	8.02104E-07	8.02863E-07	8.03452E-07	8.03924E-07	8.04310E-07	8.04631E-07
M,0,1,1	13	6.02436E-09	5.27687E-09	4.69450E-09	4.22799E-09	3.84589E-09	3.52719E-09
M,0,0,1	14	2.98631E-08	2.00374E-08	1.40907E-08	1.02828E-08	7.73242E-09	5.96046E-09
J,1,3,0	15	8.85543E-04	8.86380E-04	8.87031E-04	8.87552E-04	8.87978E-04	8.88333E-04
J,1,2,0	16	6.64780E-06	5.82250E-06	5.17951E-06	4.66443E-06	4.24254E-06	3.89065E-06
J,1,1,0	17	5.00057E-08	3.83233E-08	3.03037E-08	2.45615E-08	2.03094E-08	1.70732E-08
J,0,3,0	18	5.01417E-09	5.01891E-09	5.02259E-09	5.02554E-09	5.02795E-09	5.02996E-09
J,0,2,0	19	3.76699E-11	3.29969E-11	2.93561E-11	2.64396E-11	2.40508E-11	2.20583E-11
J,0,1,0	20	2.83358E-13	2.17183E-13	1.71753E-13	1.39223E-13	1.15133E-13	9.67977E-14
J,1,0,0	21	7.51762E-08	4.84345E-08	3.27563E-08	2.30225E-08	1.66962E-08	1.24275E-08
J,0,0,0	22	2.82101E-10	1.59033E-10	9.56038E-11	6.04751E-11	3.98703E-11	2.72037E-11
J,2,2,0	23	4.95803E-01	4.96272E-01	4.96636E-01	4.96928E-01	4.97166E-01	4.97365E-01
J,2,1,0	24	3.72201E-03	3.25994E-03	2.89993E-03	2.61155E-03	2.37534E-03	2.17832E-03
J,2,0,0	25	2.79975E-05	2.14567E-05	1.69666E-05	1.37516E-05	1.13710E-05	9.55905E-06
P,2,2,0	26	2.12487E-01	2.12688E-01	2.12844E-01	2.12969E-01	2.13071E-01	2.13156E-01
P,2,1,0	27	1.59415E-03	1.39615E-03	1.24188E-03	1.11831E-03	1.01709E-03	9.32677E-04
P,2,0,0	28	1.19767E-05	9.17819E-06	7.25718E-06	5.88176E-06	4.86331E-06	4.08820E-06
P,1,2,0	29	1.19815E-06	1.19929E-06	1.20017E-06	1.20087E-06	1.20145E-06	1.20193E-06
P,1,1,0	30	9.00255E-09	7.88604E-09	7.01619E-09	6.31940E-09	5.74869E-09	5.27269E-09
P,1,0,0	31	1.19558E-08	8.82792E-09	6.73458E-09	5.27269E-09	4.21643E-09	3.43167E-09
P,0,2,0	32	6.78424E-12	6.79065E-12	6.79564E-12	6.79963E-12	6.80289E-12	6.80561E-12
P,0,1,0	33	5.10131E-14	4.46912E-14	3.97659E-14	3.58205E-14	3.25890E-14	2.98938E-14
P,0,0,0	34	4.48649E-11	2.89865E-11	1.96568E-11	1.38509E-11	1.00687E-11	7.51153E-12

TABLE A14.12
Steady-State Probabilities for Different Crews (3 Generators)

State		Crew 1	2	3	4	5	6
				Probabilities (1 of 2)			
M,1,2,0	1	1.35479E-01	1.39041E-01	1.40256E-01	1.40869E-01	1.41238E-01	1.41485E-01
M,1,1,0	2	7.11002E-03	3.64893E-03	2.45415E-03	1.84884E-03	1.48309E-03	1.23818E-03
M,1,0,0	3	2.74263E-05	1.34534E-05	8.65137E-06	6.24126E-06	4.80228E-06	3.85182E-06
M,1,1,1	4	1.35472E-01	1.39037E-01	1.40254E-01	1.40867E-01	1.41237E-01	1.41485E-01
M,1,0,1	5	6.93487E-03	3.60483E-03	2.43476E-03	1.83805E-03	1.47623E-03	1.23342E-03
M,0,2,0	6	7.67116E-07	7.87283E-07	7.94163E-07	7.97634E-07	7.99726E-07	8.01125E-07
M,0,1,0	7	4.02633E-08	2.06657E-08	1.39006E-08	1.04732E-08	8.40221E-09	7.01545E-09
M,0,0,0	8	7.20424E-07	1.76694E-07	7.57503E-08	4.09858E-08	2.52289E-08	1.68630E-08
M,0,1,1	9	7.67076E-07	7.87264E-07	7.94152E-07	7.97626E-07	7.99721E-07	8.01121E-07
M,0,0,1	10	1.82183E-04	4.73554E-05	2.13254E-05	1.20755E-05	7.75960E-06	5.40333E-06
J,1,2,0	11	8.46916E-04	8.69182E-04	8.76778E-04	8.80609E-04	8.82919E-04	8.84464E-04
J,1,1,0	12	4.46181E-05	2.28945E-05	1.53956E-05	1.15966E-05	9.30121E-06	7.76425E-06
J,0,2,0	13	4.79545E-09	4.92152E-09	4.96453E-09	4.98623E-09	4.99931E-09	5.00805E-09
J,0,1,0	14	2.52666E-10	1.29662E-10	8.72020E-11	6.56912E-11	5.26941E-11	4.39914E-11
J,1,0,0	15	7.97707E-04	1.95675E-04	8.38996E-05	4.54015E-05	2.79510E-05	1.86851E-05
J,0,0,0	16	2.09539E-05	2.56996E-06	7.34615E-07	2.98148E-07	1.46841E-07	8.18024E-08
J,2,1,0	17	4.74176E-01	4.86642E-01	4.90895E-01	4.93041E-01	4.94334E-01	4.95199E-01
J,2,0,0	18	2.49811E-02	1.28183E-02	8.61979E-03	6.49278E-03	5.20762E-03	4.34709E-03
P,2,1,0	19	2.03219E-01	2.08562E-01	2.10384E-01	2.11304E-01	2.11858E-01	2.12228E-01
P,2,0,0	20	1.06905E-02	5.48495E-03	3.68807E-03	2.77777E-03	2.22777E-03	1.85952E-03
P,1,1,0	21	1.14590E-06	1.17602E-06	1.18630E-06	1.19148E-06	1.19461E-06	1.19670E-06
P,1,0,0	22	1.38134E-05	6.75649E-06	4.34056E-06	3.12977E-06	2.40744E-06	1.93058E-06
P,0,1,0	23	6.48835E-12	6.65892E-12	6.71711E-12	6.74646E-12	6.76416E-12	6.77599E-12
P,0,0,0	24	6.25597E-09	3.05995E-09	1.96580E-09	1.41745E-09	1.09031E-09	8.74342E-10

TABLE A14.13
Steady-State Probabilities for Different Crews (3 Generators)

State		Crew 7	8	9	10	11	12
				Probabilities (2 of 2)			
M,1,2,0	1	1.41662E-01	1.41795E-01	1.41898E-01	1.41981E-01	1.42049E-01	1.42106E-01
M,1,1,0	2	1.06271E-03	9.30808E-04	8.28044E-04	7.45722E-04	6.78293E-04	6.22051E-04
M,1,0,0	3	3.18093E-06	2.68457E-06	2.30418E-06	2.00458E-06	1.76339E-06	1.56571E-06
M,1,1,1	4	1.41662E-01	1.41795E-01	1.41898E-01	1.41981E-01	1.42049E-01	1.42105E-01
M,1,0,1	5	1.05920E-03	9.28102E-04	8.25882E-04	7.43944E-04	6.76797E-04	6.20768E-04
M,0,2,0	6	8.02126E-07	8.02878E-07	8.03464E-07	8.03933E-07	8.04317E-07	8.04637E-07
M,0,1,0	7	6.02188E-09	5.27504E-09	4.69317E-09	4.22704E-09	3.84524E-09	3.52678E-09

(Continued)

TABLE A14.13 (*Continued*)
Steady-State Probabilities for Different Crews (3 Generators)

				Probabilities (2 of 2)			
State		Crew 7	8	9	10	11	12
M,0,0,0	8	1.19365E-08	8.81467E-09	6.72503E-09	5.26555E-09	4.21092E-09	3.42729E-09
M,0,1,1	9	8.02123E-07	8.02876E-07	8.03462E-07	8.03932E-07	8.04316E-07	8.04636E-07
M,0,0,1	10	3.97767E-06	3.05001E-06	2.41278E-06	1.95628E-06	1.61809E-06	1.36060E-06
J,1,2,0	11	8.85569E-04	8.86399E-04	8.87046E-04	8.87563E-04	8.87987E-04	8.88341E-04
J,1,1,0	12	6.66313E-06	5.83549E-06	5.19071E-06	4.67422E-06	4.25120E-06	3.89837E-06
J,0,2,0	13	5.01431E-09	5.01901E-09	5.02267E-09	5.02560E-09	5.02800E-09	5.03001E-09
J,0,1,0	14	3.77567E-11	3.30704E-11	2.94195E-11	2.64951E-11	2.40998E-11	2.21020E-11
J,1,0,0	15	1.32282E-05	9.76991E-06	7.45489E-06	5.83785E-06	4.66926E-06	3.80088E-06
J,0,0,0	16	4.96391E-08	3.20791E-08	2.17581E-08	1.53347E-08	1.11501E-08	8.32006E-09
J,2,1,0	17	4.95817E-01	4.96282E-01	4.96644E-01	4.96934E-01	4.97171E-01	4.97369E-01
J,2,0,0	18	3.73059E-03	3.26721E-03	2.90621E-03	2.61703E-03	2.38018E-03	2.18264E-03
P,2,1,0	19	2.12494E-01	2.12693E-01	2.12848E-01	2.12972E-01	2.13074E-01	2.13158E-01
P,2,0,0	20	1.59570E-03	1.39741E-03	1.24293E-03	1.11920E-03	1.01786E-03	9.33341E-04
P,1,1,0	21	1.19819E-06	1.19931E-06	1.20019E-06	1.20089E-06	1.20146E-06	1.20194E-06
P,1,0,0	22	1.59411E-06	1.34523E-06	1.15454E-06	1.00438E-06	8.83513E-07	7.84461E-07
P,0,1,0	23	6.78445E-12	6.79081E-12	6.79576E-12	6.79973E-12	6.80297E-12	6.80568E-12
P,0,0,0	24	7.21957E-10	6.09244E-10	5.22883E-10	4.54877E-10	4.00137E-10	3.55278E-10

TABLE A14.14
States of Operation and Unavailability for 4 and 3 Generators

State		4 Generators	State		3 Generators
M,1,3,0	1	Operation	M,1,2,0	1	Operation
M,1,2,0	2	Operation	M,1,1,0	2	Operation
M,1,1,0	3	Operation	M,1,0,0	3	Operation
M,1,0,0	4	Operation	M,1,1,1	4	Operation
M,1,2,1	5	Operation	M,1,0,1	5	Operation
M,1,1,1	6	Operation	M,0,2,0	6	Unavailability
M,1,0,1	7	Operation	M,0,1,0	7	Unavailability
M,0,3,0	8	Unavailability	M,0,0,0	8	Unavailability
M,0,2,0	9	Unavailability	M,0,1,1	9	Unavailability
M,0,1,0	10	Unavailability	M,0,0,1	10	Unavailability
M,0,0,0	11	Unavailability	J,1,2,0	11	Operation
M,0,2,1	12	Unavailability	J,1,1,0	12	Operation
M,0,1,1	13	Unavailability	J,0,2,0	13	Unavailability
M,0,0,1	14	Unavailability	J,0,1,0	14	Unavailability
J,1,3,0	15	Operation	J,1,0,0	15	Operation
J,1,2,0	16	Operation	J,0,0,0	16	Unavailability

(Continued)

TABLE A14.14 (Continued)
States of Operation and Unavailability for 4 and 3 Generators

State	4 Generators		State	3 Generators	
J,1,1,0	17	Operation	J,2,1,0	17	Operation
J,0,3,0	18	Unavailability	J,2,0,0	18	Operation
J,0,2,0	19	Unavailability	P,2,1,0	19	Operation
J,0,1,0	20	Unavailability	P,2,0,0	20	Operation
J,1,0,0	21	Operation	P,1,1,0	21	Unavailability
J,0,0,0	22	Unavailability	P,1,0,0	22	Unavailability
J,2,2,0	23	Operation	P,0,1,0	23	Unavailability
J,2,1,0	24	Operation	P,0,0,0	24	Unavailability
J,2,0,0	25	Operation			
P,2,2,0	26	Operation			
P,2,1,0	27	Operation			
P,2,0,0	28	Operation			
P,1,2,0	29	Unavailability			
P,1,1,0	30	Unavailability			
P,1,0,0	31	Unavailability			
P,0,2,0	32	Unavailability			
P,0,1,0	33	Unavailability			
P,0,0,0	34	Unavailability			

REFERENCES

Alzbutas, R. (2003). Diesel generators reliability data analysis and testing interval optimization. *Energetika* 4:27–33.

Barbu, V.S., Karagrigoriou, A. (2018). Modeling and inference for multi-state systems, In: Lisnianski A., Frenkel I., Karagrigoriou A. (eds) *Recent Advances in Multi-state Systems Reliability: Springer Series in Reliability Engineering.* Springer, Cham, Switzerland. doi:10.1007/978-3-319-63423-4_16.

Barlow, R., Proschan, F. (1996). *Mathematical Theory of Reliability.* John Wiley & Sons, New York.

Ben-Daya, M., Duffuaa, S., Raouf, A. (2000). *Maintenance Modelling and Optimization.* Springer Science and Media, New York.

Billinton, R., Li, Y. (2007). Incorporating multi state unit models in composite system adequacy assessment. *European Transactions on Electrical Power* 17:375–386.

Brocken, E.M. (2016). *Improving the Reliability of Ship Machinery: A Step Towards Unmanned Shipping.* Delft University of Technology, Delft, the Netherlands.

Chowdhury, C. (1988). A systematic survey of the maintenance models. *Periodica Polytechnica Engineering Mechanical Engineering* 32(3–4):253–274.

Det Norske Veritas DNV (2011). *Machinery Systems General, in Rules for Classification of Ships,* Høvik, Norway.

Eryilmaz, S. (2015). Assessment of a multi-state system under a shock model. *Applied Mathematics and Computation* 269:1–8.

IEEE 1159–1995: IEEE Recommended Practice for Monitoring Electric Power Quality, 1995.

IEEE 45–2002: IEEE Recommended Practice for Electrical Installations on Shipboard, 2002.

IMO (2005). Unified Interpretations to SOLAS Chapters II-1 and XII and to the Technical Provisions for Means of Access for Inspections, London, UK. http://imo.udhb.gov.tr/dosyam/EKLER/SOLAS__BOLUM_II_1_EK(21).pdf.

IMO Study on the optimization of energy consumption as part of implementation of a Ship Energy Efficiency Management Plan (SEEMP) 2016.

IMO MEPC 1-CIRC 866 (E). (2014). Guidelines on the Method of Calculation of the Attained Energy Efficiency Design Index (EEDI) For New Ships, As Amended (Resolution Mepc.245(66), As Amended By Resolutions Mepc.263(68) And Mepc.281(70), January 2017.

Levitin, G., Lisnianski, A. (1999). Joint redundancy and maintenance optimization for multi-state series-parallel systems. *Reliability Engineering & System Safety* 64(1):33–42.

Levitin, G., Xing, L. (2018). Dynamic performance of series parallel multi-state systems with standby subsystems or repairable binary elements, In: Lisnianski A., Frenkel I., Karagrigoriou A. (eds) *Recent Advances in Multi-state Systems Reliability: Springer Series in Reliability Engineering*. Springer, Cham, Switzerland. doi:10.1007/978-3-319-63423-4_16.

Lisnianski, A., Frenkel, I., Ding, Y. (2010). *Multi State Systems Reliability and Optimization for Engineers and Industrial Managers*. Springer, London, UK.

Lisnianski, A., Elmakias, D., Laredo, D., Haim, H.B. (2012). A multi-state Markov model for a short-term reliability analysis of a power generating unit. *Reliability Engineering Systems Safety* 98:1–6.

Liu, Y., Huang, H.Z. (2010). Optimal replacement policy for multi-state system under imperfect maintenance. *IEEE Transactions on Reliability* 59(3):483–495.

Liu, Y.W., Kapur, K.C. (2006). Reliability measures for dynamic multi state non repairable systems and their applications to system performance evaluation. *IIE Transaction* 38(6):511–520.

Maes, W. (2013). *Marine Electrical Knowledge*. Antwerp Maritime Academy, Antwerp, Belgium.

MAIB (2011). Report on the investigation of the catastrophic failure of a capacitor in the aft harmonic filter room on board RMS Queen Mary 2 while approaching Barcelona on 23 September 2010. Marine Accident Investigation Branch. http://www.maib.gov.uk/publications/investigation_reports/2011/qm2.cfm (last accessed May 7, 2018).

Markopoulos, T., Platis, A. (2018). Reliability analysis of a modified IEEE 6 BUS RBTS, In: Lisnianski A., Frenkel I., Karagrigoriou A. (eds) *Recent Advances in Multi-state Systems Reliability. Springer Series in Reliability Engineering*. Springer, Cham, Switzerland. doi:10.1007/978-3-319-63423-4_16.

Mennis, E., Platis, A. (2013). Availability assessment of diesel generator system of a ship: A case study. *International Journal of Performability Engineering* 9(5):561–567.

MEPC 61/inf.18: Reduction of GHG Emissions from Ships—Marginal abatement costs and cost-effectiveness of energy-efficiency measures, October 2010.

MEPC.1-Circ.681–2: Interim Guidelines on the Method of Calculation of the Energy Efficiency Design Index for New Ships, August 2009.

MEPC.1-Circ.684: Guidelines for Voluntary Use of the Ship Energy Efficiency Operational Indicator (EEOI), August 2009.

Miller, T. (2012). Risk focus: Loss of power. http://www.ukpandi.com/fileadmin/uploads/uk-pi/Documents/Brochures/Risk%20Focus%20-%20Loss%20of% 20Power.pdf.

Mindykowski, J. (2014). Power quality on ships: Today and tomorrow's challenges. *International Conference and Exposition on Electrical and Power Engineering* (EPE 2014), Iasi, Romania.

Mindykowski, J. (2016). Case study—Based overview of some contemporary challenges to power quality in ship systems. *Inventions* 1(2):12.

Mindykowski, J., Tarasiuk, T. (2015). Problems of power quality in the wake of ship technology development. *Ocean Engineering* 107:108–117.

MSC/Circular.645-Guidelines for Vessels with Dynamic Positioning Systems-(adopted on 6 June 1994).

MTS DP Technical Committee. DP Vessel Design Philosophy Guidelines Part II.

MUNIN. D6.7: Maintenance indicators and maintenance management principles for autonomous engine room, 2014.

Nakagawa, T. (2006). *Maintenance Theory of Reliability*. Springer Science & Business Media, London, UK.

OREDA (2002). *Offshore Reliability Data Handbook*, 4th ed. OREDA, Trondheim, Norway.

Patel, M.R. (2012). *Shipboard Electrical Power Systems*. CRC Press, Boca Raton, FL.

Prousalidis, J., Styvaktakis, E., Hatzilau, I.K., Kanellos, F., Perros, S., Sofras, E. (2008). Electric power supply quality in ship systems: An overview. *International Journal of Ocean Systems Management* 1(1):68–83.

Prousalidis, J.M., Tsekouras, J.G., Kanellos, F. (2011). New challenges emerged from the development of more efficient electric energy generation units. *From Electric Ship Technologies Symposium (ESTS), IEEE*. doi:10.1109/ESTS.2011.5770901.

Shagar, V., Jayasinghe, S.G., Enshaei, H. (2017). Effect of load changes on hybrid shipboard power systems and energy storage as a potential solution: A review. *Inventions* 2:21.

Stevens, B., Dubey, A., Santoso, S. (2015). On improving reliability of shipboard power system. *IEEE Transactions on Power Systems* 30(4):1905–1906.

Trivedi, K.S. (2002). *Probability and Statistics with Reliability, Queuing and Computer Science Applications*. Wiley, New York.

Wärtsilä (2014). *WSD 42111K, Aframax Tanker for Oil and Products*. Wärtsilä Corporation, Helsinki, Finland.

Wu, Z., Yao, Y., Wang, D. (2013). The reliability modeling of marine power station. *Applied Mechanics and Materials* 427–429:404–407.

Yingkui, G., Jing, L. (2012). Multi state system reliability: A new and systematic review. *Procedia Engineering* 29:531–536.

15 Vulnerability Discovery and Patch Modeling

State of the Art

Avinash K. Shrivastava, P. K. Kapur, and Misbah Anjum

CONTENTS

15.1 INTRODUCTION

With the continual evolution of information technology (IT) infrastructures, the related vulnerabilities and exploitations are increasing because of the security issues raised during the operational phase. Today, there is no software system that is free from weaknesses or vulnerabilities, whether it is a system for personal use or for a large-scale organization. According to National Vulnerability Database (NVD), a total of 16,555 security vulnerabilities were reported in 2018 (the highest figures thus far). This statistic indicates that vulnerability assessment is the most ignored security technology today. Thus, there is a need to quantify the discovered software vulnerabilities with some mathematical models with an improvement in security without increasing penetration costs. However, some considerable work has been done on modeling the vulnerabilities with respect to time (Alhazmi & Malaiya 2005a, 2005b; Kimura 2006; Kim et al. 2007; Okamura et al. 2013; Joh and Malaiya 2014; Kapur et al. 2015; Sharma et al. 2016; Kansal et al. 2017a, 2017b; Movahedi et al. 2018). In the next section, we will discuss briefly the vulnerability life cycle followed by a literature review of vulnerability discovery models (VDMs) in Section 15.2. In Section 15.3, we provide a description of the modeling frameworks of VDMs based on a different set of assumptions followed by vulnerability patching models (VPM). Then modeling of the multi-version vulnerability discovery will be discussed in Section 15.4 followed by the conclusion and future research directions in Section 15.5.

15.1.1 VULNERABILITY

One of the best definitions of software vulnerability is given by Schultz et al. (1990) who defined it as follows: "A vulnerability is defined as a defect which enables an attacker to bypass security measures." To assess the value of vulnerability finding, we must examine the events surrounding discovery and disclosure. Schneier (2000) described the lifecycle of a vulnerability in six phases: Introduction, Discovery, Private Exploitation, Disclosure, Public Exploitation, and Fix Release. These events do not necessarily occur strictly in this order. Disclosure and Fix Release often occur together, especially when a manufacturer discovers a vulnerability and releases the announcement along with a patch (Figure 15.1).

Expectation of a more secured software system requires longer testing that results in high cost and delay in release with increased selling price. However, due to strong market competition, the release time cannot be delayed or the price of the software cannot be increased. Therefore, a trade-off between testing and launch time is required. In the existing literature, many quantitative models have been proposed by several authors. These quantitative models can help the developers in allocating the resources for security testing, scheduling, and development of security patches (Alhazmi & Malaiya 2005a, 2005b; Kimura 2006; Kim et al. 2007; Okamura et al.

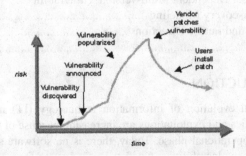

FIGURE 15.1 Lifecycle of a vulnerability.

2013; Joh et al., 2014; Kapur et al. 2015; Sharma et al. 2016; Younis et al., 2016; Kansal et al. 2017a, 2017b). In addition, developers can use VDMs to assess risk and estimate the redundancy needed in resources and procedures to deal with potential breaches. These measures help to determine the resources needed to test a specific part of software. The prime objective of this study is to understand the mathematical models pertaining to vulnerability discovery and patching phenomenon.

15.2 LITERATURE REVIEW

In past few decades, various researchers considered software security to be analogous with software reliability and developed the vulnerability discovery models on similar lines (Alhazmi & Malaiya 2005a, 2005b; Woo et al. 2006; Younis et al. 2011; Narang et al. (2017)). Anderson et al. (2002) examined and measured the security in open and closed systems by proposing a thermo-dynamic VDM. They modeled the discovery rate based on the mean time between failure (MTBF) and defined the model analogous to the software reliability growth model (SRGM). However, they concluded that the there is no difference between open and closed system because both are similar in the long run. Rescorla (2003, 2005) determined that the vulnerability finding is a better approach if it is followed by the white hat users. He evaluated the economic effectiveness of finding and fixing rediscovered vulnerabilities on the developing organizations especially when they are identified by black hat users. The author has fitted the non-homogeneous Poisson process (NHPP) reliability growth model to the observed vulnerability data to evaluate the vulnerability discovery rate over time. However, he proposed two statistical models—Rescorla exponential (RE) and Rescorla linear (RL) or Rescorla quadratic (RQ) model—that are later proved to be insignificant as they are not able to predict the behavior of all empirical data sets.

Ozment and Schechter (2006) stated that vulnerability modeling is analogous to SRGM with an aim to increase the reliability of the system regardless of their operating environments. Ozment (2007) identified the OpenBSD operating system data set and stated that some vulnerability within the data set is dependent. However, he does not apply any VDM on the data set and considered the engineering tools to measure the software security. Alhazmi (2007) attempted to develop a logistic VDM (known as the Alhazmi Malaiya Logistic (AML) model) that quantitatively evaluates the vulnerabilities trend over time. He also proposed a new metric, the vulnerability density, which is analogous to defect density. If the software has a high vulnerability density, then it is at major risk. The author divided the discovery process into three phases: linear, learning, and saturation. Later, he proposed an effort based on VDM (known as the Alhazmi Malaiya Effort [AME] based model) that exhibits the environment changes with respect to the effort instead. However, these models are solely dependent on discovery time that seems inappropriate since there are various operational factors that may influence the vulnerability discovery process. Kim et al. (2007) extended the work done by Alhazmi and others for single version by developing a new VDM for multiple versions (known as the multi-version vulnerability discovery model [MVDM]). He proved that the behavior of the vulnerability discovery rate for multiple versions is different from single-version modeling for open-source and commercial software systems. Joh et al. (2008) developed a VDM that follows the Weibull distribution and is known as the Joh-Weibull (JW) model. The model represents the asymmetric nature of the vulnerability discovery rate because of the skewness present in probabilistic density functions. Although this model is also exclusively dependent on discovery time, Bass et al. (1969) scrutinized the factors that motivate the vulnerability discoverers to spend the effort in findings. As per the study, the discoverers are more attracted toward bug bounty programs that have become the main reason for their encouragement. However, they have not modeled the vulnerability discovery process. Massacci and Nguyen (2014) proposed a methodology to validate the performance of empirical VDMs. The methodology focuses on two quantitative metrics: quality and prediction capability. The quality is measured on the basis of good fit and inconclusive fit while the predictive accuracy is measured on current and future horizon. However, he does not propose any mathematical model. Joh et al. (2014) found the relationship between performance of S-shaped vulnerability discovery models and the skewness in some vulnerability data sets and applied Weibull, Beta, Gamma, and Normal distributions. Anand et al. (2017) proposed an approach to quantify the discovered vulnerabilities using various software versions. The authors examined their approach using Windows and Windows Server Operating Systems. Zhu et al. (2017) proposed a mathematical model that predicts the software vulnerabilities and used the estimated parameters to develop a new risk model. The authors also determined the severity of vulnerability using logistic function and binomial distribution, respectively. Although this model also is dependent exclusively on discovery time, Wai et al. (2018) proposed two new algorithms—mean fit and trend fit—to predict the vulnerability discovery rate using past vulnerability data. Recently, Movahedi et al. (2018) used a clustering approach to group vulnerabilities into different clusters and then used NHPP-based software reliability models to predict the number of vulnerabilities in each cluster and then combined them together to find the total number of vulnerabilities in the system. In the next section, we will briefly discuss the VDMs proposed in the literature so far.

15.2.1 Anderson Thermodynamic Model

The pioneering work in developing a VDM was carried out by Ross Anderson (2002) resulting in a model known as the Anderson Thermodynamic (AT) model. The assumptions taken to develop this model are that (1) as soon as a vulnerability is encountered it is removed with certainty and (2) no extra vulnerabilities are added while fixing the existing vulnerability. Let $\Omega(t)$ be the remaining number of vulnerabilities left after t tests and $p(t)$ be the probability that a test fails, then according to the AT model $p(t)$ is given by:

$$p(t) = \frac{k}{\gamma t} \tag{15.1}$$

where:

k is a constant

γ is value that takes care of lower failure rate during beta testing by the users in comparison with alpha testing

On solving Equation 15.1, we get the cumulative number of vulnerabilities as follows:

$$\Omega(t) = \frac{k}{\gamma t} \ln(Ct) \tag{15.2}$$

where C is the constant of integration. This model is applicable only when $t \geq 1$.

15.2.2 ALHAZMI MALAIYA LOGISTIC MODEL

Alhazmi et al. (2005) proposed another s-shaped VDM and called it the AML model. Their model is based on the following assumption that the rate of change of the cumulative number of vulnerabilities Ω is dependent on the number of existing and leftover undetected vulnerabilities. According to them, vulnerability follows three phases: learning, linear, and saturation (shown in Figure 15.2).

Following the assumptions of AML model, we get the following differential equation:

$$\frac{d\Omega}{dt} = A\Omega(B - \Omega) \tag{15.3}$$

where Ω is the cumulative number of vulnerabilities, t is the calendar time, A and B are the empirical constants to be determined from recorded data. After solving Equation 15.3 we get:

$$\Omega(t) = \frac{B}{BCe^{-ABt} + 1} \tag{15.4}$$

where C is the constant of integration and B is total number of vulnerabilities in the system.

15.2.3 RESCORLA QUADRATIC AND RESCORLA EXPONENTIAL MODELS

Rescorla (2005) proposed two models: Rescorla Quadratic and Rescorla Exponential. These models are described in the following sections.

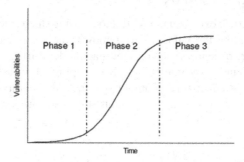

FIGURE 15.2 The basic 3-phase s-shaped model proposed by Alhazmi and Malaiya.

15.2.3.1   Rescorla Quadratic Model

According to this model the failure rate $\omega(t)$ takes the linear form that is given by:

$$\omega(t) = Bt + K \tag{15.5}$$

where B and K are constants. On integrating Equation 15.5, we get the cumulative number of vulnerabilities given by:

$$\Omega(t) = \frac{Bt^2}{2} + kt \tag{15.6}$$

At $t = 0$, $\Omega(t) = 0$ so, the constant of integration comes out to be zero.

15.2.3.2   Rescorla Exponential Model

Rescorla used exponential distribution to fit the vulnerability data which is given as:

$$\omega(t) = B\lambda e^{-\lambda t} \tag{15.7}$$

where B represents the total number of vulnerabilities in the system and λ is the rate constant. On integrating Equation 15.7, we get the cumulative number of vulnerabilities as:

$$\Omega(t) = B(1 - e^{-\lambda t}) \tag{15.8}$$

Kapur et al. (2015) applied two of the SRGMs (i.e., the Kapur & Garg (1992) Model and the Two Stage Erlang Logistic Model) on vulnerability data sets and compared their results with the AML model. They claimed that the results are equivalent to those obtained from AML model. Shrivastava et al. (2015) applied stochastic differential equation to develop a stochastic VDM using the AML model and found that results of their model are better than the AML model. The formulation of the model follows.

15.2.4   Vulnerability Discovery Model Using Stochastic Differential Equation

Shrivastava et al. (2015) extended the AML model using stochastic differential equation (SDE). Let $b(t)$ be the vulnerability removal rate per remaining vulnerabilities in the software, σ denotes the constant magnitude of irregular fluctuation and $\gamma(t)$ the Standard Gaussian White Noise. Then keeping all the assumption of AML model valid along with an extra assumption that the vulnerability discovery process follows a stochastic process with a continuous state space we get the following differential equation:

$$\frac{dN(t)}{dt} = b(t)\left[B - N(t)\right] \tag{15.9}$$

Now assuming irregular variations in $b(t)$ Equation 15.9 can be extended as the following SDE:

$$\frac{dN(t)}{dt} = \{b(t) + \sigma\gamma(t)\}\{B - N(t)\} \tag{15.10}$$

We extend the previous equation to the following SDE of an $It\hat{O}$ type:

$$dN(t) = \left\{b(t) - \frac{1}{2}\sigma^2\right\}\{B - N(t)\}dt + \sigma\left[B - N(t)\right]dW(t) \tag{15.11}$$

where $W(t)$ is called a Brownian or Wiener process. After solving Equation 15.11 using $It\hat{O}$ formula, we get:

$$N(t) = B - (B - k)\left[e^{-\int_0^t b(t)dt - \sigma W(t)}\right] \tag{15.12}$$

Therefore, the mean number of vulnerabilities will be:

$$\Omega(t) = E\left[N(t)\right] = B\left[1 - \frac{\left(\dfrac{B-k}{k}\right).e^{-\left(Bbt - \frac{1}{2}\sigma^2 t\right)}}{\left(1 + \left(\dfrac{B-k}{k}\right)e^{-Bbt}\right)}\right] \tag{15.13}$$

Using the previous equation, we can predict the number of vulnerabilities in the software.

15.2.5 Effort-Based Vulnerability Discovery Model

Alhazmi and Malaiya (2008) proposed an effort-based VDM where they measure the effort E as follows:

$$E = \sum_{i=0}^{n}(U_i - P_i) \tag{15.14}$$

Here U_i denotes the number of users working on all systems at the time period i and P_i is the percentage of the users using the system. Assuming that vulnerability detection rate is proportional to the effort and the remaining number of vulnerabilities, the effort based VDM is given as follows:

$$\Omega(t) = B(1 - e^{-\lambda_{vu}E}) \tag{15.15}$$

where λ_{vu} denotes the failure intensity.

15.2.6 USER-DEPENDENT VULNERABILITY DISCOVERY MODEL

Kansal et al. (2017a) developed a VDM considering the number of users where they assumed that vulnerability discovery is dependent on the reporting done by the users who buy the software. The notations used in this section apart from previously described are as follows:

Notations	Description
$\bar{S}$	Actual number of software buyers
$S(t)$	Cumulative number of potential software users at time t

The vulnerability intensity defined by Kansal et al. (2007a) is given as:

$$\frac{d\Omega}{dt} = \left(\frac{d\Omega}{dI}\right) \cdot \left(\frac{dI}{dS}\right) \cdot \left(\frac{dS}{dt}\right) \tag{15.16}$$

The three components on the right-hand side of Equation 15.16 are described using the following assumptions:

1. The vulnerability discovery rate is dependent on the number of instructions executed which is represented mathematically as:

$$\frac{d\Omega}{dI} = \left(x + y \cdot \frac{\Omega}{B}\right) \cdot (B - \Omega) \tag{15.17}$$

where:
$(B - \Omega)$ are the remaining vulnerabilities residing in the software
'x' is the rate with which unique vulnerabilities are detected
'y' is the rate with which the dependent vulnerabilities are detected through the support rate of $\frac{\Omega}{B}$.

2. The number of instructions executed by every user is constant which is given as:

$$\frac{dI}{dS} = k \tag{15.18}$$

3. The rate at which the number of people buys the software is given by:

$$\frac{dS}{dt} = \left(\alpha + \beta \cdot \frac{S}{\bar{S}}\right) \cdot (\bar{S} - S) \tag{15.19}$$

where:
$(\bar{S} - S)$ are the remaining number of users who have yet to buy the software
α and β are the rate with which innovators and imitators are buying the software

After solving Equation 15.19 with initial conditions, $S(t) = 0$, we get:

$$S(t) = \bar{S} \cdot \frac{1 - \exp\left(-(\alpha + \beta) \cdot t\right)}{1 + \left(\dfrac{\beta}{\alpha}\right) \cdot \exp\left(-(\alpha + \beta) \cdot t\right)} \tag{15.20}$$

Now from Equations 15.17 through 15.20, the vulnerability discovery rate becomes:

$$\frac{d\Omega}{dt} = \left(x + y \cdot \frac{\Omega}{B}\right) \cdot (B - \Omega) \cdot k \cdot \frac{dS}{dt} \tag{15.21}$$

On solving Equation 15.21 with initial conditions $\Omega(S) = 0$, $S = 0$, we get:

$$\Omega(t) = B \cdot \frac{\left(1 + h \cdot \exp\left(-(x + y) \cdot S(t)\right)\right)^{k} - \left((1 + h) \cdot \exp\left(-(x + y) \cdot S(t) \cdot k\right)\right)}{\left(1 + h \cdot \exp\left(-(x + y) \cdot S(t)\right)\right)^{k}} \tag{15.22}$$

where $h = \frac{y}{x}$, $A = x + y$

15.2.7 VULNERABILITY DISCOVERY MODEL FOR OPEN AND CLOSED SOURCE

Sharma et al. (2016) proposed a VDM using a Gamma distribution function and claimed that their model has better prediction capabilities for open and closed source software. The failure density function for gamma distribution is given as:

$$f(t) = \frac{1}{\Gamma(\alpha)\beta} \left(\frac{t}{\beta}\right)^{\alpha - 1} e^{-\left(\frac{t}{\beta}\right)}; \quad t \geq 0, \quad \alpha, \beta > 0 \tag{15.23}$$

where α, β denote the shape and scale parameters, respectively. α controls the shape of distribution. The cumulative distribution function for Gamma to perform vulnerability prediction is given by:

$$cdf\,(Gamma) = F\left(t; \alpha, \beta\right) = \int_{0}^{t} f\left(u; \alpha, \beta\right) du = \frac{\gamma\left(\alpha, \beta t\right)}{\Gamma(\alpha)} \tag{15.24}$$

So,

$$\Omega(t) = B * F(t, \alpha, \beta) \tag{15.25}$$

15.2.8   COVERAGE BASED VULNERABILITY DISCOVERY MODELING

Kansal et al. (2018) proposed a coverage based VDM in which they assumed that the vulnerability discovery rate is defined by:

$$\frac{d\Omega}{dt} = \left(\frac{d\Omega}{dC}\right) \cdot \left(\frac{dC}{dI}\right) \cdot \left(\frac{dI}{dX}\right) \cdot \left(\frac{dX}{dt}\right)$$

(15.26)

where C, I, and X are explicitly the operational coverage, executed instructions, and operational effort. The four components in the right-hand side are defined as:

1. *Component 1*: Here it was assumed that the vulnerability discovery rate is directly proportional to the operational coverage rate of the remaining vulnerabilities and inversely proportional to uncovered proportion of software and given by:

$$\frac{d\Omega}{dC} = A1 \cdot \left(\frac{c'}{p-c}\right) \cdot (B - \Omega)$$

(15.27)

   where c is the coverage rate.
2. *Component 2*: The coverage rate with respect to number of instructions executed is considered as constant and given by:

$$\frac{dC}{dI} = \phi_1$$

(15.28)

3. *Component 3*: The rate at which instructions are executed per operational effort is assumed to be constant and given by:

$$\frac{dI}{dX} = \phi_2$$

(15.29)

4. *Component 4*: Rate of operational effort is directly proportional to remaining resources where vulnerability discoverers and time are the resources that are considered as operational effort spent on vulnerability discovery and it is given by:

$$\frac{dX}{dt} = \beta(t) \cdot (\alpha - X(t))$$

(15.30)

   where $\beta(t)$ is the time dependent rate at which operational resources are consumed and α is the total amount of effort required for vulnerability discovery.

Using Equations 15.27 through 15.30, from Equation 15.26 we have:

$$\frac{d\Omega}{dt} = \left(A1 \cdot \left(\frac{c'}{p-c} \right) \cdot (B-\Omega) \right) \cdot (\phi_1) \cdot (\phi_2) \cdot \left(\frac{dX}{dt} \right) \tag{15.31}$$

On solving Equation 15.31 under $\Omega(0) = c(0) = 0$, we get:

$$\Omega(X(t)) = B \cdot \left(1 - \left(1 - \frac{c(X(t))}{p} \right)^{A1 \cdot \phi_1 \cdot \phi_2} \right) \tag{15.32}$$

In the previously described model, the authors took various effort functions $X(t)$, that is, to find the final model for vulnerability prediction. They used the Weibull and the Logistic effort functions in their model. They further took various operational coverage functions in their model. For example, if operational effort is assumed to follow Weibull distribution, then Mean Value Function (MVF) or VDM becomes:

$$\Omega(t) = B \cdot \left(1 - \left(e^{-A2 \cdot \left(\alpha \cdot \left(1 - e^{-\beta \cdot t^k} \right) \right)^h} \right)^{A1 \cdot \phi_1 \cdot \phi_2} \right) \tag{15.33}$$

15.2.9 VULNERABILITY PATCHING MODEL

VDM predicts the cumulative number of vulnerabilities against calendar time, software buyers, operational coverage, and operational effort. In contrast, vulnerability patch modeling (VPM) observes the cumulative number of patches with calendar time only. This research attempts to predict the vulnerability discovery rate, number of vulnerabilities discovered, vulnerability patching rate, and number of vulnerabilities patched. It provides an idea for quantifying security risks in terms of vulnerabilities. These quantitative models may help in understanding the behavior of software vulnerabilities under different assumptions. One of the major assumptions is that the failures are caused randomly. The tools of vulnerability modeling known as VDMs and VPMs may help consumers to estimate and predict the system risk, lower the patch development cost and time, increase productivity, and reduce exploitability. In this section, we will discuss the VPM (Kansal et al. [2016a, 2016b]) that determines the intensity with which discovered vulnerabilities are fixed or patched. It is assumed that the developed vulnerability model follows the NHPP properties to fulfill one of the considerations that the vulnerabilities are successfully removed when patches are applied.

15.2.9.1   One-Dimension Vulnerability Patching Model

Notation used in this section apart from those previously described are as follow:

Notation	Description
$\hat{\rho}(r)$	Expected number of patches released with respect to patching resources r
$\overline{A}$	Vulnerability patching rate
r	Patching resources
t	Patching time or patch release time
v	Vulnerabilities reported/discovered
d	Vulnerabilities disclosed
$\overline{B}$	Actual potential number of patches released
$\overline{C}$	Integration constant
Δ, δ	Intermediate variables

This model focuses on determining the successfully released/installed patches with time. The model is comprised of three components: directly patched vulnerabilities, indirectly patched vulnerabilities, and unsuccessfully patched vulnerabilities. The first component addresses the patches that are released by the vendors without customer interference (vulnerabilities are discovered directly by the developing team and no beta customers are involved). In this case, vendors/developers are free to use the maximum resources because of no external pressure on managers that makes the probability of success of these patches as 1. The second component addresses the patches which are developed and released corresponding to vulnerability reports. These patches are developed under pressure and resource constraints; thus, there is a possibility that these patches may fail. Thus, the probability of success of these patches is denoted as $(1-\sigma)$ where σ represents the unsuccessful patching rate that is considered being third component. The last component addresses the patches that are unavailable, especially in case of zero-day vulnerabilities whose highest probability of getting fail is denoted as σ.

Mathematically, the model can be presented as:

$$\frac{d\hat{\rho}}{dt} = A \cdot \left(B - \hat{\rho}\right) + C \cdot (1-\sigma) \cdot \frac{\hat{\rho}}{B} \cdot \left(B - \hat{\rho}\right) - \sigma \cdot \hat{\rho} \qquad (15.34)$$

where A represents the proportion of patches that are released or installed successfully without disruption. While C represents the proportion of patches that are released because of the reports submitted to vendors about vulnerabilities.

Under the initial condition $\hat{\rho}(t = 0) = 0$ and solving the above equation we get:

$$\hat{\rho} = \frac{\overline{B} \cdot \left(1 - e^{-\left(\overline{A} + \overline{C}\right)t}\right)}{1 + \dfrac{\overline{C}}{\overline{A}} \cdot \left(e^{-\left(\overline{A} + \overline{C}\right)t}\right)} \qquad (15.35)$$

where $\bar{B} = B^{\cdot(\Delta+\delta)/2\cdot C\cdot(1-\sigma)}$, $\bar{A} = \Delta - \delta/2$, $\bar{C} = \Delta + \delta/2$, $\delta = C\cdot(1-\sigma) - A - \sigma$ and $\delta = \sqrt{\delta^2 + 4\cdot C\cdot(1-\sigma)\cdot A}$ where Δ, $\delta\bar{A}$, $\bar{B}$ and $\bar{C}$ are the notations used for intermediate variables.

15.2.9.2 Two-Dimensional Vulnerability Patch Modeling

Kansal and Kapur (2019) proposed a two-dimensional VPM considering the relation between number of vulnerabilities discovered and the number of patches released. For this they used the Cobb-Douglas production function to show the relationship between the dependent (output) and independent (input) variables and defined it as:

$$r \cong v^{\alpha} \cdot t^{1-\alpha} \quad 0 \leq \alpha \leq 1 \tag{15.36}$$

where "r" refers to the patching resources, v refers to the quantifiable vulnerabilities, "t" refers to the patching time and α as the degree of impact to the vulnerability patching process. The model development is similar to what we have already defined in Section 15.3.1 where the only change is to replace "t" with "r" to obtain the final equation as:

$$\hat{\rho}(r) = \hat{\rho}(v,t) = \frac{\bar{B}\cdot\left(1 - e^{-\left(\bar{A}+\bar{C}\right)\left(v^{\alpha}\cdot t^{1-\alpha}\right)}\right)}{1 + \dfrac{\bar{C}}{\bar{A}}\cdot\left(e^{-\left(\bar{A}+\bar{C}\right)\left(v^{\alpha}\cdot t^{1-\alpha}\right)}\right)} \tag{15.37}$$

15.2.10 VULNERABILITY DISCOVERY AND PATCHING MODEL

In the operational phase, two processes, vulnerability discovery and patching, occurs simultaneously. The vulnerability discovery process is done by software users while the patching is done by the software developers. Here, the VDP process is denoted by $V(t)$ distribution and the vulnerability removal process (i.e., patching process) by $P(t)$. The intensity with which the vulnerabilities are discovered is calculated as:

$$\frac{d\Omega(t)}{dt} = \frac{v(t)}{1-V(t)}\left[B - \Omega(t)\right] \tag{15.38}$$

where $\Omega(t)$ is the number of vulnerabilities expected to be discovered until time t, $\frac{v(t)}{1-V(t)}$ is the vulnerability discovery rate, $\frac{dV(t)}{dt} = v(t)$, and B is the potential number of discovered vulnerabilities. Solving Equation 15.38 under the initial conditions $\Omega(t=0) = 0$, we get:

$$\Omega(t) = B\cdot V(t) \tag{15.39}$$

After accounting for the number of vulnerabilities discovered, the next step taken by developers is to develop patches. Hence, we have considered the vulnerability patching time in our model under the vulnerability discovery process. The intensity with which discovered vulnerabilities are patched can be calculated as:

$$\frac{d\hat{\rho}(t)}{dt} = \frac{[v * p](t)}{1 - [V \otimes P](t)}\left[\bar{B} - \hat{\rho}(t)\right]$$ (15.40)

$$\frac{[v * p](t)}{1 - [V \otimes P](t)}$$ (15.41)

is the vulnerability patching rate wherein $\frac{d[V \otimes P](t)}{dt} = [v * p](t)$.

The symbol $[v * p](t)$ denotes convolution of v and p. Another definition of convolution function that is a stieltjes convolution is represented as $[V \otimes P](t)$. Solving Equation 15.41 under the initial conditions $\hat{\rho}(t = 0) = 0$, we get:

$$\hat{\rho}(t) = \bar{B} \cdot (V \otimes P)(t)$$ (15.42)

where $\bar{B}$ is the potential number of patched vulnerabilities, $\hat{\rho}(t)$ is the number of vulnerabilities expected to be patched at time t. Equations 15.40 and 15.42 denote the generalized modeling approach in which the first step is vulnerability discovery and the other step is the vulnerability patching process. In this research paper, we have considered that $V(t)$ follows the exponential distribution which is represented as:

$$V(t) = (1 - \exp(-A \cdot t))$$ (15.43)

where A represents the vulnerability discovery rate. The cumulative vulnerability discovery model as proposed by Rescorla et al. (2002) can be derived from Equations 15.42 and 15.43. The mean value function becomes:

$$\Omega(t) = B \cdot (1 - \exp(-A \cdot t))$$

Subsequently, the VPM is formulated where $P(t)$ is represented by the logistic distribution function since patching is a more complex process than discovery. The mean value function follows:

$$P(t) = \left(\frac{1 - \exp(-A \cdot t)}{1 + C \cdot \exp(-A \cdot t)}\right)$$ (15.44)

where A represents the vulnerability patching rate with learning and C represents the shape parameter.

To obtain the simple mathematical form for the proposed model, we have assumed that the discovery rate A as in Equation 15.43 is same as the patching rate with learning as in Equation 15.44. In other words, we have considered that the discovery rate and patching rate are the same.

The stieltjes convolution as shown in Equation 15.42 is calculated as:

$$V(t) \otimes P(t) = \int P(t-x) dV(x)$$ (15.45)

Equation 5.45 shows the time delay between the vulnerability discovery and the patching process wherein the vulnerability discovery time is denoted as x and the vulnerability patching time is denoted as $t - x$. Here, the model also manifests that it is not necessary that the number of vulnerabilities discovered and patched are always same. However, at time infinity the numbers may become similar.

Thus, from Equations 15.43 and 15.44, Equation 15.45 can be re-written as:

$$V(t) \otimes P(t) = \int_0^t \left(\frac{1 - \exp(-A \cdot (t-x))}{1 + C \cdot \exp(-A \cdot (t-x))} \right) \cdot (A \cdot \exp(-A \cdot x)) \cdot dx$$ (15.46)

On solving Equation 15.46, we get:

$$\rho(t) = \bar{B} \cdot \left(1 - \exp(-A \cdot t) + (1 + C) \cdot \exp(-A \cdot t) \ln \frac{(1 + C) \cdot \exp(-A \cdot (t))}{(1 + C \cdot \exp(-A \cdot (t)))} \right)$$ (15.47)

Equation 15.47 is used further for predicting the number of vulnerabilities discovered and patched.

15.3 VULNERABILITY DISCOVERY IN MULTI-VERSION SOFTWARE SYSTEMS

Generally, we have several versions of software in which we keep trying to upgrade the previous versions by adding new functionalities. This up-gradation adds advanced features in the software and provides better user experiences. As none of the software developed is free of bugs, the trend of discovering vulnerabilities also continues in each version of the software. This phenomenon of discovering vulnerabilities in multiple versions is developed by Kim et al. (2007) by using the AML model in which they have considered that the new version is developed by keeping the previous functionalities and adding new features on the base code. Even if the base code is reduced in the new version, the vulnerability found in the common code will be counted in the older version only while predicting the number of vulnerabilities of each version (see Figure 15.3).

The cumulative number of vulnerabilities $\Omega(t)$ in each version of software is given by:

$$\Omega(t) = \frac{B}{BCe^{-ABt} + 1} + \alpha \frac{B'}{B'C'e^{-A'B'(t-\varepsilon)} + 1}$$ (15.48)

where the parameter α indicates shared components such as shared code and shared functionality, and ε denotes the time lag between the release dates of the two versions. Equation 15.48 is referred to as the multi-version vulnerability discovery

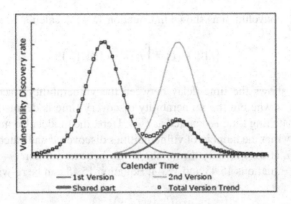

FIGURE 15.3   Multi-version software vulnerability discovery model.

model (MVDM). Equation 15.48 can be generalized to write the mathematical form of multi-version software modeling as:

$$\Omega(t) = \sum_{i=1}^{n} \alpha_i \frac{B_i'}{B_i' C_i' e^{-A_i' B_i'(t-\varepsilon_i)} + 1} \tag{15.49}$$

Following the assumptions of Kim et al. (2007), Anand et al. (2017) developed a framework for predicting the number of vulnerabilities in multi-versions of software and proposed a similar model and showed that the results are equivalent to those obtained from the model proposed by Kim et al. (2007).

15.3.1   User Dependent Multi-version Vulnerability Discovery Modeling

Narang et al. (2017) developed a user dependent model to predict the number of vulnerabilities in multi-versions of vulnerabilities. They developed the multi-version model, which is like the effort-based multi-version framework developed in software reliability literature for predicting the number of faults in multi-releases of software (Kapur et al. 2011). They assumed t1 and t2 as the time frame of the new version of software vulnerabilities, where t1 and t2 are the release times of the first and second versions, respectively. The cumulative number of vulnerabilities detected in the first version of software is given by:

$$\Omega_1(S_1(t)) = B_1.F_1(S_1(t)) \tag{15.50}$$

where $F_1(S_1(t))$ represents the user dependent vulnerability discovery function. For predicting the number of vulnerabilities in the next version, Narang et al. (2018) considered the vulnerabilities of previous version which were removed in the current version should be counted in the newer version. The mathematical form for the next version of vulnerabilities is given by:

$$\Omega_2(S_2(t)) = B_2.F_2(S_2(t-t_1)) + B_1(1-F_1(S_1(t_1)).F_2(S_2(t-t_1)) \tag{15.51}$$

where $B_1(1 - F_1(S_1(t_1)))$ are some left over vulnerabilities of the first version, and $F_1(S_1(t))$ and $F_2(S_2(t))$ are the vulnerability discovery rates of older and newer versions.

15.4 CONCLUSION AND FUTURE DIRECTIONS

In this paper, we have discussed various vulnerability discovery models proposed in the literature based on different sets of assumptions. We have tried to cover all the major contributions in quantitative assessments of software security. This field of research is becoming the topic of interest to the various researchers working in the field of software engineering due to the sensitivity and significance to real-life activities that are based on the smooth functioning of software systems. In the vulnerability data set irrespective of software versions, it is quite possible that the newer versions of software may have some vulnerabilities that are in common with previous version. Keeping this in mind, few researchers applied the VDMs to detect the number of vulnerabilities in multi-versions of software. Also, the literature on software patching models is presented in this chapter. Researchers may classify the vulnerabilities with different versions to check and compare the improvements in vulnerability discovery rates in the future. Further, the impact of incentives or the bug bounties can be analyzed in the future for vulnerability discovery processes. Due to lack of information related to patching, we were not able to explicitly calculate the total number of patched vulnerabilities w.r.t. effort. Thus, it proposes an important research question in future research studies. In this chapter, we have not covered the cost models proposed in the literature related to determination of optimal vulnerability discovery and patch release times. This research is another area in software security that is still in its initial phase. Research on vulnerability prioritization is another direction for researchers. Few attempts have been made in this direction but research in this direction is also in its early stages. Research on prioritizing vulnerability with respect to categorization could be another very interesting area.

REFERENCES

Alhazmi, O. (2007). Assessing vulnerabilities in software systems: A quantitative approach. Thesis, Colorado State University.

Alhazmi, O.H., & Malaiya, Y.K. (2005a). Modeling the vulnerability discovery process. In *16th IEEE International Symposium on Software Reliability Engineering (ISSRE'05)* (pp. 10–pp). IEEE.

Alhazmi, O.H., & Malaiya, Y.K. (2005b). Quantitative vulnerability assessment of systems software. *IEEE*, pp. 615–620.

Alhazmi, O.H., & Malaiya, Y.K. (2008). Application of vulnerability discovery models to major operating systems. *IEEE Transactions on Reliability*, 57(1), 14–22.

Anand, A., Das, S., Aggrawal, D., & Klochkov, Y. (2017). Vulnerability discovery modelling for software with multi-versions. In *Advances in Reliability and System Engineering* (pp. 255–265). Cham, Switzerland: Springer International Publishing.

Anderson, R. (2002). Security in open versus closed systems: The dance of Boltzmann, Coase and Moore. Technical report, Cambridge University.

Bass, F.M. (1969), A new-product growth model for consumer durables. *Management Science*, 15, 215–227.

Joh, H., & Malaiya, Y.K. (2014). Modeling skewness in vulnerability discovery: Modeling skewness in vulnerability discovery. *Quality and Reliability Engineering International*, 30(8), 1445–1459.

Kansal, Y., & Kapur P.K. (2019). Two-dimensional vulnerability patching model. In: Kapur, P., Klochkov, Y., Verma, A., Singh, G. (Eds.), *System Performance and Management Analytics: Asset Analytics* (Performance and Safety Management) (pp. 321–331). Singapore: Springer.

Kansal, Y., Kapur, P.K., & Kumar, U. (2018). Coverage based vulnerability discovery modeling to optimize disclosure time using multi-attribute approach. *Quality and Reliability Engineering International*, 35(1), 62–73. doi:10.1002/qre.2380.

Kansal, Y., Kapur, P.K., Kumar, U., & Kumar, D. (2017a). User-dependent vulnerability discovery model and its interdisciplinary nature. *International Journal of Life Cycle Reliability and Safety Engineering*, 6(1), 23–29.

Kansal, Y., Kapur, P.K., Kumar, U., & Kumar, D. (2017b). Effort and coverage dependent vulnerability discovery modeling In: *IEEE Xplore, International Conference on Telecommunication and Networking (TELNET)*, Noida.

Kansal, Y., Kumar, D., & Kapur, P.K. (2016a). Assessing optimal patch release time for vulnerable software systems. In *IEEE Xplore, International Conference on Innovation and Challenges in Cyber Security (ICICCS-INBUSH)*, Noida, pp. 308–314.

Kansal, Y., Kumar, D., & Kapur, P.K. (2016b). Vulnerability patch modeling. *International Journal of Reliability, Quality and Safety Engineering*, 23(6), 1640013.

Kapur, P.K., & Garg, R.B. (1992). A software reliability growth model for an error-removal phenomenon. *Software Engineering Journal*, 7(4), 291–294.

Kapur, P.K., Pham, H., Gupta, A., & Jha, P.C. (2011). *Software Reliability Assessment with OR Applications*. London, UK: Springer.

Kapur, P.K., Yadavalli, V.S.S., & Shrivastava, A.K. (2015). A comparative study of vulnerability discovery modeling and software reliability growth modeling. In *The IEEE Xplore Proceedings of International Conference on Futuristic Trends in Computational Analysis and Knowledge Management*, Amity University, Greater Noida, February 25–27, pp. 246–251.

Kim, J., Malaiya, Y.K., & Ray, I. (2007). Vulnerability discovery in multi-version software systems. In *10th IEEE High Assurance Systems Engineering Symposium. HASE'07*, pp. 141–148.

Kimura, M. (2006). Software vulnerability: Definition, modelling, and practical evaluation for e-mail transfer software. *International Journal of Pressure Vessels and Piping*, 83(4), 256–261.

Massacci, F., & Nguyen, V.H. (2014). An empirical methodology to evaluate vulnerability discovery models. *IEEE Transactions on Software Engineering*, 40(12), 1147–1162.

Movahedi, Y., Cukier, M., Andongabo, A., & Gashi, I. (2018). Cluster-based vulnerability assessment of operating systems and web browsers. *Computing*, 1–22. doi:10.1007/s00607-018-0663-0.

Narang, S., Kapur, P.K., Damodaran, D., & Shrivastava, A.K. (2017). User-based multi-upgradation vulnerability discovery model. In *6th International Conference on Reliability, Infocom Technologies and Optimization (Icrito 2017)* (Trends and Future directions) to be held during September 20–22, 2017, Amity University Uttar Pradesh.

Narang, S., Kapur, P.K., Damodaran, D., & Shrivastava, A.K. (2018). Bi-criterion problem to determine optimal vulnerability discovery and patching time. *International Journal of Quality Reliability and Safety Engineering*, 25(1), 1850002.

Okamura, H., Tokuzane, M., & Dohi, T. (2013). Quantitative security evaluation for software system from vulnerability database. *International Journal of Software Engineering & Applications*, 6(3), 15.

Ozment, A., & Schechter, S.E. (2006). Milk or wine: Does software security improve with age? *Proceedings of the 15th Conference on Usenix Security Symposium*, Berkeley, CA.

Ozment, J.A. (2007). Vulnerability discovery & software security. PhD thesis, University of Cambridge.

Rescorla, E. (2003). Security holes. Who cares? In *Proceedings of the 12th Conference on USENIX Security Symposium*, pp. 75–90.

Rescorla, E. (2005). Is finding security holes a good idea? *IEEE Security & Privacy*, 3(1), 14–19.

Schneier, B. (2000). Full disclosure and the window of vulnerability, Crypto-Gram (September 15, 2000). www.counterpane.com/cryptogram-0009.html#1.

Schultz, E.E., Brown, D.S., & Longstaff, T.A. (1990). Responding to Computer Security Incidents, Lawrence Livermore National Laboratory, 165. http://ftp.cert.dfn.de/pub/ docs/csir/ ihg.ps.gz, July 23.

Sharma, R., Sibbal, R., & Shrivastava, A.K. (2016). Vulnerability discovery modeling for open and closed source software. *International Journal of Secure Software Engineering*, 7(4), 19–38.

Shrivastava, A.K., Sharma, R., & Kapur, P.K. (2015). Vulnerability discovery model for a software system using stochastic differential equation. In *The IEEE Xplore Proceedings of International Conference on Futuristic Trends in Computational Analysis and Knowledge Management*, Amity University, Greater Noida, February 25–27, pp. 199–205.

Wai, F.K., Yong, L.W., Divakaran, D.M. & Thing, V.L.L. (2018). Predicting vulnerability discovery rate using past versions of a software. In *2018 IEEE International Conference on Service Operations and Logistics, and Informatics (SOLI)*, Singapore, pp. 220–225.

Woo, S., Alhazmi, O., & Malaiya, Y. (2006). Assessing vulnerabilities in apache and IIS HTTP servers. In *2006 2nd IEEE International Symposium on Dependable, Autonomic and Secure Computing IEEE*, pp. 103–110.

Younis, A., Joh, H., & Malaiya, Y. (2011). Modeling learningless vulnerability discovery using a folded distribution. *Proceedings of SAM*, 11, 617–623.

Younis, A., Malaiya, Y.K., & Ray, I. (2016). Assessing vulnerability exploitability risk using software properties. *Software Quality Journal*, 24, 159–202.

16 Signature Reliability Evaluations

An Overview of Different Systems

Akshay Kumar, Mangey Ram, and S. B. Singh

CONTENTS

16.1 INTRODUCTION

In recent years, substantial efforts are being made in the development of reliability theory including signature and fuzzy reliability theories and their applications to various areas of real-life problems. Barlow and Proschan (1975) discussed an important measure of the elements in a coherent system and expressed its fundamental characteristics in the fault tree. The given new important measure is a useful tool for evaluating the minimum cut sets, system reliability, and minimum cost of the fault tree system using the Monte Carlo method and life distribution. They discussed a method for computing the importance in hazard rate corresponding to series-parallel and complex systems. Owen (1975) discussed multi-linear extensions of the composite value of compounds game theory and evaluated the Banzahat value by differentiating the extension value of the game unit cube. The presidential election game and Electoral College can be computed from the proposed algorithm.

Samaniego (1985) presented the failure rate of an erratic coherent system with a lifetime element having independent identically distributed (i.i.d.) elements using the common continuous distribution F. Various examples were quoted for a coherent system including the closure theorem for k-out-of-n:F system having i.i.d. elements and obtained various characteristic of the s-coherent system. Owen (1988) defined the theory of multi-linear extensions of games and discussed its various properties in real-life situations based on the Shapley game theory. This study showed that game theory is a very useful tool for solving many real-life problems. Shapley introduced game theory in 1953, by which players could compute their utility scales and then play could be improved. Boland et al. (1990) considered a consecutive k-out-of-n:F system that consisted of n ordered elements of a coherent system and the system fails if at least k consecutive elements fail. They presented several examples for consecutive k-out-of-n:F systems applied in oil pipelines, telecommunications, and circuitry system. Also, they computed the reliability of consecutive k-out-of-n:F systems that had elements independent from each other. They developed a system having positive dependence between adjoining elements and showed the reliability of the system was less for $k \geq (n + 1)/2$. Yu et al. (1994) investigated the multi-state coherent systems (MSCS) assumed that the states of the system and its elements are totally ordered set. They discussed a new MSCS: generalized multi-state coherent system. They analyzed some properties of the MSCS generalized model and defined a new approach for computing signature of MSCS. They analyzed some properties of the MSCS generalized model and defined a new approach for evaluating the signature of MSCS. Ushakov (1986, 1994) discussed reliability engineering that plays a key role in real life. He reviewed and discussed the system reliability and applied it to engineering systems. He introduced some basic techniques applicable in cutting-edge results, probabilistic reliability ,and statistical reliability, etc. He presented various techniques and applications of reliability theory in real-life systems. Kochar et al. (1999) discussed the different techniques and properties for discussing coherent systems having i.i.d. lifetime elements. They assumed that all comparisons rely on the presentation of a system's lifetime element as a function of the system's signature. Signature of the coherent system was based on the probability of that system and failed with the ith failure element. They introduced a method for evaluating the system signature from the stochastic method, hazard rate ordering, and likelihood ordering ratio method and presented an approach to the coherent system. Levitin (2001) considered a redundancy optimization system for a multi-state system that has a fixed amount of resources for its work performance and resource generator from the subsystem. The suggested algorithm evaluated the optimal system structure and system availability. The system productivity, availability, and cost were evaluated from performance based on each element. The main goal of the study was to minimize the cost investment, total demand, and to present the demand curve based on system probability. A genetic algorithm was used for solving universal generating function (UGF) based problems, to compute the system availability, optimal structure function while the working element of the subsystem had a maximum performance rate under given demand distribution. Boland (2001) studied the characteristic of signatures having an i.i.d. lifetime element based on a coherent system. He concluded that a signature is a widely useful technique for comparing different systems properties and discussed simple and indirect majority system characteristics. Based on signature and system lifetime, the ith order statistic described the probabilities of system element and its computation for the path set and ordered cut set of the system lifetime element. Levitin (2002) proposed a new system linear multi-state sliding window system that generalized the multi-state consecutive k-out-of-r-from-n:F system. The considered system consisted of n linearly ordered multi-state elements. Each element could have two states: total failure or completely working. If the performance sum of the r consecutive element is lower than the total allocated weight, then the system called fails. The author evaluated various characteristics of the linear multi-state sliding window system with the suggested algorithm to find the order of elements and maximum system reliability. A genetic algorithm is used as the optimal solution based on a UGF technique for reliability computation. Levitin (2003a) introduced a two-state linear multi-state sliding window system which consisted of n linearly ordered multistate elements. The system performance rate was based on a given performance weight. The author presented an approach for calculating the reliability of the sliding window system (SWS) to the common supply failures (CSFs) and common supply groups (CSGs). He also described a method for comparing optimal element distributions of the CSG system reliability. The proposed study computed the optimization result with the help of the UGF technique and the genetic algorithm. Levitin (2003b) proposed multi-state a system that generalized the consecutive k-out-of-r-from-n:F system. The considered linear multi-state SWS consisted of n ordered multi-state element and every element could have two states. In this study, he evaluated the system reliability, mean time to failure (MTTF) and cost of the considered system using the extended universal moment generating function. Boland and Samaniego (2004) described the various characteristic of a system called its "signature." They defined a concept between a system's signature and other well-known system reliabilities and found that the signature was useful for comparing different systems. They provided different stochastic comparisons between systems and signature-based comparisons of a coherent system. They investigated the signature of different systems having an i.i.d. lifetime element and evaluated expected lifetime and expected cost using the system reliability function and order statistical methods. Belzunce and Shaked (2004) reviewed and studied the properties of the failure profile in the coherent system. In this study, the authors presented system reliability based on the methods of path set and cut set and discussed the relationship between elements and properties of failure profiles. They derived an expression for the independent element and density function of the lifetime distribution of a coherent system. Also, they presented the likelihood ratio of lifetimes of two systems using failure profiles and obtained bounds of failure profiles in the likelihood ratio on the lifetimes of coherent systems with independent and without identical lifetimes. Navarro and Rychlik (2007) studied the structure functions and the MTTF rate of coherent systems depending on exchangeable elements having a lifetime distribution function depending on the signature. They discussed exchangeable elements with absolutely continuous joint distribution order statistics with the weights identical to the signature based on any coherent system. They assessed expectation bounds for exchangeable exponential elements and expressed the parent marginal reliability function from reliability bounds for all the coherent with three and four exchangeable elements with exponential lifetime distribution. Navarro et al. (2007a) introduced the various properties of a coherent system with a dependent element based on the signature. They presented hyper-minimal and hyper-maximal distributions. The authors evaluated distributions, bound of series, parallel, and k-out-of-n systems. They studied the application of the coherent system in multi-variate lifetime exponential distributions. Navarro et al. (2007b) provided the survey of the ordered statistical coherent system with exchangeable lifetimes element and concept of signatures having i.i.d. elements. They discussed lifetime coherent system representation of the generalized mixture distribution for series, parallel, and k-out-of-n system. Researchers also defined the nature of the hazard rate on the basis of the series system and ordered statistical concepts. Samaniego (2007) discussed properties of series, parallel, and k-out-of-n system based on the signatures with i.i.d. lifetime elements. In this study, he evaluated system signatures, characteristic theorems, and preservation with the help of structure function and ordering statistical methods and showed the application of system signature in network reliability and reliability economics. Navarro et al. (2008) discussed the application and extension of the coherent system in various engineering problems. In this study, the authors defined and reviewed the signature-based description and conservation theorems for systems whose elements have i.i.d. lifetime based on structural reliability. They showed that the distribution of the element system's lifetime could be defined as a mixture of the distributions of k-out-of-n systems. Finally, they evaluated signatures, the expected lifetime of the binary and MSS with the help of reliability functions and the order statistic method. Bhattacharya and Samaniego (2008) reviewed and studied the optimal allocation of i.i.d. elements with reliabilities to specific locations within a given coherent system. They gave the same sufficient condition on the system structure for which the highest possible system reliability is achieved. They evaluated the optimal allocation element in series, parallel and series-parallel systems within the independent element and its reliabilities. Also, they examined long-standing interest problems in reliability theory within a coherent system having defined relevant and monotone elements and obtained many solutions of a coherent system based on signatures using order statistics and gave a sufficient condition for the optimal solution. Li and Zhang (2008) investigated the coherent systems having i.i.d. elements along with the system properties based on stochastic methods for comparing the system lifetime distribution and computed the signature of a coherent system with the help of order statistical methods. They discussed the characteristic of a coherent system for signature evaluating using reliability functions and order statistical methods consisting of i.i.d elements. Navarro and Rubio (2009) studied the signature-based coherent system and its characteristics in various engineering fields using stochastic orderings. They discussed signatures with 2, 3, and 4 system elements using the minimum path set and the order statistical approach. An algorithm was suggested for calculating signature, system moment, system reliability function, and the expected value of elements with n elements and they also compared the system in i.i.d. case. The given algorithm was based on the minimum path set. Eryilmaz et al. (2009) discussed the consecutive k-within-m-out-of-n:F system with exchangeable elements having the reliability properties based on survival function. They obtained the system-bound reliability using Monte Carlo estimator simulation and moving order statistics. The system signature was also discussed with the help of Samaniego signature simulation and defined system characteristics based on a coherent system. The multivariate Pareto distribution was used to evaluate the results of the system with exchangeable elements. Eryilmaz (2010) examined the reliability functions of the consecutive systems as a mixture of the reliability of order statistics which consisted of exchangeable lifetime elements. He also revealed that the reliability and stochastic ordering results for consecutive k-system can be computed from mixture representations. The consecutive k-systems can be applied in an oil pipeline, a system in accelerators, vacuums, telecom networks, and spacecraft relay stations. Navarro and Rychlik (2010) discussed the expected lifetime of system reliability and compared their bounds and calculated expected lifetimes of the coherent system and mixed systems based on elements with independently distributed lifetimes. They obtained better inequalities dependent on a concentration measure connected to the Gini dispersion index in case of i.i.d. The expected lifetimes of series systems of compact sizes could be derived from bounds and expected a lifetime of one unit in the case of i.i.d. lifetime distribution. Da Costa Bueno (2011) determined the importance measure of a coherent system in the presentation of its signature and described the properties of the dynamic system signature, Barlow-Proschan importance, and element importance under compensator transforms in case of deterministic compensators having i.i.d elements using lifetime distribution. Eryilmaz et al. (2011) discussed the m-consecutive-k-out-of-n:F systems with exchangeable elements based on reliability properties and evaluated the recurrence relations for the signature of the system by exact methods. They introduced order statistics and the lifetime distribution for describing system reliability metrics. They also computed the system minimum and maximum signature having i.i.d. elements and MTTF from stochastic ordering methods for the m-consecutive-k-out-of-n:F system. Lisnianski and Frenkel (2011) studied the MSS reliability evaluation on the basis of signature, optimization, and statistical inference. They discussed the advanced role of a signature in dynamic reliability and non-parametric inference for lifetime distribution. The authors defined the role of a coherent system in various engineering problems and dynamic reliability based on the signature. They also presented various methods for signature-based representation of a coherent system using order statistical, Markov process, and multiple-valued logic methods and computed MSS reliability, expected lifetime, and cost. Mahmoudi and Asadi (2011) evaluated the properties of dynamic signature for a coherent system. They reviewed and studied the concept of signature for the stochastic and advance advantage of coherent systems. They considered a coherent system and described its various characteristics and measures in real-life situations and evaluated engineering reliability based on partial information and obtained the lifetime failure probability of the coherent system. Triantafyllou and Koutras (2011) proposed a 2-within consecutive k-out-of-n:F system that consisted of exchangeable elements. The system was based on the signature and they gave some stochastic comparisons between the reliability function and the lifetime element. Researchers presented many stochastic orderings in the 2-within consecutive k-out-of-n:F system with signature. In this study, they discussed the preservation of intrinsic failure rate (IFR) property with the help of the proposed system. A 2-within consecutive k-out-of-n:F system is used in telecommunication, oil pipeline, and vacuum systems in accelerators. Balakrishnan et al. (2012)

presented an observation of the present theories relating to the signatures and their applicable use in the study of dynamic reliability, systems with i.i.d. elements and non-parametric inference for an element lifetime distribution. They introduced the various properties of the signature based on a coherent system. The authors discussed various methods, algorithms for obtaining system reliability, expected lifetime, Barlow-Proschan index, and expected cost rate using order statistics and reliability functions of a coherent systems. Eryilmaz (2012) investigated the number of elements that fail at the time of system failure. The author discussed the coherent systems such as linear consecutive k-within-m-out-of-n:F and m-consecutive-k-out-of-n:F and obtained expected lifetime, expected x value, and system reliability of considered linear consecutive k-within-m-out-of-n:F and m-consecutive-k-out-of-n:F systems using lifetime distributions and ordering statistics. Da Costa Bueno (2013) introduced the multi-state monotone system using decomposition methods and evaluated the signature of a coherent system in the classical case through exchangeability properties. The system reliability function was obtained with monotone i.i.d. elements and the Samaniego signature. The work also included the study of the signature of the binary and MSS with the help of the proposed theorem. Marichal and Mathonet (2013) evaluated that the Samaniego signature of a coherent system has i.i.d. lifetime elements using Boland's formula, which had structure function. They measured the signature of the coherent system: derivative, Barlow-Proschan index, and tail signature with lifetime distribution. For computing the signature of the coherent system with structure function, they used Owen's method. In real-life situations, various engineering problems were discussed and provided various methods and algorithms for determining system signature. Da et al. (2014) studied and discussed the signature of a k-out-of-n coherent system consisting of n elements. They computed the minimal signature and the signature of the binary coherent system and their combination of elements were derived. The authors gave several numerical examples for defining the characteristic of a coherent system with i.i.d. elements based on the minimum path set along with application in engineering fields. Also, they obtained the signature from order statistics and suggested algorithms. Eryilmaz (2014) discussed the signature of a system that is an effective tool not only for investigation of the binary coherent systems but also for application in network systems. For evaluating the system signature of series and parallel systems, he derived a simple method based on the signature and minimum signature of modules with the help of system structure functions. A simple statistical approach was given for comparing the system signature, which was dependent on a coherent system and computation of series and parallel system modules. Eryilmaz (2015) defined the representation for a mixture of the 3-state system with three state elements and reliability modeling of 3-state systems consisting of 3-state s-independent elements. The systems and its element could have three states: perfect functioning, partial performance, and complete failure. The presented study showed that survival functions of the systems were of different state subsets. Markov process was used for analyzing the signature of the 3-state consecutive-k-out-of-n:G systems consisting of s-independent elements. Lindqvist and Samaniego (2015) introduced that the signature reliability of a coherent system is a very useful tool in the study with i.i.d. lifetime elements. The signature of a coherent system in n element was a vector whose

kth failure element caused a system failure. They evaluated the dynamic signature of binary and complex systems with minimum repair called system conditional dynamic signature with the help of suggested stochastic and minimal path sets. Eryilmaz and Tuncel (2015) studied a k-out-of-n system that consisted of n linearly ordered element (linear and circular). They discussed signature with the help of simulation and that the system could have various numbers of the element. After obtaining the signature based on the expression for the structure function, MTTF, mean number of the failed element, they provided various applications in the engineering fields. Franko and Tutuncu (2016) computed the reliability of the weighted k-out-of-n:G system based on the signature with repairable i.i.d. lifetime elements. They studied the reliability and some reliability indices with the repairable weighted k-out-of-n:G system and found several uncertainties via signature. The proposed system is widely used in the engineering field such as solar field, military system, etc. They computed the system signature of the considered system depending on the weights of the element using the stochastic method and path set and calculated the Birnbaum and Barlow-Proschan element importance measures through the suggested algorithms. Chahkandi et al. (2016) discussed a repairable coherent system to examine signature and Samaniego's notation for i.i.d. lifetime elements. The Poisson process was used to calculate the failure element that has the same intensity function. They presented Samaniego for i.i.d. random variable, whereas the Poisson process could have an identical intensity function. The authors supposed that the reliability function of a coherent system depends on the mixture of the probabilities and number of repairable elements. They determined the reliability function of the series system using a stochastic order statistic algorithm. Samaniego and Navarro (2016) studied the coherent system and its properties for comparing heterogeneous elements. They used various methods for comparing coherent systems having both independent and dependent elements. In the independent case, for computing the signature in survival function, Coolen and Coolen-Maturi methods were used. Kumar and Singh (2017a, 2017b, 2017c) evaluated the signature, expected cost, MTTF, and Barlow-Proschan index of various engineering systems with the help of reliability functions and using UGF techniques. Bisht and Singh (2019) discussed the signature of complex bridge networks with binary state nodes using UGF techniques. They computed the signature of each node in series, parallel, and complex forms of the network system.

16.2 ALGORITHMS USED IN SIGNATURE RELIABILITY

16.2.1 ALGORITHM FOR COMPUTING THE SIGNATURE USING RELIABILITY FUNCTION

Step 1: Determine the signature of the system using reliability functions (see Boland, 2001).

$$A_a = \frac{1}{\binom{s}{s-a+1}} \sum_{\substack{H \subseteq [s] \\ |H|=s-a+1}} \phi(H) - \frac{1}{\binom{s}{s-1}} \sum_{\substack{H \subseteq [s] \\ |H|=s-1}} \phi(H) \qquad (16.1)$$

Calculate the reliability polynomial of SWS

$$H(P) = \sum_{j=1}^{s} C_j \binom{s}{j} P^j q^{n-j}.$$

where $C_i = \sum_{i-s-j+1}^{s} V_i, \; j = 1,2,...s.$

Step 2: Evaluate the tail signature of the system, i.e., $(s+1)$-tuple $\bar{V} = (\bar{V}_0,...,\bar{V}_s)$
with

$$\bar{V}_a = \sum_{i=a+1}^{s} V_i = \frac{1}{\binom{s}{s-a}} \sum_{|H|=s-a} \phi(H) \tag{16.2}$$

Step 3: Calculate the reliability function from a polynomial form with the help
of Taylor evolution at $v = 1$ by:

$$P(v) = v^s H\left(\frac{1}{v}\right) \tag{16.3}$$

Step 4: Compute the tail signature of the system with the help of the reliability
function using Equation 16.2 by (see Marichal and Mathonet, 2013).

$$\bar{V}_a = \frac{(s-1)!}{s!} D^a P(1), \, a = 0,1,...,s \tag{16.4}$$

Step 5: Obtain the signature from tail signature:

$$V = \bar{V}_{a-1} - \bar{V}_a, \, a = 1,2,...,s \tag{16.5}$$

16.2.2 THE ALGORITHM TO ASSESS THE EXPECTED LIFETIME OF THE SYSTEM BY USING MINIMUM SIGNATURE

Step 1: Determine the MTTF of the i.i.d. of the element of the system that have
exponentially distributed elements with the mean ($\mu = 1$).

Step 2: Assessment $E(T)$ of the system, which has i.i.d. elements (see Navarro,
2009):

$$E(T) = \mu \sum_{i=1}^{n} \frac{C_i}{i} \tag{16.6}$$

where $C = (C_1, C_2,...,C_n)$ is a vector coefficient we obtain with the help of the
minimal signature.

16.2.3 ALGORITHM FOR OBTAINING THE BARLOW-PROSCHAN INDEX FOR THE SYSTEM

Compute the Barlow-Proschan index of the i.i.d. elements with the help of the reliability function as (see Shapley, 1953; Owen, 1975, 1988).

$$I_{BP}^{(a)} = \int_0^1 \left(\partial_a H \right)(v)dv, a = 1, 2, ..., n \tag{16.7}$$

where H are reliability functions of the system.

16.2.4 ALGORITHM TO DETERMINE THE EXPECTED VALUE OF THE SYSTEM (ERYILMAZ, 2012)

Step 1: Calculate the expected value of the system elements using the signature:

$$E(X) = \sum_{i=1}^n iV_i, i = 1, 2, ..., n.$$

Step 2: Evaluate $E(X)$ and $E(X)/E(T)$ of the system.

16.2.5 ALGORITHM FOR OBTAINING THE RELIABILITY OF THE SLIDING WINDOW SYSTEM (LEVITIN, 2005)

Step 1: Estimate the UGF of the individual element, given $F = 0, U_{1-r}(z) = z^{0,b_0}$.
Step 2: Change the value of $i = 1, 2, ..., K$.
Step 3: Compute $U_{i-r+1}(z) = U_{i-r}(z) \phi U_i(z)$.
Step 4: Find all the expressions that satisfied the condition $i \geq r$ and added the terms $\alpha_f(U_{i-r+1}(z))$ to F.
Step 5: Find the reliability of SWS as $R = 1 - F$.

16.3 ILLUSTRATIONS

Case 1: Find a series system that has five elements in a series manner and reliability of the proposed system can be computed as shown in Figure 16.1 such as:

Structure function of the series system from Figure 16.1 as:

$$R(P) = \prod_{j=1}^n R_j$$

$$R(P) = R_1 R_2 R_3 R_4 R_5 \tag{16.8}$$

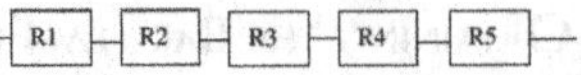

FIGURE 16.1 Series system.

In this case when elements are identically distributed ($R_j = R$), the reliability function $R(P)$ of the series system which has i.i.d. in the element can be revealed as:

$$R(P) = P^5.$$

1. **Signature of a series system**
   Using Owen's method for the system, express the reliability function in the terms of v as:

$$H(v) = v^5. \tag{16.9}$$

   With the help of Equations 16.3 and 16.9, the reliability function can be written as:

$$P(v) = v^5 H\left(\frac{1}{v}\right) = 1.$$

   Now, obtain the tail signature V of the series system by using Equation 16.4 as

$$\bar{V}_0 = 1, \bar{V}_1 = 0, \bar{V}_2 = 0, \bar{V}_3 = 0, \bar{V}_4 = 0, \bar{V}_5 = 0.$$

$$\bar{V} = (1,0,0,0,0,0).$$

   Calculate the signature V of the series system from Equation 16.5:

$$V = (1, 0, 0, 0, 0, 0).$$

2. **Barlow-Proschan index of the series system**
   From Equations 16.8 and 16.7, we obtain the Barlow-Proschan index of the series system:

$$I_{BP}^{(1)} = \int_0^1 (d_1 H) dH = \int_0^1 v^4 dv = \frac{1}{5}.$$

   Similarly, we obtain all elements Barlow-Proschan index $I_{BP}^{(K)}$ for $K = (1,2,...,5)$ given as:

$$I_{BP} = \left(\frac{1}{5}, \frac{1}{5}, \frac{1}{5}, \frac{1}{5}, \frac{1}{5}\right).$$

3. **The expected lifetime of the series system**

We have evaluated the minimal signature M of the series system from Equation 16.9:

$$\text{Minimal signature } (1, 0, 0, 0, 0).$$

Using step 3 Algorithms 16.2.2 determine the expected lifetime of the series system as:

$$E(t) = 1 \tag{16.10}$$

4. **Expected cost rate**

We have evaluated the expected value of the series system with step 1 of Algorithm 16.2.4:

$$E(X) = 1. \tag{16.11}$$

Using Equations 16.10 and 16.11, the expected cost rate is defined as:

$$= E(X) / E(t)$$

$$= 1.$$

Case 2: We find a parallel system that has five elements in a parallel manner and the reliability function of the proposed system can be evaluated as shown in Figure 16.2 such as:

Reliability function of the parallel system from Figure 16.2 defined as:

$$R(P) = 1 - \prod_{j=1}^{n} (1 - R_j)$$

$$R(P) = 1 - [(1 - R_1)(1 - R_2)(1 - R_3)(1 - R_4)(1 - R_5)]. \tag{16.12}$$

Now, $(R_i = R)$ because elements are identically distributed, the reliability function $R(P)$ of the parallel system from i.i.d. in the element can be written as:

$$R(P) = 5R - 10R^2 + 10R^3 - 5R^4 + R^5.$$

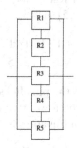

FIGURE 16.2 Parallel system.

The reliability function can be expressed in the form of P as:

$$H(P) = 5P - 10P^2 + 10P^3 - 5P^4 + P^5. \tag{16.13}$$

1. **Signature of a parallel system**

   With the help of Owen's method for the system, we have computed the signature of the parallel system from Equation 16.13 as:

   $$H(P) = 5P - 10P^2 + 10P^3 - 5P^4 + P^5. \tag{16.14}$$

   Using Equations 16.3 and 16.14, the reliability function can be expressed in term of v as:

   $$P(v) = v^5 H\left(\frac{1}{v}\right) = 1 - 5v + 10v^2 - 10v^3 + 5v^4.$$

   Now, obtaining the tail signature $\bar{V}_i$ of the proposed system by using Equation 16.4 as:

   $$\bar{V}_0 = 1, \bar{V}_1 = 1, \bar{V}_2 = 1, \bar{V}_3 = 1, \bar{V}_4 = 1, \bar{V}_5 = 0.$$

   $$\bar{V} = (1, 1, 1, 1, 1, 0).$$

   Therefore, we have evaluated the signature V of the parallel system from Equation 16.5:

   $$V = (0, 0, 0, 0, 0, 1).$$

2. **Barlow-Proschan index of the parallel system**

   From Equation 16.12 and Algorithm 16.2.3, we have calculated the Barlow-Proschan index of the parallel system by:

   $$I_{BP}^{(1)} = \int_0^1 (d_1 H) \, dH = \int_0^1 (1 - 4v + 6v^2 - 4v^3 + v^4) \, dv = \frac{1}{5}.$$

   Similarly, we compute all rest of the elements in the Barlow-Proschan index $I_{BP}^{(K)}$ for $K = (1, 2, ..., 5)$ such as:

   $$I_{BP} = \left(\frac{1}{5}, \frac{1}{5}, \frac{1}{5}, \frac{1}{5}, \frac{1}{5}\right).$$

3. **The expected lifetime of the parallel system**

   From the reliability function, we have determined the minimal signature M of the system from Equation 16.13:

   $$\text{Minimal signature} \, (1, 0, 0, 0, 0).$$

The expected lifetime of a parallel system assessed from Equation 16.6 defined as:

$$E(t) = 2.28 \tag{16.15}$$

4. **Expected cost rate**
   Assessment the expected value of the parallel system from using step 1 of Algorithm 16.2.4:

$$E(X) = 5 \tag{16.16}$$

Therefore, the expected cost rate is:

$$= E(X)/E(t)$$

$$= 2.19298.$$

Case 3: Consider an SWS that has four window elements with $n = 4$, $r = 3$, and $W = 4$ as shown in Figure 16.3. Each window having two states: complete successor and complete failure. Suppose the performance rates of the window from 1 to 4 are 1,2,3,4, respectively.
Now from UGF of the proposed system from Figure 16.3 given as:

$$U_j(z) = P_j z^j + \left(1 - P_j\right) z^0$$

where $j = 1,2,3,4$, and P_j is given the probability function and z^j, z^0 is the performance and non-performance rate.

Therefore, the UGF $U_j(z)\left(j = 1,2,3,4\right)$ of the system is given by:

$$U_1(z) = P_1 z^1 + \left(1 - P_1\right) z^0$$

$$U_2(z) = P_2 z^2 + \left(1 - P_2\right) z^0$$

$$U_3(z) = P_3 z^3 + \left(1 - P_3\right) z^0$$

$$U_4(z) = P_4 z^4 + \left(1 - P_4\right) z^0.$$

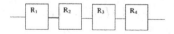

FIGURE 16.3   Sliding window system.

From the Algorithm 16.2.5 of SWS, we obtain the beginning element of the SWS as:

For $i = 1$

$$U_0(z) = \phi(U_{-1}(z), U_1(z))$$

$$U_0(z) = \phi(z^{0,(0,0,0)}, P_1 z^1 + (1 - P_1)z^0)$$

$$= P_1 z^{0,(0,0,1)} + (1 - P_1) z^{0,(0,0,0)}$$

For $i = 2$

$$U_1(z) = \phi(U_0(z), U_2(z))$$

$$= \phi(P_1 z^{0,(0,0,1)} + (1 - P_1)z^{0,(0,0,0)}, P_2 z^2 + (1 - P_2)z^0)$$

$$= P_1 P_2 z^{0,(0,1,2)} + P_1(1 - P_2)z^{0,(0,1,0)} + P_2(1 - P_1)z^{0,(0,0,2)} + (1 - P_1)(1 - P_2)z^{0,(0,0,0)}$$

For $i = 3$

$$U_2(z) = \phi(U_1(z), U_3(z))$$

$$= P_1 P_2 P_3 z^{0,(1,2,3)} + P_1(1 - P_2)P_3 z^{0,(1,0,3)} + P_2(1 - P_1)P_3 z^{0,(0,2,3)}$$

$$+ (1 - P_1)(1 - P_2)P_3 z^{0,(0,0,3)} + P_1 P_2(1 - P_3)z^{0,(1,2,0)}$$

$$+ P_1(1 - P_2)(1 - P_3)z^{0,(1,0,0)} + P_2(1 - P_1)(1 - P_3)z^{0,(0,2,0)}$$

$$+ (1 - P_1)(1 - P_2)(1 - P_3)z^{0,(0,0,0)}$$

From the condition $i \geq w$, obtained unreliability F and $U_2(z)$ are given as:

$$F = (1 - P_1)(1 - P_2)P_3 + P_1 P_2(1 - P_3) + P_1(1 - P_2)(1 - P_3)$$

$$+ P_2(1 - P_1)(1 - P_3) + (1 - P_1)(1 - P_2)(1 - P_3)$$

$$\tag{16.17}$$

For $i = 4$

$$U_3(z) = \phi(U_2(z), U_4(z))$$

$$= \phi(P_1 P_2 P_3 z^{0,(1,2,3)} + P_1(1 - P_2)P_3 z^{0,(1,0,3)} + (1 - P_1)P_2 P_3 z^{0,(0,2,3)}, P_4 z^4 + (1 - P_4)z^0)$$

$$= P_1 P_2 P_3 P_4 z^{1,(2,3,4)} + P_1(1 - P_2)P_3 P_4 z^{1,(0,3,4)} + (1 - P_1)P_2 P_3 P_4 z^{0,(2,3,4)}$$

$$+ P_1 P_2 P_3(1 - P_4)z^{1,(2,3,0)} + P_1(1 - P_2)P_3(1 - P_4)z^{1,(0,3,0)} + (1 - P_1)P_2 P_3(1 - P_4)z^{0,(2,3,0)}$$

$$F = P_1(1 - P_2)P_3(1 - P_4)$$

$$\tag{16.18}$$

Now, adding Equations 16.17 and 16.18, we obtain reliability R of SWS as:

$$R = P_2P_3 + P_1P_3P_4 - P_1P_2P_3P_4 \qquad (16.19)$$

The reliability function R of the SWS is defined as:

$$R(P) = P^2 + P^3 - P^4.$$

1. **Signature of the sliding window system**

   We obtain the reliability function in the form of v from Owen's method for the system as:

   $$H(v) = v^2 + v^3 - v^4 \qquad (16.20)$$

   Now using Equations 16.3 and 16.20, the reliability function is:

   $$P(v) = v^4 H\left(\frac{1}{v}\right) = -1 + v + v^2.$$

   Now calculate the tail signature V of the SWS from using step 4 of Algorithm 16.2.1 as:

   $$\bar{V}_0 = 1, \bar{V}_1 = \frac{3}{4}, \bar{V}_2 = \frac{1}{2}, \ \bar{V}_3 = 0, \ \bar{V}_4 = 0.$$

   The tail signature $\bar{V}$ of the SWS is:

   $$\bar{V} = \left(1, \frac{3}{4}, \frac{1}{2}, 0, 0\right).$$

   Now, find the signature of the SWS from step 5 Algorithm 16.2.1 is:

   $$V = \left(\frac{1}{4}, \frac{1}{4}, \frac{1}{2}, 0\right).$$

2. **Barlow-Proschan index of the sliding window system**

   From Equation 16.20 and Algorithm 16.2.3, we obtain the Barlow-Proschan index of the SWS by:

   $$I_{BP}^{(1)} = \int_0^1 (v^2 - v^3) \, dv = \frac{1}{12}.$$

   Similarly, we compute all the rest of the elements of the Barlow-Proschan index $I_{BP}^{(K)}$ for $K = (1, 2, ..., 4)$ such as:

   $$I_{BP} = \left(\frac{1}{12}, \frac{1}{14}, \frac{7}{12}, \frac{1}{12}\right).$$

3. **The expected lifetime of the parallel system**

From using the reliability function, we have determined the minimal signature M of the system from Equation 16.20 is:

$$\text{Minimal signature}(0,1,1,-1)$$

Then, the expected lifetime of SWS can be determined by using Equation 16.6 given as:

$$E(t) = 0.58. \tag{16.21}$$

4. **Expected cost rate**

From Algorithm 16.2.4 with step 1, we have calculated the expected value of the SWS is:

$$E(X) = 2 \tag{16.22}$$

The expected cost rate by using Equations 16.21 and 16.22 is:

$$= E(X)/E(t)$$

$$= 3.4483.$$

16.4 CONCLUSION

In this chapter, we discussed the properties of signature and its factor like a tail signature, expected cost rate, mean time to failure, and Barlow-Proschan index with the help of the reliability function and Owen's method. Also, we evaluated the reliability function by using UGF. Further, different systems such as series, parallel, and SWS and computed signature with the help of given algorithms were discussed.

REFERENCES

Balakrishnan, N., Navarro, J., & Samaniego, F. J. (2012). Signature representation and preservation results for engineered systems and applications to statistical inference. In *Recent Advances in System Reliability*, Springer, London, UK, pp. 1–22.

Barlow, R. E., & Proschan, F. (1975). Importance of system elements and fault tree events. *Stochastic Processes and Their Applications*, 3(2), 153–173.

Belzunce, F., & Shaked, M. (2004). Failure profiles of coherent systems. Naval Research Logistics (NRL), 51(4), 477–490.

Bhattacharya, D., & Samaniego, F. J. (2008). On the optimal allocation of elements within coherent systems. *Statistics & Probability Letters*, 78(7), 938–943.

Bisht, S., & Singh S. B. (2019). Signature reliability of binary state node in complex bridge network using universal generating function. *International Journal of Quality & Reliability Management*, 36(2), 186–201.

Boland, P. J. (2001). Signatures of indirect majority systems. *Journal of Applied Probability*, 38(2), 597–603.

Boland, P. J., Proschan, F., & Tong, Y. L. (1990). Linear dependence in consecutive k-out-of-n: F systems. *Probability in the Engineering and Informational Sciences*, 4(3), 391–397.

Boland, P. J., & Samaniego, F. J. (2004). The signature of a coherent system and its applications in reliability. In *Mathematical Reliability: An Expository Perspective*, Springer US, pp. 3–30.

Chahkandi, M., Ruggeri, F., & Suárez-Llorens, A. (2016). A generalized signature of repairable coherent systems. *IEEE Transactions on Reliability*, 65(1), 434–445.

Da Costa Bueno, V. (2011). A coherent system element importance under its signatures representation. *American Journal of Operations Research*, 1(3), 172.

Da Costa Bueno, V. (2013). A multistate monotone system signature. *Statistics & Probability Letters*, 83(11), 2583–2591.

Da, G., Xia, L., & Hu, T. (2014). On computing signatures of k-out-of-n systems consisting of modules. *Methodology and Computing in Applied Probability*, 16(1), 223–233.

Eryılmaz, S. (2010). Mixture representations for the reliability of consecutive-k systems. *Mathematical and Computer Modelling*, 51(5), 405–412.

Eryilmaz, S. (2012). The number of failed elements in a coherent system with exchangeable elements. *IEEE Transactions on Reliability*, 61(1), 203–207.

Eryilmaz, S. (2014). On signatures of series and parallel systems consisting of modules with arbitrary structures. *Communications in Statistics-Simulation and Computation*, 43(5), 1202–1211.

Eryilmaz, S. (2015). Mixture representations for three-state systems with three-state elements. *IEEE Transactions on Reliability*, 64(2), 829–834.

Eryilmaz, S., Kan, C., & Akici, F. (2009). Consecutive k-within-m-out-of-n: F system with exchangeable elements. *Naval Research Logistics (NRL)*, 56(6), 503–510.

Eryilmaz, S., Koutras, M. V., & Triantafyllou, I. S. (2011). Signature based analysis of m-consecutive k-out-of-n: F systems with exchangeable elements. *Naval Research Logistics (NRL)*, 58(4), 344–354.

Eryilmaz, S., & Tuncel, A. (2015). Computing the signature of a generalized k-out-of-n system. *IEEE Transactions on Reliability*, 64(2), 766–771.

Franko, C., & Tütüncü, G. Y. (2016). Signature based reliability analysis of repairable weighted k-out-of-n: G systems. *IEEE Transactions on Reliability*, 65(2), 843–850.

Kochar, S., Mukerjee, H., & Samaniego, F. J. (1999). The signature of a coherent system and its application to comparisons among systems. *Naval Research Logistics (NRL)*, 46(5), 507–523.

Kumar, A., & Singh, S. B. (2017a). Signature reliability of linear multi-state sliding window system. *International Journal of Quality & Reliability Management*, 35(10), 2403–2413.

Kumar, A., & Singh, S. B. (2017b). Computations of signature reliability of coherent system. *International Journal of Quality & Reliability Management*, 34(6), 785–797.

Kumar, A., & Singh, S. B. (2017c). Signature reliability of sliding window coherent system. In *Mathematics Applied to Engineering*, Elsevier International Publisher, London, UK, pp. 83–95.

Levitin, G. (2001). Redundancy optimization for multi-state system with fixed resource-requirements and unreliable sources. *IEEE Transactions on Reliability*, 50(1), 52–59.

Levitin, G. (2002). Optimal allocation of elements in a linear multi-state sliding window system. *Reliability Engineering & System Safety*, 76(3), 245–254.

Levitin, G. (2003a). Common supply failures in linear multi-state sliding window systems. *Reliability Engineering & System Safety*, 82(1), 55–62.

Levitin, G. (2003b). Linear multi-state sliding-window systems. *IEEE Transactions on Reliability*, 52(2), 263–269.

Levitin, G. (2005). *The Universal Generating Function in Reliability Analysis and Optimization*, Springer, London, UK, p. 442. doi:10.1007/1-84628-245-4.

Li, X., & Zhang, Z. (2008). Some stochastic comparisons of conditional coherent systems. *Applied Stochastic Models in Business and Industry*, 24(6), 541–549.

Lindqvist, B. H., & Samaniego, F. J. (2015). On the signature of a system under minimal repair. *Applied Stochastic Models in Business and Industry*, 31(3), 297–306.

Lisnianski, A., & Frenkel, I. (Eds.). (2011). *Recent Advances in System Reliability: Signatures, Multi-state Systems and Statistical Inference.* Springer Science & Business Media, London, UK.

Mahmoudi, M., & Asadi, M. (2011). The dynamic signature of coherent systems. *IEEE Transactions on Reliability*, 60(4), 817–822.

Marichal, J. L., & Mathonet, P. (2013). Computing system signatures through reliability functions. *Statistics & Probability Letters*, 83(3), 710–717.

Navarro, J., & Rubio, R. (2009). Computations of signatures of coherent systems with five elements. *Communications in Statistics-Simulation and Computation*, 39(1), 68–84.

Navarro, J., Ruiz, J. M., & Sandoval, C. J. (2007a). Properties of coherent systems with dependent elements. *Communications in Statistics: Theory and Methods*, 36(1), 175–191.

Navarro, J., & Rychlik, T. (2007). Reliability and expectation bounds for coherent systems with exchangeable elements. *Journal of Multivariate Analysis*, 98(1), 102–113.

Navarro, J., & Rychlik, T. (2010). Comparisons and bounds for expected lifetimes of reliability systems. *European Journal of Operational Research*, 207(1), 309–317.

Navarro, J., Rychlik, T., & Shaked, M. (2007b). Are the order statistics ordered? A survey of recent results. *Communications in Statistics: Theory and Methods*, 36(7), 1273–1290.

Navarro, J., Samaniego, F. J., Balakrishnan, N., & Bhattacharya, D. (2008). On the application and extension of system signatures in engineering reliability. *Naval Research Logistics (NRL)*, 55(4), 313–327.

Owen, G. (1975). Multilinear extensions and the Banzhaf value. *Naval Research Logistics Quarterly*, 22(4), 741–750.

Owen, G. (1988). Multilinear extensions of games. *The Shapley Value Essays in Honor of Lloyd S Shapley*, Cambridge University Press, New York, pp. 139–151.

Samaniego, F. J. (1985). On closure of the IFR class under formation of coherent systems. *IEEE Transactions on Reliability*, 34(1), 69–72.

Samaniego, F. J. (2007). *System Signatures and Their Applications in Engineering Reliability.* Springer Science & Business Media, London, UK, p. 110.

Samaniego, F. J., & Navarro, J. (2016). On comparing coherent systems with heterogeneous elements. *Advances in Applied Probability*, 48(1), 88–111.

Shapley, L.S. (1953). A value for n-person games. In: *Contributions to the Theory of Games*, Vol. 2. In: *Annals of Mathematics Studies*, vol. 28. Princeton University Press, Princeton, NJ, pp. 307–317.

Triantafyllou, I. S., & Koutras, M. V. (2011). Signature and IFR preservation of 2-within-consecutive k-out-of- n- F: Systems. *IEEE Transactions on Reliability*, 60(1), 315–322.

Ushakov, I. (1986) Universal generating function. *Journal of Computer Science and Systems Biology*, 24, 118–129.

Ushakov, I. A. (Ed.). (1994). *Handbook of Reliability Engineering.* John Wiley & Sons, New York.

Yu, K., Koren, I., & Guo, Y. (1994). Generalized multistate monotone coherent systems. *IEEE Transactions on Reliability*, 43(2), 242–250.

Index

Printed in the United States
by Baker & Taylor Publisher Services